Springer Proceedings in Physics

Volume 274

Indexed by Scopus

The series Springer Proceedings in Physics, founded in 1984, is devoted to timely reports of state-of-the-art developments in physics and related sciences. Typically based on material presented at conferences, workshops and similar scientific meetings, volumes published in this series will constitute a comprehensive up-to-date source of reference on a field or subfield of relevance in contemporary physics. Proposals must include the following:

- name, place and date of the scientific meeting
- a link to the committees (local organization, international advisors etc.)
- scientific description of the meeting
- list of invited/plenary speakers
- an estimate of the planned proceedings book parameters (number of pages/articles, requested number of bulk copies, submission deadline).

Please contact:

For Americas and Europe: Dr. Zachary Evenson; zachary.evenson@springer.com
For Asia, Australia and New Zealand: Dr. Loyola DSilva; loyola.dsilva@springer.com

More information about this series at https://link.springer.com/bookseries/361

Nicolae Herisanu · Vasile Marinca
Editors

Acoustics and Vibration of Mechanical Structures – AVMS-2021

Proceedings of the 16th AVMS, Timişoara, Romania, May 28–29, 2021

Springer

Editors
Nicolae Herisanu
Department of Mechanics and Strength of Materials
Polytechnic University of Timişoara
Timişoara, Romania

Vasile Marinca
Centre for Fundamental and Advanced Technical Research
Romanian Academy-Branch of Timişoara
Timişoara, Romania

ISSN 0930-8989 ISSN 1867-4941 (electronic)
Springer Proceedings in Physics
ISBN 978-3-030-96786-4 ISBN 978-3-030-96787-1 (eBook)
https://doi.org/10.1007/978-3-030-96787-1

This Springer imprint is published by the registered company Springer Nature Switzerland AG
The registered company address is: Gewerbestrasse 11, 6330 Cham, Switzerland

Preface

The Proceedings of the 16th International Conference *Acoustics and Vibration of Mechanical Structures* contains the written versions of most of the papers presented within the conference AVMS-2021 during 28–29 May 2021. The conference was hosted by the University Politehnica of Timisoara, Romania and the Romanian Academy, branch of Timisoara.

These Proceedings are structured in five parts, covering five different research directions specific to the field of acoustics and vibration. The first part is devoted to analytical approaches to vibration problems and reunites a group of 12 papers emphasizing the applicability of different new and emerging analytical approaches in the investigation of different vibration problems.

The second part is focused on environmental and occupational noise problems, including mainly experimental investigations in this research direction.

The third part is concerned with investigations developed in the field of biomechanics and bioacoustics, a very actual and important research direction. The fourth part contains a group of six papers oriented in the study of different real-life problems encountered in different industrial processes accompanied with vibration phenomena, while the fifth and last part is devoted to the study of different structural vibration problems, comprising nine papers containing theoretical and experimental results.

All papers presented within the conference were subjected to the review process, each of them being reviewed by at least two expert referees, and the accepted papers are included in this volume.

AVMS-2021 would like to express our sincere gratitude to all the authors for their contribution to this book. We are also expressing our sincere thanks to the members of the International Scientific Committee and also to the members of the Organizing Committee for ensuring the success of this Conference, which would not have been possible without their efforts.

Finally, we would like to thank SPRINGER for producing this volume.

Timişoara, Romania

Nicolae Herisanu
Vasile Marinca

Committees

Organized By

University Politehnica Timişoara—Acoustics and Vibration Laboratory
University of Nis—Noise and Vibration Laboratory
Romanian Academy—Centre for Fundamental and Advanced Technical Research
Romanian Acoustical Society

Organising Committee

Nicolae Herisanu—**Chairman,** University Politehnica Timişoara, Romania
Momir Praščevič—**Co-chairman,** University of Nis, Serbia

Members

Florin Dragan—Rector—University Politehnica Timişoara, Romania
Vasile Marinca—University Politehnica Timişoara, Romania
Darko Mihajlov—University of Nis, Serbia
Ramona Nagy—University Politehnica Timişoara, Romania
Karoly Menyhardt—University Politehnica Timişoara, Romania
Cosmina Vigaru—University Politehnica Timişoara, Romania
Marko Ličanin —University of Nis, Serbia

International Scientific Committee

Jan Awrejcewicz—Lodz University of Technology, Poland
Marco Amabili—McGill University, Montreal, Canada
Vasile Bacria–University Politehnica Timişoara, Romania
Malvina Baica—University of Wisconsin, USA
Dumitru Baleanu—Çankaya University, Turkey
Polidor Bratu—ICECON Bucharest, Romania
Florin Breaban—University d'Artois, France
Veturia Chiroiu—Institute of Solid Mechanics, Bucharest, Romania
Livija Cveticanin—University of Novi Sad, Serbia
Dragan Cvetković—University of Nis, Serbia
Gilbert-Rainer Gillich—"E. Murgu" University of Resita, Romania
Nicolae Herişanu—University Politehnica Timişoara, Romania
Metin O. Kaya—Istanbul Technical University, Turkey
Ivana Kovacic—University of Novi Sad, Serbia
J. A. Tenreiro Machado—Polytechnic Institute of Porto, Portugal
Vasile Marinca—Romanian Academy, Romania
Nuno Maia—University of Lisbon, Portugal
Emil Manoach—Bulgarian Academy of Sciences, Bulgaria
Dan B. Marghitu—Auburn University, USA
Darko Mihajlov—University of Nis, Serbia
Sotirios Natsiavas—Aristotle University of Thessaloniki, Greece
Momir Praščevič—University of Nis, Serbia
M. Mehdi Rashidi—Tongji University, Shanghai, China
Mohsen Razzaghi—Mississippi State University, USA
Zlatan Šoškić—University of Kragujevac, Serbia
Sorin Vlase—"Transilvania" University Braşov, Romania
Jerzy Warminski—Lublin University of Technology, Poland

Contents

Contributors

Simion Alexandru National Institute for Research and Development in Mine Safety and Protection to Explosion, INSEMEX Petroşani, Petroşani, Romania

Mario Ardeljan Babes-Bolyai University, Cluj-Napoca, Romania

Delicia Arsene Polytehnic University of Bucharest, Bucharest, Romania

Mihai Arsene Afico, Bucharest, Romania

Vasile Bacria University Politehnica Timisoara, Timișoara, Romania

Sandor Ianos Bernad Romanian Academy, Centre of Fundamental and Advanced Research in Engineering Sciences, Timisoara Branch, Timisoara, Romania

Maria-Luiza Beşliu-Gherghescu University of Pitesti, Pitesti, Romania

Nebojša Bogojević Faculty of Mechanical and Civil Engineering in Kraljevo, University of Kragujevac, Kraljevo, Serbia

Claudia Borda Polytehnic University of Bucharest, Bucharest, Romania

Paul Nicolae Borza University Transilvania of Brasov, Brasov, Romania

Polidor Bratu Research Institute for Construction Equipment and Technology, ICECON SA, Bucharest, Romania

Larisa Butu Polytehnic University of Bucharest, Bucharest, Romania

Angelica Calamar National Institute for Research and Development in Mine Safety and Protection to Explosion, INSEMEX Petroşani, Petroşani, Romania

Seda Vatan Can Dokuz Eylül University, İzmir, Turkey;
Manisa Celal Bayar University, Manisa, Turkey

Cristina Chilibaru-Opritescu University Politehnica Timisoara, Timișoara, Romania

Veturia Chiroiu Solid Mechanics Institute of the Romanian Academy, Bucharest, Romania

Dejan Ciric University of Nis, Nis, Serbia

Daniel Condurache Technical University of Iasi, Iasi, Romania

Andrei Craifaleanu University "Politehnica" of Bucharest, Bucharest, Romania

D. Cveticanin Novi Sad, Serbia

Livija Cveticanin University of Novi Sad, Novi Sad, Serbia;
Obuda University, Budapest, Hungary

Gigel Căpăţână Research Center for Mechanics of Machines and Technological Equipments, Engineering and Agronomy Faculty in Brăila, "Dunărea de Jos" University of Galaţi, Braila, Romania

Cornelia Florentina Dobrescu INCD URBAN—INCERC, Bucharest, Romania

Nicusor Dragan University Dunarea de Jos, Galati, Romania

Ciprian Dragne Solid Mechanics Institute of the Romanian Academy, Bucharest, Romania

Corneliu Drugă Transylvania University of Brasov, Brasov, Romania

Ştefan Dumitru University "Politehnica" of Bucharest, Bucharest, Romania

Maciej Dutkiewicz Faculty of Civil, Environmental Engineering and Architecture, Bydgoszcz University of Science and Technology, Bydgoszcz, Poland

Gilbert-Rainer Gillich Babes-Bolyai University, Cluj-Napoca, Romania

Delia Gârleanu Polytehnic University of Bucharest, Bucharest, Romania

Gabriel Gârleanu Polytehnic University of Bucharest, Bucharest, Romania

Codruta Oana Hamat Babes-Bolyai University, Cluj-Napoca, Romania

Nicolae Herisanu Department of Mechanics and Strength of Materials, Politehnica University Timisoara, Timisoara, Romania

Calin Itu University Transilvania of Brasov, Brasov, Romania

Maria-Cristina Ioncica Department of Mechanics and Strength of Materials, Politehnica University of Timisoara, Timisoara, Romania

Irina Ioniţă Transylvania University of Brasov, Brasov, Romania

Snežana Ćirić Kostić Faculty of Mechanical and Civil Engineering in Kraljevo, University of Kragujevac, Kraljevo, Serbia

Jaroslaw Latalski Department of Applied Mechanics Faculty of Mechanical Engineering, Lublin University of Technology, Lublin, Poland

Marko Licanin University of Nis, Nis, Serbia

Toth Lorand National Institute for Research and Development in Mine Safety and Protection to Explosion, INSEMEX Petroşani, Petroşani, Romania

Ş.İpek Lök Dokuz Eylül University, İzmir, Turkey

Marcela R. Machado Department of Mechanical Engineering, University of Brasilia, Brasilia, Brazil

Levent Malgaca Dokuz Eylül University, İzmir, Turkey

Tatian-Cristian Malin Babeş-Bolyai University, Cluj-Napoca, Romania

Dan B. Marghitu Department of Mechanical Engineering, Auburn University AL, Auburn, AL, USA

Bogdan Marinca Politehnica University Timişoara, Timişoara, România

Vasile Marinca University Politehnica Timisoara, Timisoara, Romania

Marinela Marinescu Polytehnic University of Bucharest, Bucharest, Romania

Kovacs Marius National Institute for Research and Development in Mine Safety and Protection to Explosion, INSEMEX Petroşani, Petroşani, Romania

Karoly Menyhardt Politehnica University Timisoara, Timişoara, Romania

Matache Mihai INMA, Bucharest, Romania

Darko Mihajlov University of Nis, Nis, Serbia

Cristina Modrogan Polytehnic University of Bucharest, Bucharest, Romania

L. Munteanu Romanian Academy, Institute of Solid Mechanics, Bucharest, Romania

Gina Diana Muscă Anghelache Engineering and Agronomy Faculty in Brăila, Research Centre for Machines and Technological Equipment's Mechanics, "Dunărea de Jos" University of Galai, Galai, Romania

Ramona Nagy Politehnica University Timisoara, Timişoara, Romania

Dorian Nedelcu Babeş-Bolyai University, Cluj-Napoca, Romania

N. Nedelcu Romanian Academy, Institute of Solid Mechanics, Bucharest, Romania

Adrian Eracle Nicolescu University Ovidius, Constanta, Romania

I. Ninkov Obuda University, Budapest, Hungary

Nicolae Pandrea University of Pitesti, Pitesti, Romania

Roxana Alexandra Petre University "Politehnica" of Bucharest, Bucharest, Romania

Cardei Petru INMA, Bucharest, Romania

Laurentiu Picu Dunarea de Jos, Galati University, Galati, Romania

Mihaela Picu European Centre of Excellence for the Environment, "Dunărea de Jos" University of Galati, Galati, Romania

Marius-Vasile Pop Babes-Bolyai University, Cluj-Napoca, Romania

Bianca Popa Politehnica University Timisoara, Timişoara, Romania

Aurora Potîrniche Research Center for Mechanics of Machines and Technological Equipments, Engineering and Agronomy Faculty in Brăila, "Dunărea de Jos" University of Galați, Braila, Romania

Momir Prascevic University of Nis, Nis, Serbia

Elena-Larisa Predel University of Pitesti, Pitesti, Romania

K. Rabiei Department of Mathematics and Statistics, Mississippi State University, Starkville, USA

Branko Radičević Faculty of of Mechanical and Civil Engineering in Kraljevo, University of Kragujevac, Kraljevo, Serbia

Miomir Raos Faculty of Occupational Safety in Niš, University of Niš, Niš, Serbia

M. Razzaghi Department of Mathematics and Statistics, Mississippi State University, Starkville, USA

Ileana Rosca Faculty of Product Design and Environment, Transylvania University of Brasov, Brasov, Romania

Fanel Scheaua Engineering and Agronomy Faculty of Braila, "Dunarea de Jos" University of Galati, MECMET Research Center, Galati, Romania

Ionel Şerban Transylvania University of Brasov, Brasov, Romania

Vladimir Sinđelić Faculty of Mechanical and Civil Engineering in Kraljevo, University of Kragujevac, Kraljevo, Serbia

Simion Sorin National Institute for Research and Development in Mine Safety and Protection to Explosion, INSEMEX Petroşani, Petroşani, Romania

Zlatan Šoškić Faculty of Mechanical and Civil Engineering in Kraljevo, University of Kragujevac, Kraljevo, Serbia

A. Stan Romanian Academy, Institute of Solid Mechanics, Bucharest, Romania; University of Pitesti, Piteşti, Romania

Alin-Florentin Stan University of Pitesti, Pitesti, Romania;
IMSAR, Bucharest, Romania

N. Stan Romanian Academy, Institute of Solid Mechanics, Bucharest, Romania

Anca Stanciu Transylvania University of Brasov, Brasov, Romania

Andreea Stoia Department of Mechanics and Strength of Materials, Politehnica University Timisoara, Timisoara, Romania

Dan Ioan Stoia Department of Mechanics and Strength of Materials, Politehnica University Timisoara, Timisoara, Romania

Nicolae-Doru Stănescu University of Pitesti, Pitesti, Romania

Zofia Szmit Department of Applied Mechanics Faculty of Mechanical Engineering, Lublin University of Technology, Lublin, Poland

Jelena Tomić Faculty of Mechanical and Civil Engineering in Kraljevo, University of Kragujevac, Kraljevo, Serbia

Alin-Florin Totorean Department of Mechanics and Strength of Materials, Politehnica University of Timisoara, Timisoara, Romania

Eugen Tudor Politehnica University Timişoara, Timişoara, România

Mehmet Uyar Bayburt University, Bayburt, Turkey;
Dokuz Eylül University, İzmir, Turkey

Ovidiu Vasile University Politehnica Bucharest, Bucharest, Romania

Cosmina Vigaru Department of Mechanics and Strength of Materials, Politehnica University Timisoara, Timisoara, Romania

Sorin Vlase University Transilvania of Brasov, Brasov, Romania

Jerzy Warminski Department of Applied Mechanics Faculty of Mechanical Engineering, Lublin University of Technology, Lublin, Poland

Analytical Approaches to Vibration Problems

The Sample Entropy of Two Spatial Links

Dan B. Marghitu(✉)

Department of Mechanical Engineering, Auburn University AL, Auburn, AL 36349, USA
marghitu@auburn.edu

Abstract. The nonlinear equations of motion for two spatial links are developed using Gibbs–Appell method. The system is analyzed for different initial conditions. The results are analyzed with return maps and sample entropy.

Keywords: Gibbs–Appell equations of motion · Spatial two links · Sample entropy

1 Introduction

In classical mechanics, the equations of motion are developed using Newton, Lagrange, or Hamilton procedures [1,2]. The notion of quasi-coordinates was utilized by Euler and Lagrange [2]. Kane's dynamical equations of motion take into consideration the generalized inertia forces and the generalized speeds [3–7]. Kane's method was used to study a wheeled mobile manipulator [8] and a robot arm [9]. The Gibbs–Appell function is methodical and straight forward to apply using symbolical MATLAB. The Gibbs–Appell method uses the generalized velocities and generalized forces [3].

The entropy estimates the level of regularity in a time series. Chaotic or random processes have high entropy, and the regularity of a system is represented by low entropy. The complexity of a nonlinear system can be determined with the entropy. Shannon described the entropy as the information content of a system [10]. Pincus used the approximate entropy for time series [11]. Richman and Moorman defined Sample Entropy (SampEn) as an improvement of approximate entropy [12]. The permutation entropy and the multi-scale entropy were studied in [13,14]. The computation of the SampEn for physiological time series was examined in [15].

For this research, the Gibbs–Appell equations are developed for a spatial two link system with revolute joints. The symbolic MATLAB is used to obtain the nonlinear equations of motion, and `ode45` function was employed to numerically solve the equations. The SampEn was calculated for different initial conditions of the links.

N. Herisanu and V. Marinca (eds.), *Acoustics and Vibration of Mechanical Structures – AVMS-2021*, Springer Proceedings in Physics 274,
https://doi.org/10.1007/978-3-030-96787-1_1

2 Equations of Motion

2.1 Kinematics

The system shown in Fig. 1 has a slender rod 1 of mass m_1 and is rotating about the fixed revolute joint at O. At the other end of the rod A, there is another revolute joint that connects a straight link 2 (ABD) of mass m_2. The link 2 rotates about the symmetry axis of link 1. The centroid of bar 1 is at C_1 and the centroid of link 2 is at A. The following length is given $OC_1 = C_1A = 2\,l_1$ and $BD = l_2$. A Newtonian frame of reference has the unit vectors $[\imath_0, \jmath_0, \mathbf{k}_0]$ as depicted in the Fig. 1. A rotating frame of reference attached to rod 1 has the unit vectors $[\imath_1, \jmath_1, \mathbf{k}_1]$ and a rotating frame of reference attached to link 2 has the unit vectors $[\imath_2, \jmath_2, \mathbf{k}_2]$.

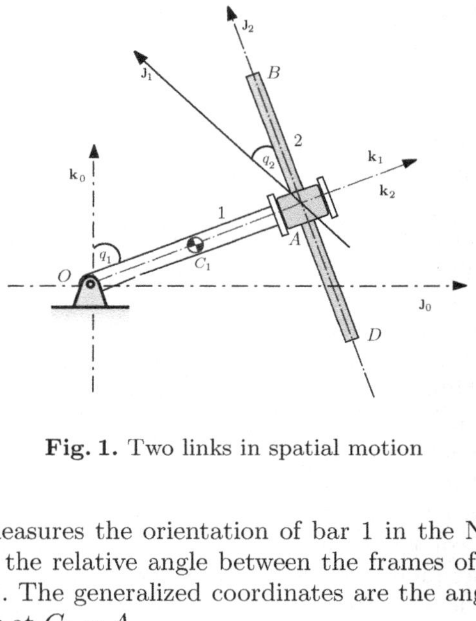

Fig. 1. Two links in spatial motion

The angle q_1 measures the orientation of bar 1 in the Newtonian frame of reference and q_2 is the relative angle between the frames of reference attached to link 2 and bar 1. The generalized coordinates are the angles q_1 and q_2. The centroid of link 2 is at $C_2 = A$.

The central inertia dyadic of rod 1 is

$$I_{C_1} = (I_{1x}\,\imath_1)\imath_1 + (\mathbf{I_{1y}}\,\jmath_1)\jmath_1 + (\mathbf{I_{1z}}\,\mathbf{k}_1)\mathbf{k}_1,$$

where $I_{1x} = I_{1y} = m_1\,(2\,l_1)^2/12$ and $I_{1z} = 0$. The central inertia dyadic of round link 2, with the radius R, is

$$I_{C_2} = \bar{I}_A = (I_{2x}\,\imath_2)\imath_2 + (I_{2y}\,\jmath_2)\jmath_1 + (I_{2z}\,\mathbf{k}_2)\mathbf{k}_2,$$

where $I_{2x} = I_{2z} = m_2\,(l_2^2 + 3\,R^2)/12$ and $I_{2y} = m_2\,R^2/2$. The central principle axes of links 1 and 2 are parallel to the frames of reference attached to links 1 and 2.

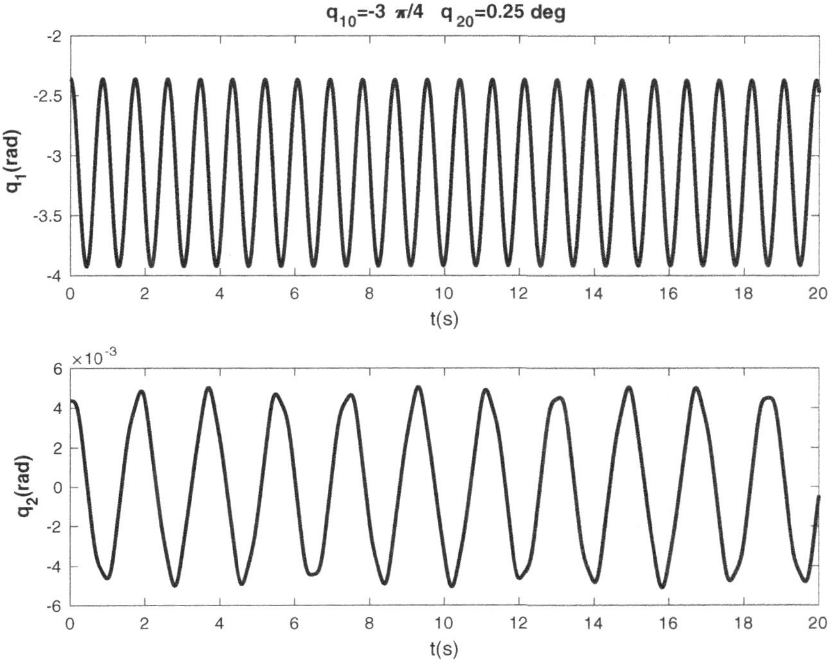

Fig. 2. Generalized coordinates for $q_1(0) = -3\,\pi/4$ and $q_2(0) = 0.25°$

The transformation from rod 1 to the Newtonian frame of reference is given by

$$\begin{bmatrix} \imath_1 \\ \jmath_1 \\ \mathbf{k}_1 \end{bmatrix} = \begin{bmatrix} 1 & 0 & 0 \\ 0 & c_1 & s_1 \\ 0 & -s_1 & c_1 \end{bmatrix} \begin{bmatrix} \imath_0 \\ \jmath_0 \\ \mathbf{k}_0 \end{bmatrix}, \tag{1}$$

and the transformation from link 2 to rod 1 is expressed as

$$\begin{bmatrix} \imath_2 \\ \jmath_2 \\ \mathbf{k}_2 \end{bmatrix} = \begin{bmatrix} c_2 & s_2 & 0 \\ -s_2 & c_2 & 0 \\ 0 & 0 & 1 \end{bmatrix} \begin{bmatrix} \imath_1 \\ \jmath_1 \\ \mathbf{k}_1 \end{bmatrix}, \tag{2}$$

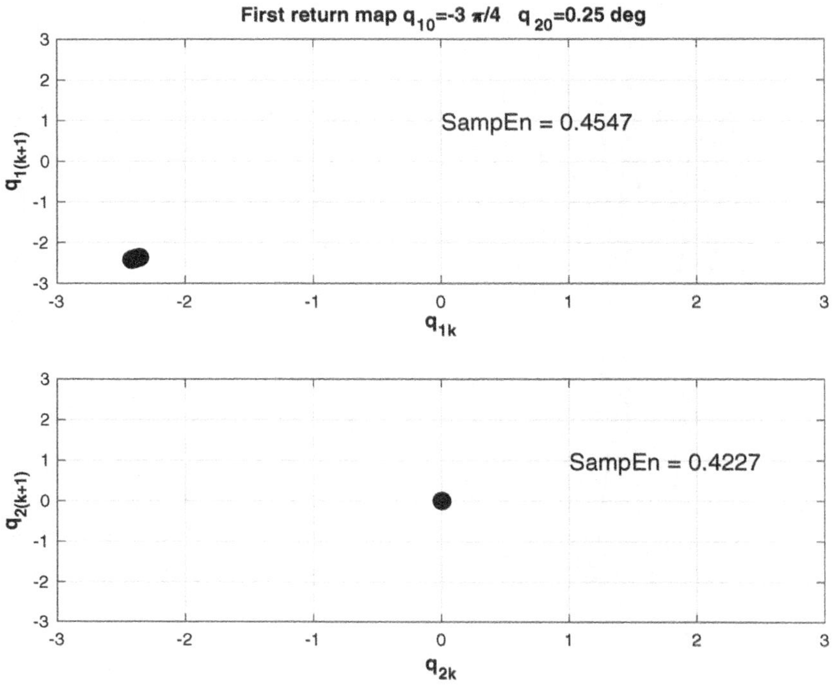

Fig. 3. First return map for $q_1(0) = -3\,\pi/4$ and $q_2(0) = 0.25°$

where $s_1 = \sin q_1, c_1 = \cos q_1, s_2 = \sin q_2$, and $c_2 = \cos q_2$. The generalized velocities are $u_i = \dot{q}_i$, $i = 1, 2$. The angular velocity of rod 1 in the Newtonian frame of reference is

$$\boldsymbol{\omega}_{10} = \dot{q}_1\, \mathbf{1}_0 = \dot{q}_1\, \mathbf{1}_1 = u_1\, \mathbf{1}_1. \tag{3}$$

The relative angular velocity of link 2 with respect to rod 1 is

$$\boldsymbol{\omega}_{21} = \dot{q}_2\, \mathbf{k}_1 = \dot{q}_2\, \mathbf{k}_2 = u_2\, \mathbf{k}_2. \tag{4}$$

The link 2 has an absolute angular velocity in the fixed frame of reference given by

$$\boldsymbol{\omega}_{20} = \boldsymbol{\omega}_{10} + \boldsymbol{\omega}_{21} = \dot{q}_1\, \mathbf{1}_1 + \dot{q}_2\, \mathbf{k}_1. \tag{5}$$

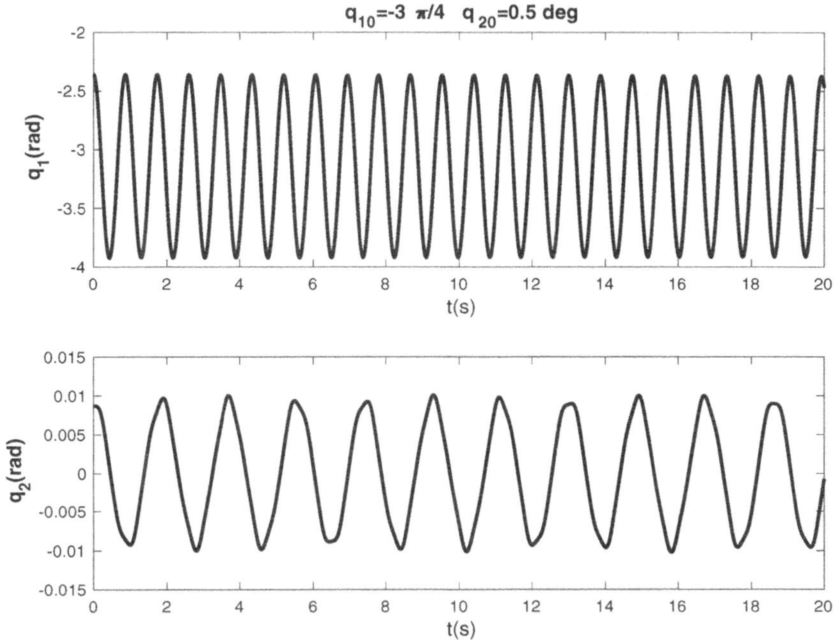

Fig. 4. Generalized coordinates for $q_1(0) = -3\,\pi/4$ and $q_2(0) = 0.5°$

The link 1 has an absolute angular acceleration in the fixed frame of reference expressed as

$$\boldsymbol{\alpha}_{10} = \ddot{q}_1\, \imath_1 = \dot{u}_1\, \imath_1. \tag{6}$$

The absolute angular acceleration of link 2 in the fixed frame of reference is

$$\boldsymbol{\alpha}_{20} = \frac{d}{dt}\boldsymbol{\omega}_{20} = \frac{^{(2)}d}{dt}33_{20} + 33_{20} \times 33_{20} = \frac{^{(2)}d}{dt}33_{20}, \tag{7}$$

where $\frac{^{(2)}d}{dt}$ is the partial derivative with respect to time in the frame of reference attached to link 2.

The centroid of rod 1 has the position vector

$$\mathbf{r}_{C_1} = l_1\, \mathbf{k}_1, \tag{8}$$

and the centroid of link 2 ($C_2 = A$) has the position vector

$$\mathbf{r}_{C_2} = \mathbf{r}_A = 2\, l_1\, \mathbf{k}_1. \tag{9}$$

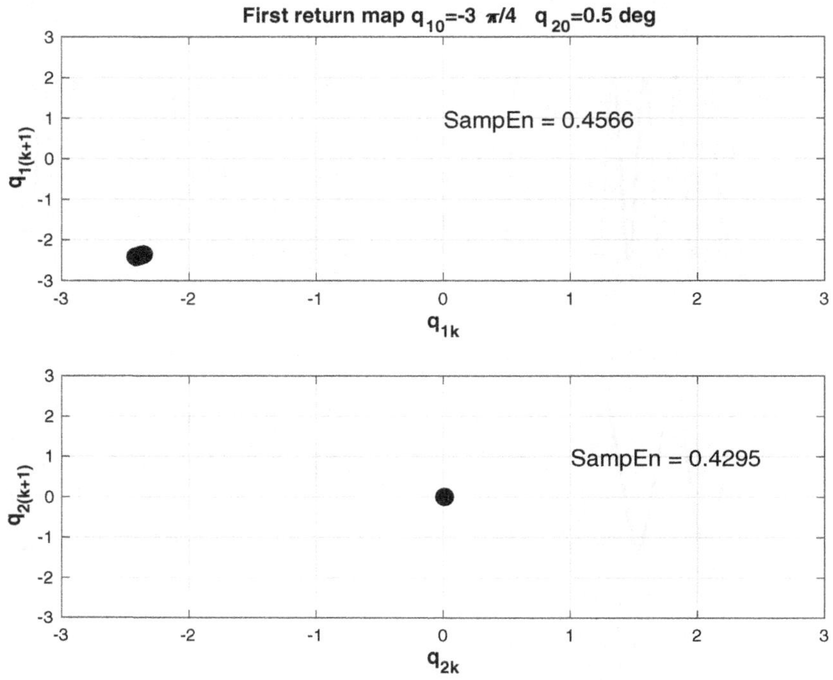

Fig. 5. First return map for $q_1(0) = -3\,\pi/4$ and $q_2(0) = 0.5°$

The velocities and the accelerations of the centroids in the Newtonian frame of reference are

$$
\begin{aligned}
\mathbf{v}_{C_1} &= \frac{d}{dt}\mathbf{r}_{C_1} = \frac{^{(1)}d}{dt}\mathbf{r}_{C_1} + 33_{10} \times \mathbf{r}_{C_1},\\
\mathbf{v}_{C_2} &= \frac{d}{dt}\mathbf{r}_{C_2} = \frac{^{(1)}d}{dt}\mathbf{r}_{C_2} + \boldsymbol{\omega}_{10} \times \mathbf{r}_{C_2},\\
\mathbf{a}_{C_1} &= \frac{d}{dt}\mathbf{v}_{C_1} = \frac{^{(1)}d}{dt}\mathbf{v}_{C_1} + 33_{10} \times \mathbf{v}_{C_1},\\
\mathbf{a}_{C_2} &= \frac{d}{dt}\mathbf{v}_{C_2} = \frac{^{(1)}d}{dt}\mathbf{v}_{C_2} + 33_{10} \times \mathbf{v}_{C_2}.
\end{aligned} \tag{10}
$$

2.2 Generalized Forces

The generalized forces, Q_r, are

$$Q_r = \frac{\partial \mathbf{v}_{C_1}}{\partial u_r} \cdot \mathbf{G}_1 + \frac{\partial \mathbf{v}_{C_2}}{\partial u_r} \cdot \mathbf{G}_2, \quad r = 1, 2, \tag{11}$$

where the gravitational forces are

$$\mathbf{G}_1 = -m_1\, g\, \mathbf{k}_0 \quad \text{and} \quad \mathbf{G}_2 = -m_2\, g\, \mathbf{k}_0. \tag{12}$$

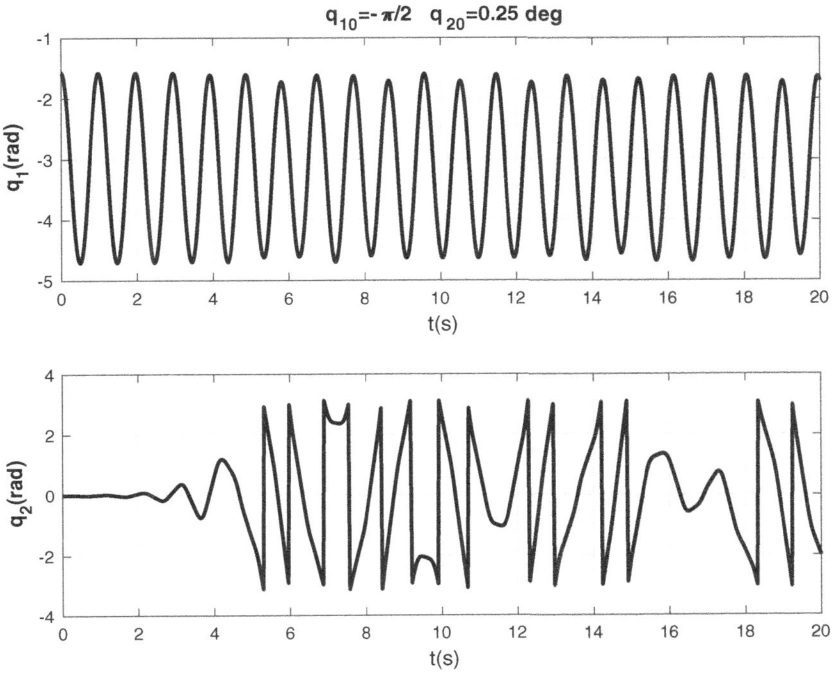

Fig. 6. Generalized coordinates for $q_1(0) = -\pi/2$ and $q_2(0) = 0.25°$

2.3 Gibbs–Appell Equations of Motion

The Gibbs function for the rod 1 is [3,16]

$$\begin{aligned} S_1 = & \frac{1}{2} m_1 \mathbf{a}_{C_1} \cdot \mathbf{a}_{C_1} + \frac{1}{2} \boldsymbol{\alpha}_{10} \cdot \bar{I}_{C_1} \cdot \boldsymbol{\alpha}_{10} + \boldsymbol{\alpha}_{10} \cdot (\boldsymbol{\omega}_{10} \times \bar{I}_{C_1} \cdot \boldsymbol{\omega}_{10}) \\ & + \frac{1}{2} \omega_{10}^2 \, \boldsymbol{\omega}_{10} \cdot \bar{I}_{C_1} \cdot \boldsymbol{\omega}_{10}, \end{aligned} \tag{13}$$

and the Gibbs function for the link 2 is

$$\begin{aligned} S_2 = & \frac{1}{2} m_2 \mathbf{a}_{C_2} \cdot \mathbf{a}_{C_2} + \frac{1}{2} \boldsymbol{\alpha}_{20} \cdot \bar{I}_{C_2} \cdot \boldsymbol{\alpha}_{20} + \boldsymbol{\alpha}_{20} \cdot (\boldsymbol{\omega}_{20} \times \bar{I}_{C_2} \cdot \boldsymbol{\omega}_{20}) \\ & + \frac{1}{2} \omega_{20}^2 \, \boldsymbol{\omega}_{20} \cdot \bar{I}_{C_2} \cdot \boldsymbol{\omega}_{20}. \end{aligned} \tag{14}$$

The equations of motion using Gibbs–Appell method are

$$\frac{\partial^2 (S_1 + S_2)}{\partial \ddot{q}_r} = Q_r, \quad r = 1, 2. \tag{15}$$

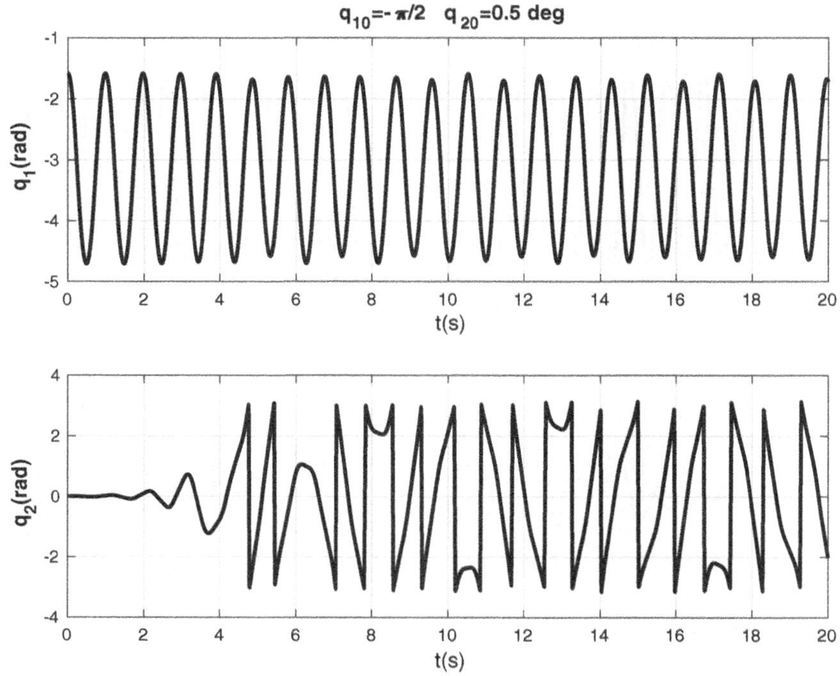

Fig. 7. Generalized coordinates for $q_1(0) = -\pi/2$ and $q_2(0) = 0.5°$

3 Sample Entropy

A time series $u_1, u_2, \ldots, u_N$ is given and vectors $U_m(k) = \{u(k+i), i = 0, \ldots, m-1\}$ are constructed, where $k = 1, \ldots, N-m+1$ [17–19]. Vectors of length m are constructed as $U_m(k) = \{u(k+i), i = 0, \ldots, m-1\}$, where $k = 1, \ldots, N-m+1$. The Euclidean norm can be the distances between vectors. The record A_i represents the number of vectors with the distance less than a tolerance r. The count of different vectors is defined as

$$A^m(r) = \frac{1}{N-m} \sum_{i=1}^{N-m} A_i^m(r). \tag{16}$$

The sample entropy (SampEn) is introduced as

$$\text{SampEn}(m, r, N) = -\ln \left[\frac{A^{m+1}(r)}{A^m(r)} \right]. \tag{17}$$

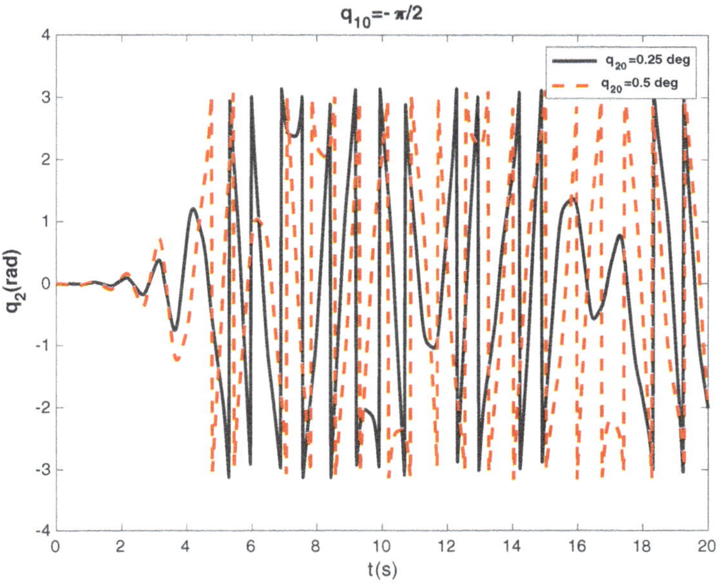

Fig. 8. Generalized coordinate, $q_2(t)$, for $q_1(0) = -\pi/2$, $q_2(0) = 0.25°$ and $q_2(0) = 0.5°$

4 Results

The input data for the numerical application of the mechanical system are: $m_1 = 0.01$ kg, $m_2 = 0.1$ kg, $l_1 = 0.075$ m, $l_2 = 0.2$ m, $R = 0.05$ m, $g = 9.81$ m/s^2. For the first set of Simulation, the following initial conditions, at $t = 0$, are used: $q_1(0) = -3\,\pi/4$, $\dot{q}_1(0) = \dot{q}_2(0) = 0$. Figure 2 shows the evolution in time of the angles $q_1(t)$ and $q_2(t)$ when the initial condition for $q_2(t)$ is $q_2(0) = 0.25°$. The system behaves in a regular way. The same configuration is studied with first return maps. The instant of maximum values for the angles q_{1max} and q_{2max} was selected for inserting the Poincare section. The Poincare map for the relative angular position q_i at the instant of maximum value can now be obtained by plotting the values of q_i at the kth cycle versus the values at $(k+1)$-th. The discrete map that represents the dynamics of the nonlinear is shown in Fig. 3. The Poincare section is approximately reduced to a single point. The sample entropy is SampEn = 0.4547 for $q_1(t)$ and SampEn = 0.4227 for $q_2(t)$.

Similar results are obtained if the initial condition for $q_2(t)$ is changed from 0.25° to $q_2(0) = 0.5°$. Figure 4 depicts the angles $q_1(t)$ and $q_2(t)$ versus time, and Fig. 5 shows the Poincare section and the sample entropies, SampEn = 0.4566 for $q_1(t)$ and SampEn = 0.4295 for $q_2(t)$. When the initial angle is $q_1(0) = -3\,\pi/4$, the system has a regular function for $q_2(0) = 0.25°$ and $q_2(0) = 0.5°$. The values for the sample entropy are close.

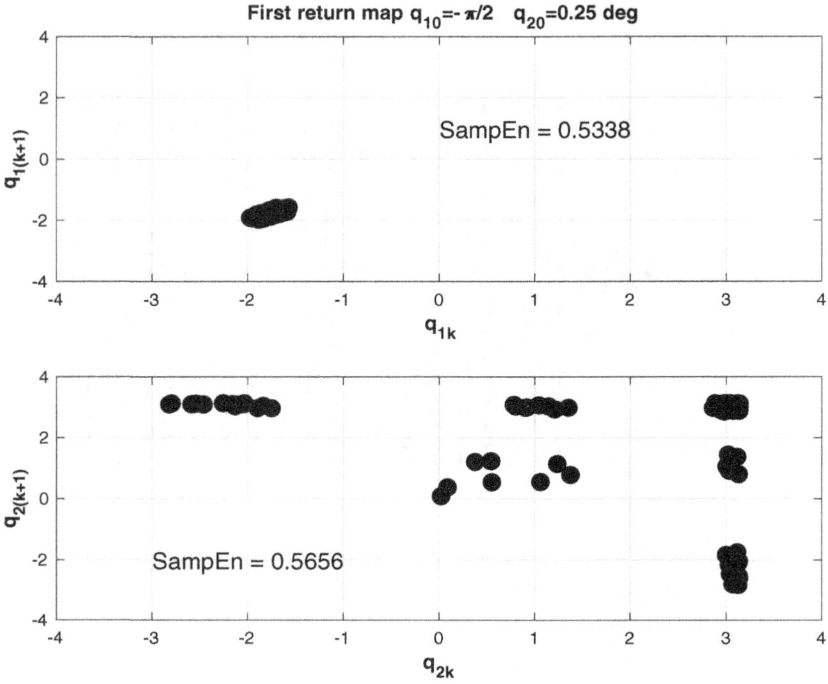

Fig. 9. First return map for $q_1(0) = -\pi/2$ and $q_2(0) = 0.25°$

The next set of simulation is for the initial conditions $q_1(0) = -\pi/2$, $\dot{q}_1(0) = \dot{q}_2(0) = 0$. Figure 6 shows the evolution in time of the angles $q_1(t)$ and $q_2(t)$ when the initial condition for $q_2(t)$ is $q_2(0) = 0.25°$, and Fig. 7 depicts the angles for $q_2(0) = 0.5°$. The angle $q_2(t)$ has boundary conditions from $-\pi$ to π. For this set of initial conditions, the motion is chaotic, a small variation in the initial value of the angle q_2 will result in different results as shown in Fig. 8. A small change of $0.25°$ in the initial conditions produces in divergent outcomes.

The discrete maps for the new initial conditions are shown in Figs. 9 and 10. The maps with a scatter of dots indicates that the time series was not periodic or quasi-periodic. The sample entropy for $q_2(t)$ is SampEn $= 0.5656$ for $q_2(0) = 0.25°$ and SampEn $= 0.5812$ for $q_2(0) = 0.5°$. If the attractor is a limit cycle, the signals have low entropy (initial condition $q_1(0) = -3\,\pi/4$). More complicated chaotic signals have higher entropy (initial condition $q_1(0) = -\pi/2$).

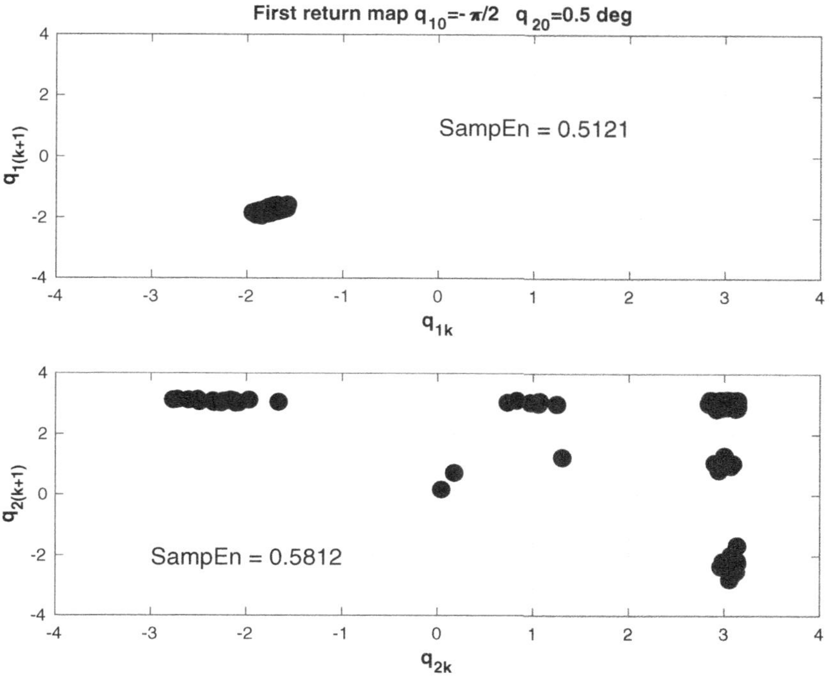

Fig. 10. First return map for $q_1(0) = -\pi/2$ and $q_2(0) = 0.5°$

5 Conclusions

Sample entropy can quantify the tendency of time series to repeat. As we would expect, the regular response of the system has lower Entropy, and the chaotic signal has higher entropy. Sample entropy can be used to characterize the dynamics of a nonlinear mechanical system. To validate our results, the experimental simulation will be needed.

References

1. Pars, L.A.: A Treatise on Analytical Dynamics. Ox Bow Press, Woodbridge, Connecticut (1965)
2. Whittaker, E.T.: Analytical Dynamics. Cambridge University Press, London (1937)
3. Kane, T.R., Levinson, D.A.: Dynamics: Theory and Applications. McGraw-Hill, New York (1985)
4. Kane, T.R.: Mechanical Demonstration of Mathematical Stability and Instability. International Journal of Engineering Education **2**(4), 45–47 (1974)
5. MotionGenesis^{TM} Kane Tutorial, www.MotionGenesis.com (2016)
6. Kane, T.R., Likins, P.W., Levinson, D.A.: Spacecraft Dynamics. McGraw-Hill, New York (1983)

7. Gillespie, R.B.: Kane's equations for haptic display of multibody systems. Haptics-e Electron. J. Haptics Res. **2**(3) (2003)
8. Tanner, H.G., Kyriakopoulos, K.J.: Modeling of a mobile manipulator with the use of Kane's dynamical equations. In: IFAC Intelligent Autonomous Vehicles, Madrid, Spain, pp. 611– 617 (1998)
9. Purushotham, A., Anjeneyulu, M.J.: Kane's method for robotic arm dynamics: a novel approach. J. Mech. Civ. Eng. **6**, 7–13 (2013)
10. Shannon, C.: A mathematical theory of communication. Bell Syst. Tech. J. **27**(3), 379–423 (1948)
11. Pincus, S.M.: Approximate entropy as a measure of system complexity. Proc. Natl. Acad. Sci. USA **88**, 2297–2301 (1991)
12. Richman, J.S., Moorman, J.R.: Physiological time-series analysis using approximate entropy and sample entropy. Am. J. Physiol. Heart Circ. Physiol. (2000)
13. Bandt, C., Pomp, B.: Permutation entropy: a natural complexity measure for time series. Phys. Rev. Lett. **88**, 174102 (2002)
14. Costa, M., Goldberger, A.L., and Peng, C.K.: Multiscale entropy analysis of complex physiologic time series. Phys. Rev. Lett. (2002)
15. Gan, C.C., Learmonth, G.: Comparing Entropy with Tests for Randomness as a Measure of Complexity in Time Series. 1512.00725, arXiv, stat.ME (2015)
16. Mata, V., Provenzano, S., Cuadrado, P., Valero, F.: An O(n) algorithm for solving the inverse dynamic problem in robots by using the gibbs-Appell formulation. Tenth World Congress IFToMM Oulu **3**, 1208–1215 (1999)
17. Alcaraz, R., Abasolo, D., Hornero, R., Rieta, J.J.: Optimal parameters study for sample entropy-based atrial fibrillation organization analysis. Comput. Methods Programs Biomed. **99**(1), 124–32 (2010)
18. Semmlow, J.L., Griffel, B.: Biosignal and Medical Image Processing. CRC Press, Taylor & Francis Group, Boca Raton (2014)
19. Semmlow, J.L.: Circuits, Signals, and Systems for Bioengineers. Academic Press, London (2018)
20. Lake, D.E., Richman, J.S., Griffin, P.M., Moorman, J.R.: Sample entropy analysis of neonatal heart rate variability. Am. J. Physiol. - Regul. Integr. Comput. Physiol. **283**(3), 789–797 (2002)

Self-driving Car: Challenges in Vibration

Livija Cveticanin[1,2(✉)], I. Ninkov[2], and D. Cveticanin[3]

[1] University of Novi Sad, Trg D. Obradovica 6, Novi Sad, Serbia
cveticanin@uns.ac.rs
[2] Obuda University, Nepszinhaz u. 7, Budapest, Hungary
[3] B. Nusica 15, Novi Sad, Serbia
dragan.cveticanin@remming.co.rs

Abstract. Recently, there is the intensive investigation in a self-driving car as a comfort vehicle for passenger voyages. The pleasant travel requires low level of vibration. In this paper, the vibration suppression with mechanical metastructure in the seat is suggested. The basic unit of the metastructure is the mass-in-stiffness system. The metastructure available the vibration energy from the basic structure to be transmitted to the added system giving the option for basic structure to stop to move. Using the energy harvesting system, the absorbed energy of motion of metastructure is transformed into electric one, which is appropriate for energy supply in sensors and other micro-electro-mechanical systems in a self-driving car. The special case of powering of a LiDAR by energy harvester is tested. The efficiency of the system is proved.

Keywords: Mechanical metamaterial · Energy harvester · Seat vibration suppression

1 Introduction

In 2016, the Society of Automotive Engineers SAE made the classification in autonomous cars (J3016 201,609) by introducing six levels [28]: Level 0—No automation, Level 1 ("hands on")—Driver assistance is necessary, Level 2 ("hands off")—Partial automation, Level 3 ("eyes off")—Conditional automation, Level 4 ("mind off")—High automation, and Level 5 ("steering wheel optional")—Full automation. For Level 5, no human intervention is required at all. The car is known as a self-driving car (SDC), and also as autonomous vehicle (AV or auto), or driverless car, or robot-car. SDC is a cyber-physical system (CPD) capable of sensing its environment and moving safely without human input [1]. The person in the car needs not to operate as the driver. The passenger may relax, rest, read, write, sleep, call, or work during travel. According to these requirements, the comfort of the passengers in car has to be improved. The feeling of comfort is connected with level of vibration [2]. However, the human response to vibration is quite different from person to person. It depends on human biology, anatomy and physiology, psychology and biomechanics response but also vibration properties (frequency, intensity, wave form, direction, etc.). The frequency in the interval of 1 to

N. Herisanu and V. Marinca (eds.), *Acoustics and Vibration of Mechanical Structures – AVMS-2021*, Springer Proceedings in Physics 274,
https://doi.org/10.1007/978-3-030-96787-1_2

50 Hz has the negative effect. The low-frequency vibration causes lethargy, nausea, loss of well-being and sometimes vomiting. As the side effect of driving motion is the sickness. Investigation in elimination of this feeling during sitting in the vehicle seat is done [3]. It is interesting to be mentioned that it is obtained that the special vibration helps to reduce the sickness and because of that algorithms for acceleration have to be developed to minimize the sickness. In [4], the laboratory study on the influence of road conditions on vehicle vibration and comfort of passengers is tested. Four types of roads are investigated: asphalt, concrete, gravel, and bump road. The whole body vibration is measured to give the appropriate estimation. It was found that for the same acceleration, especially at relative great one which is higher than 1.5 m/s^2, vibration containing more high-frequency components (higher than 20 Hz) cause greater discomfort than those containing less high-frequency components. It is concluded that for increasing of the comfort, the high components of vibration have to be eliminated. In addition, the type of vibration differs from the structure of the vehicle and transfer mechanism of the car [5].

According to this consideration, it is concluded that in SDC, the reduction of the vibration at certain frequencies and suppression of vibration is necessary. Usually, for vibration elimination in the car, the shock absorbers in the suspension system are mounted. However, in this paper, the concept of vibration suppression in the seat is considered. The mechanical metastructures, which are recently developed for vibration elimination, are suggested to be incorporated into the seat construction. Three types of mechanical metastructures are already investigated: those with mass-in-mass subsystem [6–8], mass-in-spring system [9], and auxetic hexagonal units [10–12]. The first one is shown to be convenient for vibration elimination at the certain frequency [13]. The second type of metastructures acts like vibration isolator (vibration is decreased in the wide range of frequencies) [14]. The third type is convenient for the elimination of vibration shock and impact. The principle of vibration suppression in metastructure is to stop the motion of the basic body by transmitting the energy of vibration onto the added mass-spring assembly. The main disadvantage is that the waste energy is transformed into heat. (It is found that the total energy loss in the car, including vibrations, is up to 78.5% [15]). To reduce the energy waste in the car, recently, the energy harvesting systems are developed [15]. The vibration harvesting system has the aim to transform the vibration energy into electric one. Harvesting systems are located in the suspension system.

Already, a significant number of papers dealing with energy harvesting devices in vehicle suspensions are published. Various concepts and design of harvesters are considered. Two groups of harvesters are developed: those with electromagnet and the other with piezoelement. One of the simplest concepts of harvesting is with magnet moving-coil mechanism [16]. Two types of harvesters are designed: the electromagnetic linear harvesters and electromagnetic rotary harvesters. In electromagnetic linear harvester, the kinetic energy of the vertical oscillations is directly transformed by electromagnetic induction into electric energy [17, 18]. The harvester, also called 'magnetic spring' device, contains a casing, coils, and three magnets (two of them are fixed rings on the top and on the bottom of the device, while the third magnet is moving up and down) [19]. According to the action of the excitation force, the magnet is levitated between two sections of coil on the harvester casing. The kinetic energy is converted into electric one, through the coil. In electromagnetic rotating harvester, the kinetic energy of the perpendicular vibration of the car suspension is harvested into electric energy using a rotary electromagnetic motor. Perpendicular vibration is transformed into a rotational motion through a transmission mechanism, which drives the rotary motion. The transmission mechanism is mechanical or hydraulic-based [20–22]. The mechanical mechanism has a rack-pinion [23] or a ball screw assembly [24, 25]. In [26], an inertial rotary energy harvester device is suggested. It contains a twist-driving system (which converts the vibration into rotating motion of a disc), a pawl-ratchet clutch system (which delivers and store the kinetic energy of vibration and keep rotating inertially during the interval between two impacts) and a magnet-coil system (which transforms the kinetic energy into electricity). In [27], the piezoelectric material is applied for energy harvesting in the vehicle, too. The excitation, due to road roughness, is considered. In the suspension system, the piezoelement is added. In paper [28], it is stated that the effect of harvesting depends not only on the effect of the road class but also on the frequency domain of vibration, vehicle velocity, and vehicle center of gravity location. It is concluded that the most harvesting effect is reached for vibration with natural resonant frequency. The amount of the harvested energy depends on the peak values of the electro power and energy density.

In spite of the fact that the harvested energy in the car is relatively small, it is obtained that this produced electric energy is enough to prolong the traveling range of the electric vehicle for 7 km [29] and to give energy supply for sensors.

In general, it can be concluded that the usual way for vibration absorption in the car is with shock absorbers in sustainable system, which are sometimes connected with energy harvesters. The aim of this paper is to improve the effect of vibration suppression and to increase the electric energy production.

As it is already mentioned, the SDC is a cyber-physical system (CPS) [30], which represents an assembly of physical and a highly sophistic artificial computing system, which is able to make perception, collect data, plan, realize, and control the driving process [31]. Devices that construct a CPS include sensors (for detection of physical values) and from simple hardware to high-end work computers for data managing and control to the most complex hardware for overall of the system. To percept and sense the environment, the vehicle must be supplied with a system of various sensors, which would detect the position of vehicle among other objects. Based on these information, the path of motion is planned and the navigation paths are identified. Sensors in SDC are: radars, lidars, ultrasonic sensors, video, thermal and far infra-red cameras, and inertial measurement units (IMUs). All of these sensors need small energy supply.

In this paper, the mechanical metastructure—harvesting assembly for vibration suppression and electric energy supply for sensors in SDC is investigated. The paper has four sections. After Introduction in Sect. 2, the mathematical model of the mechanical metastructure with mass-in-spring unit is developed. Vibration properties of the unit are analytically analyzed. In Sect. 3, the linear harvesting system is connected with the metastructure unit. The coupled electro-mechanical system is obtained. The electric power production is calculated. The paper ends with conclusion.

2 Mathematical Model of the Mechanical Metastructure with Mass-in-Spring Unit

In Fig. 1, the model of the mass-in-spring unit of the metastructure suggested to be connected with the car's seat is shown. The seat is assumed to be massless beam supported with stiffness $k_1/2$.

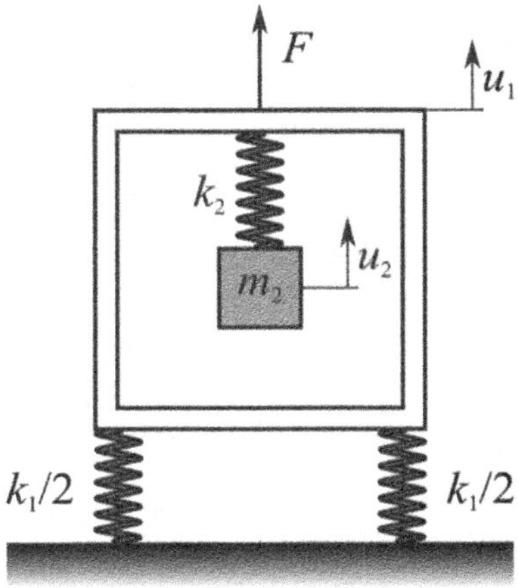

Fig. 1. Model of the mass-in-spring unit.

On the massless beam, a excitation force $F = F_0\exp(i\Omega t)$, with amplitude F_0 and frequency Ω, acts. It causes vibration. The unit system has two degrees of freedom and the motion of the mass is u_2 and of the massless beam u_1. Mathematical model of the mass-in-spring system is

$$\begin{aligned} m_2\ddot{u}_2 = k_2(u_2 - u_1) = 0 \\ (k_1 + k_2)u_2 - k_2u_2 = F \end{aligned} \tag{1}$$

The system of a second-order differential equation and a linear algebraic Eq. (1) has the closed-form solution

$$u_1 = Aexp(i\Omega t), u_2 = Bexp(i\Omega t), \tag{2}$$

where the amplitudes of vibration A and B are

$$A = \frac{F_0(k_2 - m_2\Omega^2)}{(k_1 + k_2)(k_2 - m_2\Omega^2) - k_2^2}, \quad B = \frac{F_0k_2}{(k_1 + k_2)(k_2 - m_2\Omega^2) - k_2^2} \tag{3}$$

The beam stops to vibrate if the amplitude A is zero. For parameter values of the added mass-spring system that satisfies the relation

$$\sqrt{\frac{k_2}{m_2}} = \omega_2 = \Omega \tag{4}$$

where ω_2 is the frequency of the added system, the vibration absorption for the beam occurs. The added mass m_2 vibrates with amplitude

$$B = -\frac{F_0}{k_2}. \tag{5}$$

The amplitude of vibration depends on the intensity of excitation and stiffness of the added spring: the amplitude of vibration is higher for smaller value of stiffness parameter of the added spring.

For mathematical reasons, let us introduce the equivalent stiffness in the form

$$\frac{1}{k} = \frac{1}{k_1} + \frac{1}{k_2} \tag{6}$$

and the frequency of vibration

$$\omega = \sqrt{\frac{k}{m_2}}. \tag{7}$$

Comparing ω and ω_2, it is

$$\omega^2 = \omega_2^2 \frac{1}{1 + \frac{k_2}{k_1}}. \tag{8}$$

The frequency ratio ω/ω_2 decreases with increase in stiffness ratio k_2/k_1 i.e. stiffness of the added spring k_2. Using (7) and after some modification of (3), the following amplitude–frequency relations are obtained

$$A = \frac{F_0}{k_1} \frac{1}{\left(1 - \frac{\Omega^2}{\omega^2}\right)} \left(1 - \frac{\Omega^2}{\omega^2} \frac{k}{k_2}\right), \quad B = \frac{F_0}{k_1} \frac{1}{\left(1 - \frac{\Omega^2}{\omega^2}\right)}. \tag{9}$$

Dividing the relations (9), the amplitude ratio follows as

$$\frac{A}{B} = 1 - \frac{\Omega^2}{\omega^2} \frac{k}{k_2}. \tag{10}$$

Analyzing (9) and (10), it is obvious that for the theoretical case when damping is omitted, both amplitudes A and B are indefinitely high for $\Omega/\omega = 1$. In addition, the amplitude of vibration A is zero for

$$\frac{\Omega^2}{\omega^2} = \frac{k_2}{k}. \tag{11}$$

The frequency relation (11) depends on the stiffness relation k_2/k. In Fig. 2, the amplitude–frequency diagrams (9) for various values of k/k_2 are plotted.

Analyzing diagrams in Fig. 2 and Eq. $(9)_2$ it is seen that the amplitude of vibration B is independent on k/k_2. However, it is not the case for the amplitude A. Namely, for small values of frequency relation Ω/ω, the amplitude is almost constant and has the

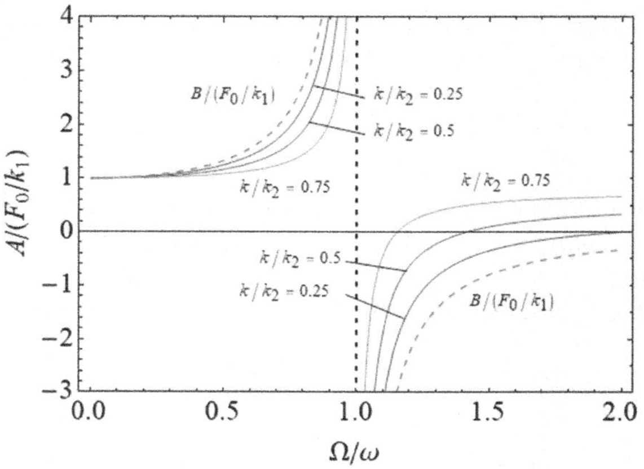

Fig. 2. Amplitude $A/(F_0/k_1)$—frequency (Ω/ω) diagrams for various values of k/k_2 (full line) and amplitude $B/(F_0/k_1)$—frequency (Ω/ω) diagram of the added mass-spring system (dotted line).

value $A = F_0/k_1$, i.e. it is independent on the stiffness of the added spring. The effect of added mass-spring system is more evident for the frequency ratio Ω/ω, which tends to 1 and is higher than 1. Specially, the amplitude A is zero for higher value of excitation if the stiffness ratio k/k_2 is smaller. According to (6), it is

$$\frac{k}{k2} = \frac{1}{1 + \frac{k_2}{k_1}}. \tag{12}$$

Using (11) and (12), the frequency value for the zero amplitude of vibration is higher for higher values of k_2/k_1.

For $\Omega/\omega >> \sqrt{k_2/k}$, the amplitude of vibration tends to $A = F_0k/(k_1k_2)$. After substituting (9), the amplitude value is $A = F_0/(k_1 + k_2)$. Analyzing the suggested relation, it is seen that for the body without added mass-spring system, the amplitude of vibration is almost equal to the initial value F_0/k_1. For the added spring with stiffness k_2, the amplitude tends to the value of $F_0/(k_1 + k_2)$. The higher is the stiffness of the added spring the amplitude of vibration A is smaller. Thus, if the stiffness of the added spring is extremely high and tends to infinity, the amplitude of vibration tends to zero in the wide frequency range.

2.1 Effective Stiffness

Let us introduce an equivalent stiffness k_{eff} which for the excitation force F gives the same motion u1 as it is already discussed mass-in-spring unit, i.e.

$$k_{eff}u_1 = k_1u_1 + k_2u_2. \tag{13}$$

Substituting (2) into (13), the effective stiffness follows as

$$k_{eff} = k_1 + k_2\frac{B}{A}. \tag{14}$$

For (10), we obtain

$$k_{eff} = k_1 + k_2 \frac{1}{1 - \frac{\Omega^2}{\omega^2} \frac{k}{k_2}}. \tag{15}$$

The frequency relation (8) with (15) yields

$$k_{eff} = k_1 + k_2 \frac{1}{1 - \frac{\Omega^2}{\omega_2^2}}. \tag{16}$$

In Fig. 3, effective stiffness–frequency diagrams for various values of stiffness ratio k_2/k_1 are plotted. Analyzing (16) and Fig. 3, we obtain that the effective stiffness coefficient k_{eff} is negative for $\omega_2 < \Omega < \omega_2(1 + \frac{k_2}{k_1})$. In this frequency gap, the amplitude of vibration of the beam is obtained to be almost zero and the vibration is absorbed. For $\Omega > \omega_2\left(1 + \frac{k_2}{k_1}\right)$, the effective stiffness is positive and tends to a constant value k_1 when the excitation frequency tends to infinity.

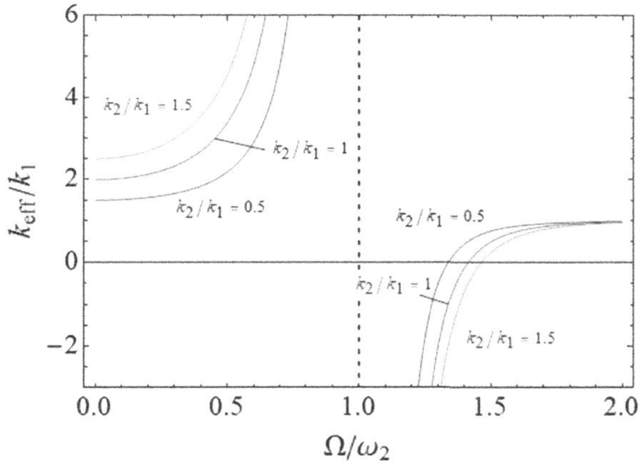

Fig. 3. Effective stiffness–frequency diagrams for various values of k_2/k_1.

It is concluded that the mass-in-stiffness unit acts as a vibration isolator in a wide range of frequencies, i.e. $\Omega \epsilon (\omega_2(1 = \frac{k_2}{k_1}), \Omega)$.

3 Model of the Harvesting System

In the basic unit of the metastructure, the electromagnetic harvester is incorporated (see Fig. 4).

The added mass of the metastructure, made of magnet, is oscillating inside a coil inducing the electric current [19]. The scheme of the electric circuit of the harvester contains the coil with inductance Le, resistance R, and external resistance Re. The generated electrical voltage U has to satisfy the balance equation [28]

$$\frac{L_e}{R_e}\dot{U} + \frac{R + R_e}{R} U = h(u_1 - u_2) \tag{17}$$

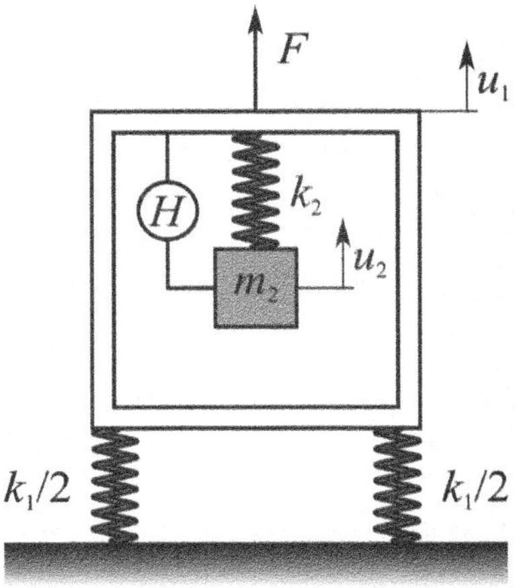

Fig. 4. Mass-in-spring unit with energy harvester H.

where h is the constant of harvester. The relation (17) directly depends on u_1—u_2. Substituting (2) into (17), a linear first-order non-homogenous differential equation is obtained

$$\frac{L_e}{R_e}\dot{U}+\frac{R+R_e}{R}U=h(A-B)\sin(\Omega t) \tag{18}$$

where A and B are the amplitudes of vibration (9) and

$$A-B=\frac{F_0}{k_1}\frac{(\frac{\Omega}{\omega_2})^2}{1-(\frac{\Omega}{\omega_2})^2\left(1+\frac{k_2}{k_1}\right)}. \tag{19}$$

In general, the solution of (18) is

$$U=Kexp(-t/\lambda)+Csin(\Omega t+\varphi) \tag{20}$$

where K is an arbitrary constant and

$$\lambda=\frac{L_e}{R_e}\frac{R}{R+R_e},\ C=\frac{h(A-B)}{\frac{R+R_e}{R}-\frac{L_e}{R_e}\Omega^2\lambda}\sqrt{1+\lambda^2\Omega^2},\ \varphi=tan^{(-1)}(-\lambda\Omega). \tag{21}$$

For the initial condition $U(0)=0$, the voltage–time relation is

$$U=C\left(sin(\Omega t+\varphi)-exp\left(-\frac{t}{\lambda}\right)\sin(\varphi)\right). \tag{22}$$

Analyzing (22), it is obtained that the steady—state voltage is a periodical time function, i.e.

$$U=Csin(\Omega t+\varphi), \tag{23}$$

with parameters (21).

3.1 Power Harvesting

The power of the electric energy harvested from the vibration energy is defined as U^2/R. Substituting (23), the harvested power is

$$P = \frac{C^2 sin^2(\Omega t + \varphi)}{R}. \tag{24}$$

After elimination of the transient motion and averaging over the period of vibration, it is obtained

$$P_{avl} = \frac{1}{T}\int_0^T P(t)dt = \frac{U^2}{2R} = \frac{h^2}{2R}\frac{(\frac{F_0}{k_1})^2}{(\frac{R+R_e}{R})^2 + (\frac{L_e}{R_e})^2\Omega^2}\left(\frac{\frac{\Omega^2}{\omega_2^2}}{1 - \frac{\Omega^2}{\omega_2^2}\left(1 + \frac{k_2}{k_1}\right)}\right)^2. \tag{25}$$

According to the relation (25), it is seen that the distribution of the power produced in the harvester depends also on the mechanical parameters of the system: stiffness ratio and frequency ratio. For certain $\frac{\Omega^2}{\omega_2^2}$, the averaged power is maximal if the stiffness ratio $\frac{k_2}{k_1}$ satisfies the following relation

$$\frac{k_2}{k_1} = \frac{1}{\frac{\Omega^2}{\omega_2^2}} - 1 \geq 0. \tag{26}$$

Nevertheless, at the frequency (11) the harvested energy in the system without damping is indefinitely high. In the neighborhood of this region, the power is smaller but seems to be enough for sensor supply.

3.2 Power Consumption of the LiDAR

In the self-driving car, a significant number of sensors are installed. However, for the safety and convenience, one of the most important is the light detection and ranging (LiDAR) sensor, which measures the distance between the car and an object and help the self-driving car to detect the surrounding environment [32]. LiDAR helps in accurate positioning and gives the exact distance to an object. The LiDAR need not high but continual power supply. Due to results of experiments [33] made for the scanning LiDAR, power consumption is varying dependently on the working mode. In the red waveform, the average power consumption of 6.971 W is obtained. The blue waveform shows the variation from 7.442 W to 7.154 W in the sleep mode of LiDAR, which occurs when the surrounding environment does not change for a certain time period, and the power consumption of sensor is minimized.

For the electromagnetic harvester considered in [28] with following parameters R = 10 Ω, Re = 10 Ω, Le = 9.4 10^{-4} H, the averaged harvested power (25) is

$$P_{avl} = 1.9369\left(\frac{F_0}{k_1}\right)^2\left(\frac{\frac{\Omega^2}{\omega_2^2}}{1 - \frac{\Omega^2}{\omega_2^2}\left(1 + \frac{k_2}{k_1}\right)}\right)^2 \tag{27}$$

where h = i kc/r = 12.448 m^{-1}, gear ratio is i = 22.6, generator constant is kc = 0.018, and equivalent rotational inertia radius is r = 3.268 10^{-2} m.

Comparing the maximal value of the power consumption of LiDAR and Eq. (27), it is obvious that the requirement for power supply is fulfilled if

$$\left| \frac{F_0}{k_1} \frac{\frac{\Omega^2}{\omega_2^2}}{1 - \frac{\Omega^2}{\omega_2^2}\left(1 + \frac{k_2}{k_1}\right)} \right| \geq 1.9602. \tag{28}$$

Using (26), it is seen that for certain frequency and stiffness ratio, the relation (28) is satisfied independently on $\frac{F_0}{k_1}$. However, Eq. (28) and the energy supply requirement are fulfilled even for other frequency and stiffness ratio, but the excitation amplitude $\frac{F_0}{k_1}$ has to be appropriate.

4 Conclusion

One of the main requirements for comfort in the self-driving car is elimination of vibration, which may disturb the work, read, write, sleep, call, rest, etc. of the persons in the vehicle. In this paper, it is concluded that the vibration of the body of passenger can be eliminated by suppressing of vibration in the seat and not only in the suspension mechanism. In the seat of the car, it is suggested the mechanical metastructure for vibration elimination to be built-in. The metastructure with mass-in-stiffness unit is developed. Mathematical description of the metastructure is developed. It is obtained that the added mass-spring system acts as vibration absorber for certain frequency range. However, for excitation with the frequency higher than the resonant frequency of metastructure, the vibration tends to decrease to a small amplitude of vibration. In this frequency area, the metastructure acts as a vibration isolator: the vibration exists, but the amplitude of vibration is quite small. Using the electromagnetic energy harvester, the vibration energy, transmitted to the added mass-spring system, is transformed into electric energy. The voltage level of the harvester depends not only on the harvester's parameters but also on the coefficients of the metastrucutre. The maximal value of electric power is obtained for the resonant case of metastructure when the excitation frequency of the external force is equal to natural frequency of the added mass-spring system in the metastructure unit. The obtained power is suitable for energy supply of sensors like LiDAR in the self-driving car.

Acknowledgements. Investigation is supported by the Faculty of Technical Sciences in Novi Sad (Project Number 054/2021).

References

1. Vdovin, D.S., Khrenov, I.O.: Systems of the self-driving vehicle. In: IOP Conferences Series: Materials Science and Engineering, vols. 534, 012016, pp. 6 (2019)
2. Mansfield, N.: Vibration and shock in vehicles: new challenges, new methods, new solutions. In: 1st International Comfort Congress, Proceedings, Salerno, 7–8 June, 2017, pp. 7 (2017)
3. Lucas, G., Kemeny, A., Paillot, D., Colombert, F.: A simulation sickness study on a driving simulator equipped with a vibration platform. Transp. Res. Part F **68**, 15–22 (2020)
4. Huang, Y.: D, Li, Subjective discomfort model of the micro commercial vehicle vibration over different road conditions. Appl. Acoust. **145**, 385–392 (2019)
5. Burdzik, R.: Identification of vibration transfer to car-body from road roughness by driving car. Vibroengineering PROCEDIA **1**, 27–30 (2013)
6. Cveticanin, L., Zukovic, M.: Negative effective mass in acoustic metamaterial with nonlinear mass-in-mass subsystems. Commun. Nonlinear Sci. Numer. Simul. **51**, 89–104 (2017)
7. Cveticanin, L., Zukovic, M., Cveticanin, D.: On the elastic metamaterial with negative effective mass. J. Sound Vib. **436**, 295–309 (2018)
8. Cveticanin, L., Zukovic, M., Cveticanin, D.: Influence of nonlinear subunits on the resonance frequency band gaps of acoustic metamaterial. Nonlinear Dyn. **93**(3), 1341–1354 (2018)
9. Pierce, C.D., Willey, C.L., Chen, W.V., Hardin, J.O., Berrigan, J.D., Juhl, A.T., Matlack, K.H.:. Adaptive elastic metastructures from magneto-active elastomers. Smart Mater. Struct. **29**, Article ID 065004, 1–11, 2020.
10. Mir, M., Ali, M.N., Sami, J., Ansaari. U.: Review of mechanics and applications of auxetic structures. Adv. Mater. Sci. Eng. Article ID 753496, 17 (2014)
11. Santulli, C., Langella, C.: Study and development of of concept of auxetic structures in bio-inspired design. Int. J. Sustain. Des. **3**(1), 20–37 (2016)
12. Fan, H., Yang, L., Tian, Y., Wang. Z.: Design of metastructures with quasi-zero dynamic stiffness for vibration isolation. Compos. Struct. **243**, Article ID 112244, 1–13 (2020)
13. Zhu, R., Liu, X.N., Hu, G.K., Sun, C.T., Huang, G.L.: A chiral elastic metamaterial beam for broadband vibration suppression. J. Sound Vib. **333**(10), 2759–2773 (2014)
14. Hobeck, J.D., Laurent, C.M.V., Inman, D.J:. 3D printing of metastructures for passive broadband vibration suppression. In: 20th International Conferences on Composite Materials, Copenhagen, 19–24 July, 1–8, 2015
15. Abdelkarem, M.A.A., Xu, L., Ali, M.K.A., Elagouz, A., Mi, J., Guo, S., Liu, Y., Zuo, L.: Vibration energy harvesting in automotive suspension system: a detailed review. Appl. Energy **229**, 572–699 (2018)
16. Fathabadi, H.: Recovering waste vibration energy of an automobile using shock absorbers included magnet moving –coil mechanism and adding to overall efficiency using wind turbine, Energy **189**, 116274, 9 (2019)
17. Zuo, I., Scully, B., Shestani, J., Zhou, Y.: Design and characterization of an electromagnetic energy harvester for vehicle suspension. Smart Mater. Struct. **19**, 045003 (2010)
18. Sapinski, B., Krupa, S.: Efficiency improvement in a vibration power generator for a linear MR damper: numerical study. Smart Mater. Struct. **22**, 045011 (2013)
19. Nguyen, H.T., Genov, D.A., Bardawell, H.: Vibration energy harvesting using magnetic spring based nonlinear oscillators: design strategy and insights. Appl. Energy **269**, 115102, 16 p. (2020)
20. Li C., Tse, P.W.: Fabrication and testing of an energy-harvesting hydraulic damper. Smart Mater. Struct. **22**, 065024 (2013)
21. Fang, Z., Guo, X., Xu, I., Zhang, H.: Experimental study of damping and energy regeneration characteristics of a hydraulic electromagnetic shock absorber. Adv. Mech. Eng. **3**, 943528 (2013)

22. Li, C., Zhu, R., Liang, M., Yang, S.: Integration of shock absorption and energy harvesting using a hydraulic rectifier. J. Sound Vib. **333**, 3904–3916 (2014)
23. Li, Z., Zuo, I., Kuang, J., Luhrs, G.: Energy harvesting shock absorber with a vertical motion rectifier. Smart Mater. Struct. **22**, 025008 (2013)
24. Wang, X.: Frequency Analysis of Vibration Energy Harvesting System. Elsevier Science & Technology Books (2016)
25. Xie, L., Li, J., Li, X., Huang, L., Cai, S.: Damping-tunable energy –harvesting vehicle damper with multiplie controlled generators: design, modeling and experiments. Mech. Syst. Signals Process. **99**, 859–872 (2018)
26. Luo, A., Zhang, Y., Dai, X., Wang, Y., Xu, W., Lu, Y., Wang, M., Fan, K., Wang, F.: An inertial rotary energy harvester for vibrations at ultra-law frequency with high energy conversion efficiency. Appl. Energy **279**, 115762, 15 (2020)
27. Morangueira, Y.L.A., Pereira, J.C.C.: Energy harvesting assessment with a coupled full car and piezoelectric model. Energy **210**, 118668, 13 (2020)
28. Zhang, R., Zhao, L., Qin, X., Zhang, H., Wang, X.: A comprehensive comparison of the vehicle vibration energy harvesting abilities of the regenerative shock absorbers predicted by the quarter, half and full vehicle suspension system models. Appl. Energy **272**, 115180, 17 (2020)
29. Aria, M.: A survey of self-driving urban vehicles development. IOP Conf. Series: Mater. Sci. Eng. **662**, 042006, 6 (2019)
30. Putnik, G.D., Ferreira, L., Lopes, N., Putnik, Z.: What is a cyber-physical system: definitions and models spectrum. FME Trans. **47**, 663–674 (2019)
31. Madden, J.: Security Analysis of a Cyber Physical System: A Car Example. Missouri University of Science and Technology, Master Theses (2013)
32. Lee, S., Lee, D., Choi, P., Park, D.: Accuracy – power controllable LiDAR sensors system with 3D object recognition for autonomous vehicle. Sensors **20**, 5706, 20 (2020)

The Numerical Solution of the Fractional Bagley–Torvik Equation by the Boubaker Wavelets

K. Rabiei and M. Razzaghi(✉)

Department of Mathematics and Statistics, Mississippi State University, Starkville MS, 39762, USA
razzaghi@math.msstate.edu

Abstract. In this paper, a new numerical scheme to solve the fractional Bagley–Torvik equation is presented. The technique is based upon wavelets approximation. The properties of Boubaker wavelets are given. The exact value of the Riemann–Liouville fractional integral operator (RLFIO) for Boubaker wavelets is introduced. Then this operator is utilized to transform the fractional Bagley–Torvik differential equation into a set of algebraic equations. Two examples are solved using the present method to demonstrate the exactness of the technique.

Keywords: Fractional Bagley–Torvik equation · Boubaker wavelets · Caputo derivative · Numerical solution

1 Introduction

The fractional differential equations (FDEs) are widely applied in different fields of science and engineering [1–6]. To obtain the analytic solution of this kind of equation is complex. Therefore, finding numerical solutions to FDEs becomes an interesting subject [7,8]. The authors of [9] introduced the Bagley–Torvik equation for the first time and mentioned its important role in motion modeling of a rigid plate immersed in a Newtonian fluid [10]. This equation has been considered both analytically and numerically in [11]. Some methods given to find the solution of this kind of equation are pseudo-spectral scheme [12], Taylor collocation method [13], Bessel collocation method [14], Haar wavelet [15], enhanced homotopy perturbation method [16], and hybrid functions approximation [17]. The references [18] and [19], are devoted to the existence and numerical solution of the Bagley–Torvik. In this paper, Boubaker wavelets are applied to solve the fractional Bagley–Torvik differential equation with initial conditions. We obtain the exact value for the RLFIO of the Boubaker wavelets using the hypergeometric functions. Then the error of approximation is discussed. Also, we solve two examples to show the validity and accuracy of our method.

N. Herisanu and V. Marinca (eds.), *Acoustics and Vibration of Mechanical Structures – AVMS-2021*, Springer Proceedings in Physics 274,
https://doi.org/10.1007/978-3-030-96787-1_3

2 Basic Notation

The organization of this article is as follows. In Sect. 2, we describe the basic definitions of fractional calculus. Section 3 is devoted to the definition of Boubaker wavelets. In Sect. 4, the RLFIO of these functions is obtained by using hypergeometric functions. In Sect. 5, the numerical technique is given to approximate the solution of Bagley–Torvik DE. In Sect. 6, we provide error analysis for the proposed scheme. In Sect. 7, two numerical examples are included.

In this part, some notations from fractional calculus and Boubaker wavelets are presented.

Definition 1. The RLFIO of order $\beta \geq 0$ of a function $f(t)$ is as follows [20] :

$$I^{\beta} f(t) = \begin{cases} \frac{1}{\Gamma(\beta)} \int\limits_0^t \frac{f(s)}{(t-s)^{1-\beta}} ds, & \beta > 0,\ t > 0, \\ f(t), & \beta = 0. \end{cases} \tag{1}$$

Definition 2. The Caputo derivative of order β of $f(t)$ is given as [20]:

$$D^{\beta} f(t) = \frac{1}{\Gamma(n-\beta)} \int\limits_0^t \frac{f^{(n)}(s)}{(t-s)^{\beta+1-n}} ds, \tag{2}$$

for $n-1 < \beta \leq n$, $n \in \mathbb{N}$, $t > 0$.

The operators I^{β} and D^{β} have the following properties:

1. $I^{\beta}(\lambda_1 f(t) + \lambda_2 g(t)) = \lambda_1 I^{\beta} f(t) + \lambda_2 I^{\beta} g(t)$, for constant λ_1 and λ_2.
2. $I^{\beta} t^{\alpha} = \frac{\Gamma(\alpha+1)}{\Gamma(\alpha+\beta+1)} t^{\alpha+\beta}, \alpha > -1$,
3. $I^{\beta} D^{\beta} f(t) = f(t) - \sum_{i=0}^{n-1} f^{(i)}(0) \frac{t^i}{i!}$.

Definition 3. The hypergeometric function ${}_2F_1(a, b, c; z)$ for $|z| < 1$ is given by [21]

$$ {}_2F_1(a, b, c; z) = \sum_{k=0}^{\infty} \frac{(a)_k (b)_k}{(c)_k} \cdot \frac{z^k}{k!} \tag{3}$$

where

$$(q)_k = \begin{cases} 1, & \text{if } k = 0, \\ q(q+1)\cdots(q+k-1), & \text{if } k > 0. \end{cases}$$

In Eq. (3), a, b, c, and z are real numbers. The hypergeometric function is represented in the integral form [21] as

$$ {}_2F_1(a, b, c; z) = \frac{\Gamma(c)}{\Gamma(b)\Gamma(c-b)} \int\limits_0^1 t^{b-1} (1-t)^{c-b-1} (1-zt)^{-a} dt.$$

Let c be replaced by $b+1$, so

$$ {}_2F_1\left(a,b,b+1;z\right) = b\int_0^1 t^{b-1}\left(1-zt\right)^{-a}dt. $$

Letting $t=\frac{s}{c}$, we have

$$ {}_2F_1\left(a,b,b+1;z\right) = \frac{b}{c^b}\int_0^c s^{b-1}\left(1-\frac{zs}{c}\right)^{-a}ds. $$

The relationship between the operator I^{β} and the function ${}_2F_1$ is obtained as follows. The unit step function $\mu_c(t)$ is defined by

$$ \mu_c(t) = \begin{cases} 1 & ,t\geq c, \\ 0 & ,t<c. \end{cases} $$

If $t<c$, then $I^{\beta}\left(t^{\alpha}\mu_c\left(t\right)\right)=0$. Assume that $t\geq c$. Then by using Eq. (1), we have

$$ \begin{aligned} I^{\beta}\left(t^{\alpha}\mu_c\left(t\right)\right) &= \frac{1}{\Gamma(\beta)}\int_0^t \frac{s^{\alpha}\mu_c\left(s\right)}{\left(t-s\right)^{1-\beta}}ds = \frac{1}{\Gamma(\beta)}\int_c^t \frac{s^{\alpha}\left(s\right)}{\left(t-s\right)^{1-\beta}}ds \\ &= \frac{1}{\Gamma(\beta)}\int_0^t \frac{s^{\alpha}\left(s\right)}{\left(t-s\right)^{1-\beta}}ds - \frac{1}{\Gamma(\beta)}\int_0^c \frac{s^{\alpha}\left(s\right)}{\left(t-s\right)^{1-\beta}}ds \\ &= \frac{\Gamma\left(\alpha+1\right)t^{\alpha+\beta}}{\Gamma\left(\alpha+\beta+1\right)} - \frac{t^{\beta-1}c^{\alpha+1}}{\Gamma\left(\beta\right)\left(\alpha+1\right)}{}_2F_1\left(1-\beta,\alpha+1,\alpha+2;\frac{c}{t}\right). \end{aligned} \tag{4} $$

3 The Boubaker Wavelets

The Boubaker wavelets $b_{n,m}(t)=b(k,n,m,t)$ are defined over $[0,1)$ in [22]

$$ b_{n,m}(t) = \begin{cases} \frac{2^{\frac{k+1}{2}}}{\sqrt{w_m}}B_m(2^{k+1}t-4n+2), & \frac{n-1}{2^{k-1}}\leq t<\frac{n}{2^{k-1}}, \\ 0, & otherwise, \end{cases} \tag{5} $$

where $m=0,1,2,\ldots,M-1$; $n=1,2,\ldots,2^{k-1}$. In this formula, M is the order of the Boubaker polynomials, k is a positive integer, and w_m is the following normality coefficient:

$$ w_m(t) = \begin{cases} 4, & m=0, \\ \sum\limits_{r=0}^{\lfloor m/2\rfloor}\frac{2^{2m-4r+2}\alpha_{m,r}^2}{2m-4r+1} + 2\sum\limits_{k=0}^{\lfloor m/2\rfloor-1}\sum\limits_{s=k+1}^{\lfloor m/2\rfloor}\frac{2^{2m-2(k+s)+2}\alpha_{m,k}\alpha_{m,s}}{2m-2(k+s)+1}, & m\geq 1, \end{cases} \tag{6} $$

where

$$\alpha_{i,r} = (-1)^r \binom{i-r}{r} \frac{i-4r}{i-r}.$$

The definition of Boubaker polynomials over $[-2, 2]$ is

$$B_0(t) = 1,$$

$$B_i(t) = \sum_{r=0}^{\lfloor i/2 \rfloor} \alpha_{i,r} t^{i-2r}, \quad i \geq 1.$$

4 RLFIO of the Boubaker Wavelets

We now give a formula for calculating the RLFIO of Boubaker wavelets. Let

$$\Psi(t) = [\psi_{1,0}(t), \ldots, \psi_{2^{k-1},0}(t), \ldots, \psi_{1,M-1}(t), \ldots, \psi_{2^{k-1},M}(t)]^T. \tag{7}$$

Denote

$$I^\beta \Psi(t) = \overline{\Psi}(t, \beta). \tag{8}$$

To obtain the elements of vector $\overline{\Psi}(t, \beta)$ we write

$$\psi_{n,m}(t) = \frac{2^{\frac{k+1}{2}}}{\sqrt{w_m}} \mu_{\left(\frac{n-1}{2^{k-1}}\right)}(t) B_m\left(2^{k+1}t - 4n + 2\right) - \frac{2^{\frac{k+1}{2}}}{\sqrt{w_m}} \mu_{\left(\frac{n}{2^{k-1}}\right)}(t) B_m\left(2^{k+1}t - 4n + 2\right). \tag{9}$$

Then applying I^β on both sides, we have

$$\begin{aligned} I^\beta \{\psi_{n,m}(t)\} = {} & I^\beta \left(\frac{2^{\frac{k+1}{2}}}{\sqrt{w_m}} \mu_{\left(\frac{n-1}{2^{k-1}}\right)}(t) B_m\left(2^{k+1}t - 4n + 2\right) \right) \\ & - I^\beta \left(\frac{2^{\frac{k+1}{2}}}{\sqrt{w_m}} \mu_{\left(\frac{n}{2^{k-1}}\right)}(t) B_m\left(2^{k+1}t - 4n + 2\right) \right). \end{aligned} \tag{10}$$

Now for $m \geq 1$, by considering the following relations given in [22]

$$B_m\left(2^{k+1}t - 4n + 2\right) = \sum_{r=0}^{\lfloor m/2 \rfloor} (-1)^r \binom{m-r}{r} \frac{m-4r}{m-r} \left(2^{k+1}t - 4n + 2\right)^{m-2r},$$

and

$$\left(2^{k+1}t - 4n + 2\right)^{m-2r} = \sum_{j=0}^{m-2r} \binom{m-2r}{j} (2-4n)^{m-2r-j} 2^{(k+1)j} t^j,$$

and substituting the above relations in Eq. (10), we have

$$I^{\beta}\left\{\psi_{n,m}(t)\right\} = \frac{2^{\frac{k+1}{2}}}{\sqrt{w_m}} \sum_{r=0}^{\lfloor m/2 \rfloor} \sum_{j=0}^{m-2r} (-1)^r \binom{m-r}{r} \frac{m-4r}{m-r} \binom{m-2r}{j} (2-4n)^{m-2r-j} 2^{(k+1)j}$$
$$\left(I^{\beta}(\mu_{\left(\frac{n-1}{2^{k-1}}\right)}(t)t^j) - I^{\beta}(\mu_{\left(\frac{n}{2^{k-1}}\right)}(t)t^j)\right).$$

Hence, by using (4), for $m \geq 1$, we obtain

$$I^{\beta}\psi_{n,m}(t) = \begin{cases} 0, & t \in \left[0, \left(\frac{n-1}{2^{k-1}}\right)\right) \\ F(t), & t \in \left[\left(\frac{n-1}{2^{k-1}}\right), \left(\frac{n}{2^{k-1}}\right)\right), \\ F(t) - G(t), & t \in \left[\left(\frac{n}{2^{k-1}}\right), 1\right), \end{cases}$$

where

$$F(t) = \frac{2^{\frac{k+1}{2}}}{\sqrt{w_m}} \sum_{r=0}^{\lfloor m/2 \rfloor} \sum_{j=0}^{m-2r} (-1)^r \binom{m-r}{r} \frac{m-4r}{m-r} \binom{m-2r}{j} (2-4n)^{m-2r-j} 2^{(k+1)j}$$
$$\left[t^{j+\beta} \frac{\Gamma(j+1)}{\Gamma(j+\beta+1)} - \frac{t^{\beta-1}}{\Gamma(\beta)(j+1)} \left(\frac{n-1}{2^{k-1}}\right)^{(j+1)} {}_2F_1\left(1-\beta, j+1, j+2; t\left(\frac{n-1}{2^{k-1}}\right)\right)\right],$$

and

$$G(t) = \frac{2^{\frac{k+1}{2}}}{\sqrt{w_m}} \sum_{r=0}^{\lfloor m/2 \rfloor} \sum_{j=0}^{m-2r} (-1)^r \binom{m-r}{r} \frac{m-4r}{m-r} \binom{m-2r}{j} (2-4n)^{m-2r-j} 2^{(k+1)j}$$
$$\left[t^{j+\beta} \frac{\Gamma(j+1)}{\Gamma(j+\beta+1)} - \frac{t^{\beta-1}}{\Gamma(\beta)(j+1)} \left(\frac{n}{2^{k-1}}\right)^{(j+1)} {}_2F_1\left(1-\beta, j+1, j+2; t\left(\frac{n}{2^{k-1}}\right)\right)\right].$$

Because the formula of Boubaker polynomials is valid for $m \geq 1$, we should make the same process for $m = 0$, separately. Since

$$\psi_{n,0}(t) = 2^{\frac{k-1}{2}} \mu_{\left(\frac{n-1}{2^{k-1}}\right)}(t) - 2^{\frac{k-1}{2}} \mu_{\left(\frac{n}{2^{k-1}}\right)}(t),$$

we have

$$I^{\beta}\left\{\psi_{n,0}(t)\right\} = 2^{\frac{k-1}{2}} \left(I^{\beta}(\mu_{\left(\frac{n-1}{2^{k-1}}\right)}(t)) - I^{\beta}(\mu_{\left(\frac{n}{2^{k-1}}\right)}(t))\right).$$

Therefore, for $m = 0$, we obtain

$$I^{\beta}\psi_{n,0}(t) = \begin{cases} 0, & t \in \left[0, \left(\frac{n-1}{2^{k-1}}\right)\right) \\ F_0(t), & t \in \left[\left(\frac{n-1}{2^{k-1}}\right), \left(\frac{n}{2^{k-1}}\right)\right), \\ F_0(t) - G_0(t), & t \in \left[\left(\frac{n}{2^{k-1}}\right), 1\right), \end{cases}$$

where

$$F_0(t) = 2^{\frac{k-1}{2}} \left[t^{\beta} \frac{1}{\Gamma(\beta+1)} - \frac{t^{\beta-1}}{\Gamma(\beta)} \left(\frac{n-1}{2^{k-1}} \right) {}_2F_1 \left(1-\beta, 1, 2; t \left(\frac{n-1}{2^{k-1}} \right) \right) \right],$$

and

$$G_0(t) = 2^{\frac{k-1}{2}} \left[t^{\beta} \frac{1}{\Gamma(\beta+1)} - \frac{t^{\beta-1}}{\Gamma(\beta)} \left(\frac{n}{2^{k-1}} \right) {}_2F_1 \left(1-\beta, 1, 2; t \left(\frac{n}{2^{k-1}} \right) \right) \right].$$

Hence by using the above formulas we get $I^{\beta}\Psi(t) = \overline{\Psi}(t, \beta)$, given in equation (8).

5 Problem Statement and Numerical Method

Here, the following fractional Bagley–Torvik equation is considered [17]:

$$AD^{(2)}f(t) + BD^{\frac{3}{2}}f(t) + Cf(t) = g(t). \tag{11}$$

The initial conditions for this equation are as follows:

$$f(0) = f_0, \; f'(0) = f'_0. \tag{12}$$

For solving this problem, by using the property 3 of definition 2 and (11), we have

$$I^2(AD^{(2)}f(t) + BD^{\frac{3}{2}}f(t) + Cf(t)) = I^2(g(t)),$$
$$A(f(t) - f'_0 t - f_0) + B(I^{\frac{1}{2}}(f(t) - f'_0 t - f_0)) + CI^2 f(t) = I^2(g(t)), \tag{13}$$

to solve Eq. (13), let

$$f(t) \simeq \sum_{m=0}^{M} \sum_{n=1}^{2^{k-1}} w_{nm} b_{nm}(t) = W^T \Psi(t), \tag{14}$$

where $\Psi(t)$ is introduced in Eq. (7) and W is the following unknown coefficients vector which should be calculated.

$$W = [w_{1,0}, \ldots, w_{2^{k-1},0}, \ldots, w_{1,M-1}, \ldots, w_{2^{k-1},M}]^T.$$

Then we approximate $I^2 f(t)$ and $I^{\frac{1}{2}} f(t)$ as

$$I^2 f(t) = W^T \overline{\Psi}(t, 2), \qquad I^{\frac{1}{2}} f(t) = W^T \overline{\Psi}(t, \frac{1}{2}). \tag{15}$$

Substituting Eqs. (14) and (15) in Eq. (13), we get

$$A(W^T\Psi(t) - f'_0 t - f_0) + B(W^T\overline{\Psi}(t, \frac{1}{2}) - I^{\frac{1}{2}}(f'_0 t + f_0)) + CW^T\overline{\Psi}(t, 2) = I^2(g(t)). \tag{16}$$

Next, Eq. (16) is collocated at the collocating nodes t_i, given by

$$t_i = \frac{2i-1}{2^k(M+1)}, \quad i = 1, \ldots, 2^{k-1}(M+1). \tag{17}$$

These equations give $2^{k-1}(M+1)$ algebraic equations, which can be solved for the unknown vector W^T.

6 Estimation Error

The following proposition can be considered as a generalization of Taylor's expansion.

Proposition 1. *Let f be a function in $C(0,h]$ such that $D^{k\beta}f \in C(0,h]$ for every $k = 0,\dots,M+1$. Then $\forall t \in (0,h]$, $\exists \xi \in (0,t]$ such that*

$$f(t) = \sum_{i=0}^{M} \frac{D^{i\beta}f(0^+)}{\Gamma(i\beta+1)} t^{i\beta} + \frac{D^{(M+1)\beta}f(\xi)}{\Gamma\left((M+1)\beta+1\right)} t^{(M+1)\beta}. \tag{18}$$

Proof. See [23]. □

Theorem 1. *Suppose that $f \in L^2[0,h]$ is a function such that $D^{k\beta}f \in C[0,h]$ for $k = 0,\dots,M+1$ and $C^T\Psi_{k,M}$ is the best approximation of $f(t)$ out of*

$$\mathcal{O}_{k,M} = span\{\psi_{n,m} \mid n = 1,\dots,2^{k-1},\ m = 0,\dots,M\}$$

with respect to the L^2-norm. Then

$$||f - C^T\Psi_{k,M}||_{L^2[0,h]} \leq \frac{h^{(M+1)\beta+\frac{1}{2}}}{\Gamma((M+1)\beta+1)\sqrt{2(M+1)\beta+1}} \sup_{\xi\in[0,h]} \left|D^{(M+1)\beta}(\xi)\right|.$$

Proof. Divide the interval $[0,h)$ to the 2^{k-1} smaller intervals $I_n = \left[\frac{n-1}{2^{k-1}}, \frac{n}{2^{k-1}}\right)$ for $n = 1,\dots,2^{k-1}$. Applying the generalized fractional-order Taylor's expansion in Eq. (18), we get

$$\left|f(t) - \tilde{f}(t)\right| \leq \frac{t^{(M+1)\beta}}{\Gamma((M+1)\beta+1)} \sup_{\xi\in[0,h]} \left|D^{(M+1)\beta}(\xi)\right|,$$

where $\tilde{f}(t) = \sum\limits_{i=0}^{M} \frac{D^{i\beta}(0^+)}{\Gamma(i\beta+1)} t^{i\beta}$. On each interval I_n, the wavelet $\Psi_{k,M}$ coincides with a polynomial at t of degree at most M. The function $C^T\Psi_{n,m}$ is the best approximation of $f(t)$ from $\mathcal{O}_{k,M}$ over the interval $[0,h)$, as is on each interval I_n. Thus, since $\tilde{f} \in \mathcal{O}_{k,M}$, we have

$$\begin{aligned} ||f - C^T\Psi_{k,M}||_{L^2 I_n} &\leq \left\|f - \tilde{f}\right\|_{L^2 I_n} \\ &\leq \frac{1}{\Gamma((M+1)\beta+1)} \sup_{\xi\in[0,h]} \left|D^{(M+1)\beta}(\xi)\right| \cdot ||t^{(M+1)\beta}||_{L^2 I_n}. \end{aligned}$$

Hence

$$\begin{aligned} ||f - C^T\Psi_{k,M}||^2_{[0,h]} &= \sum_{n=1}^{2^{k-1}} ||f - C^T\Psi_{k,M}||^2_{L^2 I_n} \\ &\leq \left(\frac{1}{\Gamma((M+1)\beta+1)} \sup_{\xi\in[0,h]} \left|D^{(M+1)\beta}(\xi)\right|\right)^2 \cdot ||t^{(M+1)\beta}||^2_{L^2[0,h]} \\ &= \left(\frac{h^{(M+1)\beta+\frac{1}{2}}}{\Gamma((M+1)\beta+1)\sqrt{2(M+1)\beta+1}} \sup_{\xi\in[0,h]} \left|D^{(M+1)\beta}(\xi)\right|\right)^2. \end{aligned}$$

The theorem is then proved.

7 Numerical Examples

In this part, we give two examples. In the first example, we obtain the exact solution and in the second one our numerical findings are better than the existing results.

7.1 Example 1

Consider Eqs. (11) and (12) with conditions [13]

$$f(0) = 1, \quad f'(0) = 1, \quad g(t) = 1 + t, \quad A = B = C = 1. \tag{19}$$

$f(t) = t + 1$ is the exact solution of this problem. We choose $M = 1,\ k = 1$, and set

$$f(t) = A^T B(t) = a_{10}b_{10}(t) + a_{11}b_{11}(t), \tag{20}$$

and write

$$I^{\frac{1}{2}} f(t) = a_{10}I^{\frac{1}{2}}b_{10}(t) + a_{11}I^{\frac{1}{2}}b_{11}(t), \tag{21}$$

$$I^2 f(t) = a_{10}I^2 b_{10}(t) + a_{11}I^2 b_{11}(t). \tag{22}$$

Substituting Eqs. (19)–(22) in Eq. (13), we have

$$a_{10}b_{10}(t)+a_{11}b_{11}(t)-t-1+a_{10}I^{\frac{1}{2}}b_{10}(t)+a_{11}I^{\frac{1}{2}}b_{11}(t)-I^{\frac{1}{2}}(t+1)+a_{10}I^2 b_{10}(t)+a_{11}I^2 b_{11}(t) = I^2 g(t), \tag{23}$$

collocating Eq. (23) and then solving the obtained set of equations, we find

$$a_{10} = \frac{3}{2} \quad a_{11} = \frac{1}{2\sqrt{3}}. \tag{24}$$

By replacing these coefficients in Eq. (20), the exact $f(t)$ is achieved.

7.2 Example 2

In this example, we consider [17]

$$g(t) = \begin{cases} 8,\ 0 \leq t \leq 1, \\ 0,\ t > 1. \end{cases} \quad f(0) = f_0 = 0, \qquad f'(0) = f_0' = 0. \tag{25}$$

The analytic solution given in [15] is

$$f(t) = 8(y_U(t) - y_U(t-1)), \qquad if \quad g(t) = 8(U(t) - U(t-1)),$$

where

$$y_U(t) = U(t)\left(\frac{1}{A}\sum_{r=0}^{\infty}\frac{(-1)^r}{r!}\left(\frac{C}{A}\right)^r t^{2(r+1)}E^{(r)}_{\frac{1}{2},\frac{3r}{2}+3}\left(\frac{-B}{A}t^{\frac{1}{2}}\right)\right), \tag{26}$$

in Eq. (26), $E_{\lambda,\mu}(z)$ is the following Mittag–Leffler function with two parameters $\lambda,\ \mu > 0$ and

$$E^{(r)}_{\lambda,\mu}(y) \equiv \frac{d^r}{dy^r} E_{\lambda,\mu}(y) = \sum_{j=0}^{\infty} \frac{(j+r)!y^j}{j!\Gamma(\lambda j + \lambda r + \mu)}, \quad (r = 0, 1, 2, ..),$$

and $U(t)$ is unit step function. For solving this example, using Eqs. (25) and (13), we write

$$Af(t) + BI^{\frac{1}{2}}f(t) + CI^2 f(t) = I^2 g(t). \tag{27}$$

We set

$$f(t) = Q^T \Psi(t), \tag{28}$$

and have

$$I^{\frac{1}{2}}f(t) = Q^T \overline{\Psi}(t, \frac{1}{2}), \qquad I^2 f(t) = Q^T \overline{\Psi}(t, 2). \tag{29}$$

Substituting Eqs. (28) and (29) in Eq. (27), we get

$$AQ^T B(t) + BQ^T \overline{\Psi}(t, \frac{1}{2}) + CQ^T \overline{\Psi}(t, 2) = I^2 g(t), \tag{30}$$

For finding Q in Eq. (28), we collocate Eq. (30) at the nodes t_i, given in Eq. (17). $A = 1$, $B = 0.5$, $C = 0.5$, are considered in this example. With $k = 2$ and $M = 6$, we compare our results with approximation for Riemann–Liouville fractional derivative [10], generalized Taylor series [13], and Bernoulli hybrid functions [17] in Table 1. It is noted that the best existing findings reported in [17] are obtained by using $N = 8, M_1 = 3$, (32 basis functions), while we get better results with $k = 2$ and $M = 6$, (32 basis functions). In this Table, M_1 shows the degree of Bernoulli polynomials.

Also, the graphs of the exact and approximate solutions of $f(t)$ are shown in Fig. 1a and the absolute errors are shown in Fig. 1b with $k = 2$ and $M = 6$.

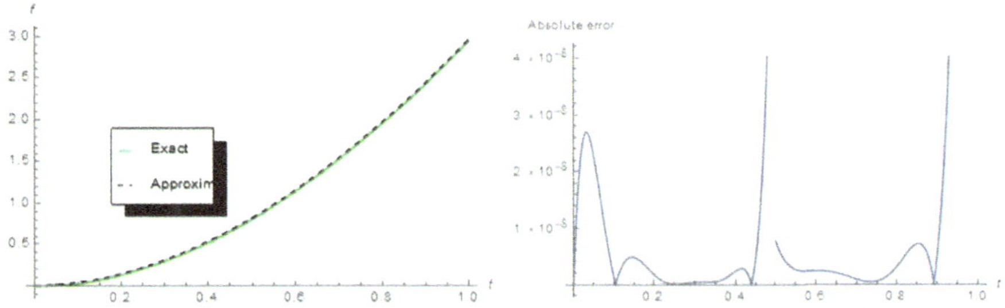

(a) Approximate and exact $f(t)$ (b) Absolute error of $f(t)$

Fig. 1. The graphs of the numerical, exact and absolute errors of $f(t)$ with $k = 2, M = 6$, Example 2.

Table 1. The values of $f(t)$ for different methods, Example 2.

t	Methods [10]	Methods [13]	Methods [17] $M_1 = 3,\quad N = 8$	Present method $M = 6,\quad k = 2$	Exact
0.1	–	0.036485547	0.0364875	0.0364875	0.036487479
0.2	–	0.140634716	0.1406398	0.1406397	0.140639621
0.3	–	0.307476229	0.3074848	0.3074846	0.307484627
0.4	–	0.533271294	0.5332842	0.5332841	0.533284109
0.5	–	0.814735609	0.8147568	0.8147568	0.814756949
0.6	–	1.148805808	1.1488372	1.1488373	1.148837422
0.7	–	1.532521264	1.5325655	1.5325654	1.532565426
0.8	–	1.962974991	1.9630293	1.9630293	1.963029254
0.9	–	2.437455982	2.4373338	2.4373339	2.437333970
1.0	2.95173	2.954070000	2.9525839	2.9525839	2.952583880

8 Conclusion

In this study, the exact RLFIO for Boubaker wavelets is obtained by using hypergeometric functions. Then, using this operator a novel method for the solution of fractional Bagley–Torvik equation is given. The numerical results are given to show the validity of our method.

References

1. Barkai, E., Metzler, R., Klafter, J.: From continuous time random walks to the fractional Fokker-Planck equation. Phys. Rev. E. **61**, 132–138 (2000)
2. Benson, D.A., Wheatcraft, S.W., Meerschaert, M.M.: Application of a fractional advection-dispersion equation. Water Resour. Res. **36**, 1403–1412 (2000)
3. Hilfer, R.: Applications of Fractional Calculus in Physics. World Scientific, Singapore (2000)
4. Mainardi, F.: Fractional calculus: some basic problems in continuum and statistical mechanics. In: Carpinteri, A., Mainardi, F. (Eds.) *Fractals and Fractional Calculus in Continuum Mechanics*. Springer, New York (1997)
5. Miller, K.S., Ross, B.: An Introduction to the Fractional Calculus and Fractional Differential Equations. Wiley, New York (1993)
6. Saichev, A.I., Zaslavsky, G.M.: Fractional kinetic equations: solutions and applications. Chaos **7**, 753–764 (1997)
7. Ding, H., Li, C.: Numerical algorithms for the fractional Diffusion-Wave equation with reaction term. Abstr. Appl. Anal. (2013); (Article ID 493406)
8. El-Sayed, A.A., Baleanu, D., Agarwal, P.: A novel Jacobi operational matrix for numerical solution of multi-term variable-order fractional differential equations. J. Taibah Univ. Sci. **14**(1), 963–974 (2020)
9. Torvik, P.J., Bagley, R.L.: On the appearance of the fractional derivative in the behavior of real materials. J. Appl. Mech. ASME Trans. **51**, 294–298 (1984)

10. Atanackovic, T.M., Zorica, D. (2013) On the Bagely-Torvik equation. J. Appl. Mech. **80**
11. Podlubny, I.: Fractional Differential Equations. Academic, San Diego, CA, USA (1999)
12. Esmaeili, S., Shamsi, M.: A pseudo-spectral scheme for the approximate solution of a family of fractional differential equations. Commun. Nonlinear Sci. Numer. Simulat. **16**, 3646–3654 (2011)
13. Cenesiz, Y., Keskin, Y., Kurnaz, A.: The solution of the Bagley-Torvik equation with the generalized Taylor collocation method. J. Frankl. Inst. **347**, 452–466 (2010)
14. YÃŒzbas, S.: Numerical solution of the Bagley-Torvik equation by the Bessel collocation method. Math. Methods Appl. Sci. **36**, 300–312 (2013)
15. Ray, S.S.: On Haar wavelet operational matrix of general order and its application for the numerical solution of fractional Bagley Torvik equation. Appl. Math. Comput. **218**, 5239–5248 (2012)
16. Zolfaghari, M., Ghaderi, R., SheikholEslami, A., Ranjbar, A., Hosseinnian, S.H., Momani, S., Sadati, J.: Application of the enhanced homotopy perturbation method to solve the fractional-order Bagley-Torvik differential equation. Phys. Scripta **136**, 7 (2009)
17. Mashayekhi, S., Razzaghi, M.: Numerical solution of the fractional Bagley-Torvik equation by using hybrid functions approximation. Math. Meth. Appl. Sci. **39**, 353–365 (2016)
18. Ray, S.S., Bera, R.K.: Analytical solution of the Bagley-Torvik equation by Adomian decomposition method. Appl. Math. Comput. **168**, 398–410 (2005)
19. Diethelm, K., Ford, J.: Numerical solution of the Bagley-Torvik equation. BIT Numer. Math. **42**, 490–507 (2002)
20. Oldham, K.B., Spanier, J.: The Fractional Calculus. Academic Press, New York (1974)
21. Andrews, G.E., Askey, R., Roy, R.: Special functions. In: Encyclopedia of Mathematics and its Applications, vol. 71. Cambridge University Press, Cambridge (1999). ISBN 978-0-521-62321-6. MR 1688958
22. Rabiei, K., Ordokhani, Y.: A new operational matrix based on Boubaker wavelet for solving optimal control problems of arbitrary order. Trans. Inst. Meas. Control **42**(10), 1858–1870 (2020)
23. Odibat, Z.M., Shawagfeh, N.T.: Generalized Taylors formula. Appl. Math. Comput. **186**, 286–293 (2007)

Impact of Piecewise Energy Harvester System of Bistable Beam with Hertzian Contact Force

Vasile Marinca and Nicolae Herisanu(✉)

University Politehnica Timisoara, Timisoara, Romania
nicolae.herisanu@upt.ro

Abstract. An investigation is developed of the nonlinear behavior of an impact-excited energy harvesting system of bistable beam by means of bilayer structure which is composed of a buckled piezoelectric beam and a rubber plate with spatial separation. The electromechanical response of this system under the standard low-frequency and nonharmonic excitation is studied using an approximate analytical approach. The periodic motion of the energy harvesting system with Hertzian contact force is illustrated. The results obtained by our procedure are useful to optimize the energy harvesting structure.

Keywords: Impact · Energy harvesting system · OAFM

1 Introduction

Vibration energy harvesting is one of the most challenging tasks for engineers nowadays and represents a major field of research, taking into account the actual need to exploit renewable energy, especially as an alternative for micro powering for small electronics energy harvesting from the renewable ambient energy interests in recent decades. The piezoelectric energy harvester can be fabricated more easily and with a simpler structure and in consequence, the piezoelectric energy harvesting system has received more attention compared to other transduction procedures.

Erturk et al. [1] investigated a piezoaeroelastic model with a focus on the generated electrical power and its effect on the aeroelastic response. The model and the experiments are given for a modified typical section undergoing self-sustained oscillations at the neutral stability condition. The effect of piezoelectric power generation on the linear flutter speed is discussed and a useful consequence of having nonlinearities in the system is addressed. Stanton et al. [2] explored linear dynamics for broadband energy harvesting, suggesting an intrinsic suitability for efficient performance in realistic vibration environments. They studied a variety of multiple attractors that may exist across a broad frequency range. The energy harvester is composed of a discontinuously laminated piezoelectric beam with a nonlinear boundary condition imposed by repelling permanent magnets. An experimental study of a Duffing oscillator exhibiting piezoelectric electromechanical coupling is presented by Sebald et al. [3]. The fast burst perturbation technique which consists of a fast voltage burst is applied to the piezoelectric element.

N. Herisanu and V. Marinca (eds.), *Acoustics and Vibration of Mechanical Structures – AVMS-2021*, Springer Proceedings in Physics 274,
https://doi.org/10.1007/978-3-030-96787-1_4

It is shown that the resonator may jump from the low solution to the high solution at a very small energy cost. Masana and Daqaq [4] explored an axially loaded clamped–clamped piezoelectric beam which functions as an energy harvester n mono-stable and bistable configurations. By numerical and experimental results, they demonstrated that the shape of the potential function plays an essential role in conjunction with the magnitude of the base excitation to determine whether the bistable harvester can outperform the mono-stable one and for what ranges of frequencies.

Litak et al. [5] analyzed an energy harvesting solution based on two magnetoelastic beams delivering power into an electrical circuit and driven by harmonic excitation. They showed that the systems with relative mistuning in the stiffness of the harvesting oscillators is vital to provide a broadband response but significantly complicates the resulting analysis. A perturbation approach to find an approximate analytical solution is proposed by Ilink et al. [6] to a vibrating model of an energy harvesting device, including the nonlinearities in the model of the piezoelectric coupling and nonideal excitation. The State Dependent Ricatti Equation linearizes the system around a point, changing the energy of the oscillator. Tang and Yang [7] considered a magnetic coupled piezoelectric energy harvester in which the magnetic interaction is introduced by a magnetic oscillator. Silva et al. [8] demonstrated that the inclusion of nonlinear terms in the energy harvester model can significantly influence the performance of vibrations of energy harvesting system, especially under resonant conditions. Kumar et al. [9] investigated a vertical isotropic nonlinear cantilever Bernoulli–Euler beam with a torsional spring at one end and a tip mass at the free end as an energy harvester using the piezoelectric patch. Shakki and Zand [10] examined the effects of dimensionless input frequency and dimensionless damping parameters for piezoelectric and piezomagnetoelectric configurations in MEMS energy harvesters. Warminski and Kecik [11] and Kecik and Perlikovski [12] studied auto parametric vibrations of a nonlinear system with pendulum and nonlinear dynamics and bifurcation of a vibration absorber-harvester with pendulum, respectively. Additionally, the smart elements can be used to eliminate or move the unstable regions and ,on the other hand, an active element can introduce a new bifurcation, branch point or period doubling. Xie et al. [13] explored the energy scavenging performance of a bistable buckled beam with a driving rubber plate. The electromechanical response of the nonlinear energy harvesting system under the standard low-frequency and nonharmonic gait excitation is studied by means of numerical simulations. Xie et al. [14] developed a mathematical model for a tapered cantilever beam energy harvester in width and thickness directions to harvest energy from ambient vibration. This harvester has a higher energy harvesting efficiency since a maximum collected power at each piezoelectric patch on the cantilever can be achieved.

In the present work, we will investigate the nonlinear behavior of impact-excited energy harvesting system by means of the Optimal Auxiliary Functions Method (OAFM). The periodic motion of the energy harvesting system with Hertzian contact force is studied. Our results are compared with numerical results obtained using a Runge–Kutta approach. Our technique introduces the so-called convergence-control parameters, whose optimal values allow a fast convergence of the solutions using only one iteration.

2 Description of Nonlinear Piecewise Energy Harvester

Figure 1 illustrates the system under study, which consists of a rubber plate (RP), a thimble (T), a steel beam (SB), two symmetric piezoelectric layers at both sides of a steel beam, and a threaded rod.

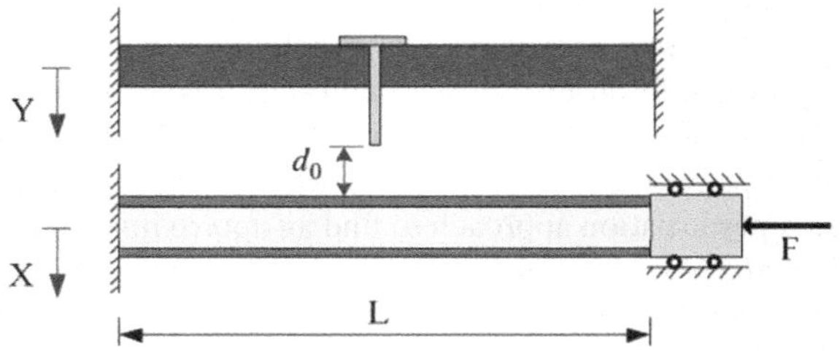

Fig. 1. Schematic representation of the impact of a rubber plate and steel beam

The steel beam is compressed by a static axial loading F, and a rubber plate is submitted to a transverse dynamic excitation. The gap between the buckled beam and the thimble is d_0. The buckled beam would vibrate after the impact excitation and a dynamic strain is generated in the piezoelectric patches, in return, a voltage V can be obtained across a connected resistance R. The nonlinear model is described by the following equation [13]:

$$\begin{aligned} &M_b\ddot{X} + C_b\dot{X} + \frac{dU}{dX} - G(X, \dot{X}, Y, \dot{Y}, d_0)H(X - Y - d_0) + \theta_0 V = 0 \\ &M_r\ddot{Y} + Cr\dot{Y} + K_r Y + G(X, \dot{X}, Y, \dot{Y}, d_0)H(X - Y - d_0) = F\sin\omega_0\bar{t} \\ &C_p\dot{V} + \frac{V}{R} - \theta_0\dot{X} = 0 \end{aligned} \tag{1}$$

where M_b, M_r are the equivalent mass, X, Y, V are the displacements and the voltage, C_b, C_r are the equivalent damping of the buckled beam and the rubber plate, U is the potential energy function, K_r is the equivalent stiffness of the rubber plate, C_p is the equivalent capacitance of the piezoelectric layer, θ_0 is the equivalent electromechanical coupling coefficient, Fsinωt is the base excitation, $G(X, \dot{X}, Y, \dot{Y}, d_0)$ is the Hertzian force, and $H(X–Y–d_0)$ is the Heaviside function defined as

$$G = \begin{cases} 0 & \textit{if } Y - X - d_0 \le 0 \\ \frac{3}{2}a_1K_c(Y - X - d_0)^{3/2}(\dot{Y} - \dot{X}) + K_c(Y - X - d_0)^{3/2} & \textit{if } Y - X - d_0 > 0 \end{cases} \tag{2}$$

$$H(X - Y - d_0) = \begin{cases} 0 & \textit{if } X - Y - d_0 \le 0 \\ 1 & \textit{if } X - Y - d_0 > 0 \end{cases} \tag{3}$$

The dot denotes the derivative with the respect to time t and

$$\frac{dU}{dX} = K_1 X + K_2 X^3 \tag{4}$$

Using the following transformations and notations

$$\begin{aligned}
&x = \frac{X}{L}\,;\ y = \tfrac{Y}{L}\,;\ d = \tfrac{d_0}{L}\,;\ \omega_n = \sqrt{\tfrac{K_r}{M_r}}\,;\ V = \tfrac{\theta_0 L}{C_p}v\,;\ t = \omega_n \bar{t}\,;\ \omega = \tfrac{\omega_0}{\omega_n}\,;\\
&\xi_r = \frac{c_r}{M_r\omega_n}\,;\ \xi_b = \tfrac{c_b}{M_b\omega_n}\,;\ v_b = \tfrac{K_1}{M_b\omega_n^2}\,;\ v_r = \tfrac{K_r}{M_r\omega_n^2}\,;\ \beta = \tfrac{K_2L^2}{M_b\omega_n^2}\\
&\xi_b = \frac{3a_1}{2M_b\omega_n}\,;\ K_{cb} = \tfrac{K_c\sqrt{L}}{M_b\omega_n^2}\,;\ \theta = \frac{\theta_0^2}{c_pM_b\omega_n^2}\,;\ \xi_{cr} = \tfrac{3a_1}{2M_r\omega_n}\\
&K_{cr} = \tfrac{K_c\sqrt{L}}{M_r\omega_n^2}\,;\ f = \tfrac{F}{M_r\omega_n^2L^2}\,;\ \lambda = \tfrac{1}{Rc_p\omega_n}
\end{aligned} \tag{5}$$

The dimensionless system of equations was derived as

$$\begin{aligned}
&\ddot{x} + \xi_b\dot{x} + v_b x + \beta x^3 - (\xi_{cb}k_{cb}(y-x-d)^{3/2}(\dot{y}-\dot{x}) +\\
&+k_{cb}(y-x-d)^{3/2})H(y-x-d) + \theta v = 0\\
&\ddot{y} + \xi_r\dot{y} + v_r y + (\xi_{cr}k_{cr}(y-x-d)^{3/2}(\dot{y}-\dot{x}) +\\
&+k_{cr}(y-x-d)^{3/2})H(y-x-d) = f\sin(\omega t)\\
&\dot{v} + \lambda v - \dot{x} = 0
\end{aligned} \tag{6}$$

The initial conditions for Eqs. (6) are

$$x(0) = A\,;\ \dot{x}(0) = 0\,;\ y(0) = B\,;\ \dot{y}(0) = 0\,;\ V(0) = 0 \tag{7}$$

In what follows, to solve Eqs. (6) and (7), we use the Optimal Auxiliary Functions Method (OAFM) [15–20].

3 Application of OAFM to Analyze the Impact of Piecewise Energy Harvester System

It is observed that the Heaviside function changed in the points t_i for which $y(t_i)$-$x(t_i)$-d = 0, $t_i < t_j$ if i < j, i, j = 1,2,... On the domain $D_1 = [0,t_1)$, the system (6) can be rewritten as

$$\begin{aligned}
&\ddot{x} + \xi_b\dot{x} + v_b x + \beta x^3 + \theta v = 0\\
&\ddot{y} + \xi_r\dot{y} + v_r y = f\sin\omega t\\
&\dot{v} + \lambda v - \dot{x} = 0
\end{aligned} \tag{8}$$

On the domain $D_2 = [t_1,t_2)$, the system, in this case, is a system (6) with H(y-x-d) = 1. For $\xi_b = \xi_r = \xi$, and $v_b = v_r = V$, the linear operators for Eq. (8) is $L(x) = \ddot{x} + \xi\dot{x} + V$. The approximate solution of the system (6) can be expressed as

$$\overline{x}(t) = x_0(t) + x_1(t)\,;\ \overline{v}(t) = v_0(t) + v_1(t) \tag{9}$$

The initial approximation $x_0$is obtained from the linear equation

$$L(x_0) = 0;\ x_0(0) = A, \dot{x}_0(0) = 0 \tag{10}$$

with the solution

$$x_0(t) = Ae^{-0.5\xi t}(\cos pt + \frac{0.5\xi}{p}\sin pt);\ p = \sqrt{v - 0.25\xi^2} \tag{11}$$

The nonlinear operator calculated for x_0, corresponding to Eq. (8) becomes

$$N(x_0) = \frac{\beta A^3 e^{-1.5t}}{4}(\alpha_1 \cos pt + \alpha_2 \sin pt + \alpha_3 \cos 3pt + \alpha_4 \sin 3pt) \tag{12}$$

where α_1, α_2, α_3, and α_4 are known. Following that, for the first approximation $x_1(t)$, we have the solution

$$x_1(t) = e^{-0.5\xi t}[c_1(\cos 3pt - \cos pt) + c_2(\sin pt - 3\sin pt)];\ x_1(0) = \dot{x}_1(0) = 0 \tag{13}$$

The approximate solution of Eq. (8) becomes

$$\begin{aligned} \overline{x}(t) &= Ae^{-0.5\xi t}(\cos pt + \frac{0.5\xi}{p}\sin pt) + \\ &+ e^{-0.5\xi t}[c_1(\cos 3pt - \cos pt) + c_2(\sin pt - 3\sin pt)];\ t \in D_1 \end{aligned} \tag{14}$$

The linear differential Eq. (9) has the solution (for B = 0 into Eq. 7)

$$\begin{aligned} \overline{y}(t) &= \frac{fe^{-0.5\xi t}}{(v-\omega^2)^2 + \xi^2\omega^2}(\xi\omega\cos pt + \frac{\omega^3 + 0.5\xi^2\omega - v\omega}{p}\sin pt) + \\ &+ \frac{f(v-\omega^2)\sin\omega t - f\xi\omega\cos\omega t}{(v-\omega^2)^2 + \xi^2\omega^2} \end{aligned} \tag{15}$$

If Ω is the frequency of the system (6) on the domain $D_2 = [t_1, t_2]$, and if one inserts the independent variable $\tau = \Omega t$ into (6), then this becomes

$$\begin{aligned} &\Omega^2(x'' + x) + \xi\Omega x' + (v - \Omega^2)x + \beta x^3 - \\ &- [\xi_{cb}k_{cb}(y - x - d)^{3/2}(\dot{y} - \dot{x}) + k_{cb}(y - x - d)^{3/2}) + \theta v = 0 \end{aligned} \tag{16}$$

$$\begin{aligned} &\Omega^2(y'' + y) + \xi\Omega y' + (v - \Omega^2)y + \xi_{cb}k_{cb}(y - x - d)^{3/2}(\dot{y} - \dot{x}) + \\ &+ k_{cr}(y - x - d)^{3/2}) = f\sin(\omega t) \end{aligned} \tag{17}$$

The first term of the Hertzian force can be developed in a power series

$$\begin{aligned} (y - x - d)^{3/2} &\approx \frac{x^3 - y^3}{54} + \frac{xy^2 - x^2y}{18} + \frac{x^2 + y^2}{6} + \frac{d+1}{18}(y - x)^2 + \\ &+ \frac{24 - (d+1)^2}{18}(y - x) - d + \frac{d+1}{6} + \ldots \end{aligned} \tag{18}$$

The linear operators L for Eqs. (16) and (17) are

$$L(x) = \Omega^2(x'' + x)\,,\; L(y) = \Omega^2(y'' + y) \tag{19}$$

The initial approximate solutions of Eqs. (16) and (17) are obtained from $L(x_0) = 0$, such that

$$x_0(t) = A\cos\tau\,,\; y_0(t) = B\cos\tau \tag{20}$$

Substituting Eq. (20) into the nonlinear operators corresponding to Eqs.(16) and (17), we can deduce that the first approximations $x_1(\tau)$ and $y_1(\tau)$ can be written as

$$x_1(\tau) = c_5(\cos\tau - \cos 3\tau) + c_6(\cos\tau - \cos 5\tau) + c_7(\cos\tau - \cos 7\tau) \tag{21}$$

$$y_1(\tau) = c_8(\cos\tau - \cos 3\tau) + c_9(\cos\tau - \cos 5\tau) + c_{10}(\cos\tau - \cos 7\tau) \tag{22}$$

We mention that the unknown convergence-control parameters C_1, C_2 from Eq. (14) have other values on the domain D_2, such that these are regarded as C_3 and C_4. From Eqs. (14), (15), (21), and (22), we obtain the approximate solution of the system (16) and (17)

$$\begin{aligned}\bar{x}(t) &= x(0)e^{-0.5\xi t}(\cos pt + \frac{0.5\xi}{p}\sin pt) + e^{-1.5\xi t}[C_3(\cos 3pt - \cos pt)+ \\ &+ C_4(\sin 3pt - 3\sin pt)] + c_5(\cos\tau - \cos 3\tau) + c_6(\cos\tau - \cos 5\tau) + c_7(\cos\tau - \cos 7\tau)\end{aligned} \tag{23}$$

$$\begin{aligned}\bar{y}(t) = & \frac{fe^{-0.5\xi t}}{(v-\omega^2)^2 + \xi^2\omega^2}(\xi\omega\cos pt + \frac{\omega^3 + 0.5\xi^2\omega - v\omega}{p}\sin pt)+ \\ & + \frac{f(v-\omega^2)\sin\omega t - f\xi\omega\cos\omega t}{(v-\omega^2)^2 + \xi^2\omega^2} + c_8(\cos\tau - \cos 3\tau)+ \\ & + c_9(\cos\tau - \cos 5\tau) + c_{10}(\cos\tau - \cos 7\tau)\end{aligned} \tag{24}$$

4 Numerical Example

For $\xi_b = \xi_r = 0.013$; $v_b = v_r = 1$; $\beta = 1/3$; $\xi_{cb} = \xi_{cr} = 0.015$; $k_{cb} = 0.014$; $\xi = 0.283$; $k_{cr} = 100$; $f = 0.2$; $\omega = 1$; $\lambda = 0.0078$; $d = 0.2$; $A = 0.3$; $B = 0$, using the collocation method on the first domain, we obtain

$$t_1 = 1.85\,;\; c_1 = 0.00228595\,;\; c_2 = -0.00380915 \tag{25}$$

On the domain D_2, the values of the unknown parameters are

$$\begin{aligned}& t_2 = 1.43;\; c_3 = 0.0967719; c_4 = -0.0409609; c_5 = 0.138517;\; c_6 = 0.00490425; \\ & c_7 = 0.00934685;\; c_8 = 0.0203771;\; c_9 = 0.106097;\; c_{10} = 0.0449993\end{aligned} \tag{26}$$

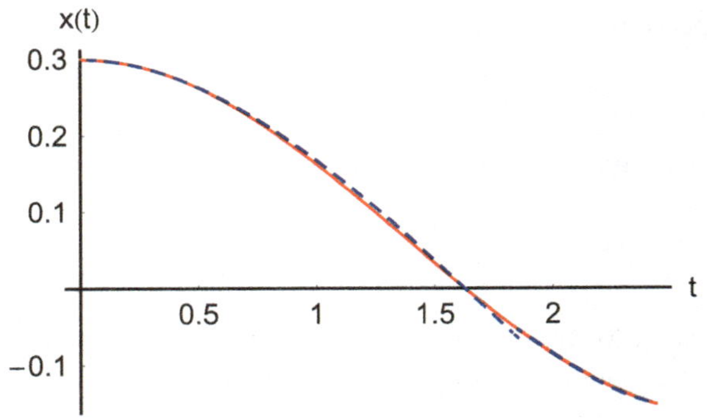

Fig. 2. Comparison between the numerical and analytical solution x: ——numerical – – – –analytical solution

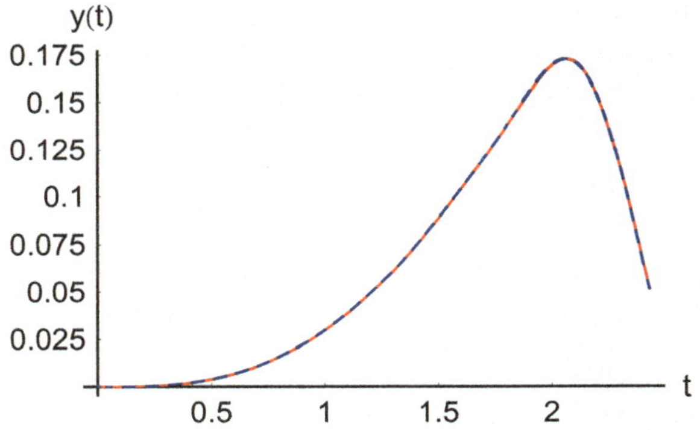

Fig. 3. Comparison between the numerical and analytical solution y: ——numerical – – – –analytical solution

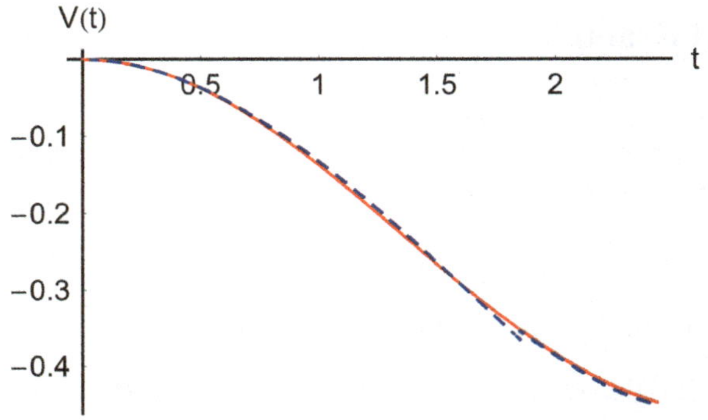

Fig. 4. Comparison between the numerical and analytical solution V: ——numerical – – – –analytical solution

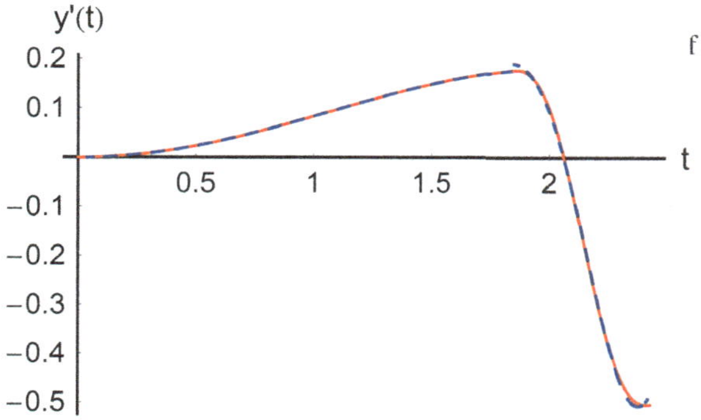

Fig. 5. Comparison between the numerical and analytical solution y': ——numerical ----analytical solution

It should be emphasized that on the other domains D_3, D_4,..., the approximate solutions (23) and (24) are valid but with corresponding values of the convergence-control parameters C_i.

The accuracy of the obtained solutions results by comparing the approximate solutions (23) and (24) with the parameters given in Eqs. (25) and (26) with the numerical integration results. Figure 2, 3, 4, and 5 present a comparison of the present solutions (25) and (26) with numerical results obtained using a Runge–Kutta approach.

From Fig. 5, it is clear that in the point $t_1 = 1.85$ appears a jump point caused by the impact.

5 Conclusions

This paper proposed an approach to analyze impact piecewise energy harvester system based on the buckled beam and a rubber plate. The electromechanical coupling equations were established by using the Hertzian contact collision model. The main advantage of the proposed approximate technique is given by the presence of the auxiliary functions combined with the convergence-control parameters which lead to high accuracy of the results.

References

1. Erturk, A., Vieira, W.G.R., De Marqui, C., Inman, D.J.: On the energy harvestingpotential of piezoaeroelastic systems. Appl. Phys. Lett. **96**, 184103 (2010)
2. Stanton, S.C., McGehee, C.C., Mann, B.P.: Nonlinear dynamics for broadband energy harvesting. Investigation of a bistable piezoelectric inertial generator. Phys. D **239**, 640–653 (2010)
3. Sebald, G., Kuwano, H., Guyomar, D., Ducharne, B.: Experimental Duffing oscillator for broadband piezoelectric energy harvesting. Smart Mater. Struct. **20**, 102001 (2011)
4. Masana, R., Duqaq, M.F.: Comparing the performance of a nonlinear energy harvester in mono- and bi-stable potential. In: Proceedings of ASME 2011 International Design Engineering Technical Conference and Computer Information in Engineering Conference IDETC/CIE, Washington (2011)
5. Litak, S.G., Friswell, M.I., Kwuimy, C.A.K., Adkikari, S., Borowiec, M.: Energy harvesting by two magmetopiezoelectric oscillators with mistuning. Theor. Appl. Mech. Lett. **2**, 043009 (2012)
6. Iliuk, I., Balthazar, J.M., Tusset, A.M., Felix, J.L.P., de Pontes Jr., B.R.: Nonlinear dynamics and control strategies. On a energy harvester vibrating system with a linear form to non-ideal motor torque. In: MATEC Web of Conferences, vol. 1, p. 08003 (2012)
7. Tang, L., Yang, Y.: A nonlinear piezoelectric energy harvester with magnetic oscillator. Appl. Phys. Lett. **101**, 094102 (2012)
8. Silva, L.L., Savi, M.A., Monteiro Jr., P.C.C., Netto, T.A.: On the nonlinear behaviour of vibration-based energy harvester. In: 22nd International Congress of Mechanical Engineering, Brasil (2013)
9. Kumar, P., Dwivedy, S.K., Kumari, P.: Nonlinear dynamic analysis of piezoelectric based energy harvester. In: Proceedings of ICTACEM 2014, International Conference on Theoretical, Applied, Computational and Experimental Mechanics (2014)
10. Shakki, S., Zand, M.M.: Studying piezoelectric and piezomagnetoelectric configurations for different excitation frequencies in MEMS energy harvesters. JCAMECH **47**(2), 241–246 (2016)
11. Warminski, J., Kecik, K.: Autoparametric vibrations of a nonlinear system with pendulum. Math. Probl. Eng. **2006**, 80705 (2006)
12. Kecil, K., Perlikowski, P.: Nonlinear dynamics and bifurcations of a vibration absorber-harvester. In: ECCOMAS Thematic Conferences on Multibody Dynamics, Barcelona (2015)
13. Xie, Z., Kwumy, Huang, W.B.: Energy harvesting from human motion by impact excited bistable buckled beams. In: 2016 Symposium on Piezoelectricity, Acoustic waves and Device Applications (SPAWDA), Xi'an, China (2016)
14. Xie, X.D., Carptinteri, A., Wang, Q.: A theoretical model for a piezoelectric energy harvester with a tapered shape. Eng. Struct. **144**, 19–25 (2017)

15. Herisanu, N., Marinca, V.: An effective analytical approach to nonlinear free vibration of elastically actuated microtubes. Meccanica **56**(4), 813–823 (2021)
16. Herisanu, N., Marinca, V.: An efficient analytical approach to investigate the dynamics of a misaligned multirotor system. Mathematics **8**, 1083 (2020)
17. Marinca, V., Herisanu, N.: Construction of analytic solution to axisymmetric flow and heat transfer on a moving cylinder. Symmetry **12**, 1335 (2020)
18. Marinca, V., Herisanu, N.: Optimal auxiliary functions method for a pendulum wrapping on two cylinders. Mathematics **8**, 1364 (2020)
19. Herisanu, N., Marinca, V., Madescu, G.: Application of the optimal auxiliary functions method to a permanent magnet synchronous generator. Int. J. Nonlinear Sci. Numer. Simul. **20**(3–4), 399–406 (2019)
20. Marinca, B., Marinca, V., Bogdan, C.: Dynamics of SEIR epidemic model by optimal auxiliary functions method. Chaos Solitons and Fractals **14**, 110999 (2021)

Higher-Order Accelerations Field with Multidual Algebra

Daniel Condurache(✉)

Technical University of Iasi, D. Mangeron Street no.59, 700050 Iasi, Romania
daniel.condurache@tuiasi.ro

Abstract. Using the property of the hypercomplex multidual numbers algebra, this paper proposes a novel computing method for higher-order accelerations vector field in the case of rigid body motion and multibody systems. The equations that allow the determination of higher-order accelerations are given for the spatial serial kinematic chains. The results are in a closed-form and coordinate-free. The properties for velocities, accelerations, jerks, and jounces fields are offered for few particular cases.

Keywords: Higher-order kinematics · Multidual algebra · Lie group

1 Introduction

Finding the pose, the fields of the velocities, accelerations, and higher-order accelerations (jerk, jounce, snap, crackle, pop, etc.) represent a fundamental problem in the kinematics of rigid bodies motion. Of multibody systems, various methods of finding the pose, the fields of the velocities, and accelerations are available. The isomorphism between screw theory and the Lie algebra of the Special Euclidean group of rigid body displacements provide results and techniques from differential geometry and the method of Lie group theory [1–3, 5, 7, 8, 12–15]. The new developments in high precision robotic systems, artificial vision systems, and molecular dynamics require the new procedures of space docking to calculate higher-order accelerations. In this paper, we aim to analyze and study these characteristics that determine the structure of motion of a system of macroscopic bodies. The key to the proposed procedure starts with the property of rigid body displacements group of forming a Lie group, accompanied by its Lie algebra. A previous result [14] proposes isomorphic representations between the Lie group SE (3) with the Lie group of the orthogonal dual tensors and the Lie algebra se (3) with the Lie algebra of dual vectors. This result opens the way of systematizing many empirical procedures used in spatial kinematics, by hypercomplex dual numbers algebra. In engineering literature, the sets of numbers called hypercomplex numbers are in fact categories rigorously studied in modern algebra, being in reality elements of some commutative or noncommutative linear algebras. The results obtained using dual algebras completely solve the problem of finding the field of n-th order accelerations, using a set of results obtained by the previous papers [4, 14]. The results can be extended for the

N. Herisanu and V. Marinca (eds.), *Acoustics and Vibration of Mechanical Structures – AVMS-2021*, Springer Proceedings in Physics 274,
https://doi.org/10.1007/978-3-030-96787-1_5

multidual commutative algebra [15]. The paper proposes a novel computing method for studying the higher-order accelerations fields for rigid body motion and serial kinematic chain, using the calculus with the dual and the multidual numbers algebra.

1.1 Higher-Order Kinematics of the Rigid Body. An Algebraic Approach

Let a rigid body that moves with respect to a fixed reference frame denote with $\{\mathbb{R}^0\}$. Consider the reference frame $\{\mathbb{R}\}$ originated at a point Q and attached to the rigid body. Let $\mathbf{V}_3$ be the set of free vectors from the three-dimensional Euclidean space and $S\mathbb{O}_3$ the Lie group of proper-orthogonal Euclidian tensors [1]. Also, let denote by $\boldsymbol{\rho}$ and $\mathbf{r}$ the position vectors of an arbitrary point P of the rigid body with respect to the reference frames $\{\mathbb{R}^0\}$ and $\{\mathbb{R}\}$. On may write:

$$\boldsymbol{\rho} = \boldsymbol{\rho}_Q + \mathbf{R}\mathbf{r} \tag{1}$$

where $\mathbf{r} \in \mathbf{V}_3$ is the relative position vector of point P with respect to the reference frame $\{\mathbb{R}\}$, $\boldsymbol{\rho}_Q \in \mathbf{V}_3$ is the position vector of point Q with respect to the reference frame $\{\mathbb{R}^0\}$ and $\mathbf{R} = S\mathbb{O}_3$ is the rotation tensor.

The parametric equations of the rigid body motion are the following:

$$\begin{cases} \boldsymbol{\rho}_Q = \boldsymbol{\rho}_Q(t) \\ \mathbf{R} = \mathbf{R}(t) \end{cases}, t \in \mathbf{I} \subseteq \mathbb{R}, \tag{2}$$

where $\boldsymbol{\rho}_Q = \boldsymbol{\rho}_Q(t) \in \mathbf{V}_3, \mathbf{R} = \mathbf{R}(t) \in S\mathbb{O}_3$, $t \in \mathbf{I} \subseteq \mathbb{R}$. Functions $\boldsymbol{\rho}_Q$ and $\mathbf{R}$ are indefinitely derivable.

The higher-order acceleration of order $n \in \mathbb{N}^*$, of that point of the rigid body with the position vector $\boldsymbol{\rho}$ is denoted by $\mathbf{a}_{\boldsymbol{\rho}}^{[n]}$, $n \in \mathbb{N}^*$

$$\mathbf{a}_{\boldsymbol{\rho}}^{[n]} \overset{\text{def}}{=} \frac{d^n \boldsymbol{\rho}}{dt^n} = \boldsymbol{\rho}^{(n)} = \mathbf{a}_Q^{[n]} + \boldsymbol{\phi}_n[\boldsymbol{\rho} - \boldsymbol{\rho}_Q], n \in \mathbb{N}^*. \tag{3}$$

In Eq. (3), $\mathbf{a}_Q^{[n]}$ is the higher-order acceleration of the fixed point Q from the body and $\boldsymbol{\phi}_n = \mathbf{R}^{(n)}\mathbf{R}^T$ represents the nth order accelerations tensor. The tensor denoted by $\tilde{\boldsymbol{\omega}} = \boldsymbol{\phi}_1 = \dot{\mathbf{R}}\mathbf{R}^T$ is the skew-symmetric tensor corresponding to the instantaneous angular velocity of a rigid body, denoted by $\boldsymbol{\omega}$ [1]. Let $\mathbf{v} = \dot{\boldsymbol{\rho}}_Q - \boldsymbol{\omega} \times \boldsymbol{\rho}_Q$ the velocity of a imaginary point of the rigid body which passes through the origin of the reference frame $\{\mathbb{R}^0\}$ at a given moment of time. We'll define the spatial twist of a rigid body by a pair of vectors noted as $(\boldsymbol{\omega}, \mathbf{v})$.

We will consider the following theorem:

Theorem 1 ([14]) *The higher-order accelerations field of the rigid body motion is given by*

$$\mathbf{a}_{\boldsymbol{\rho}}^{[n]} = \mathbf{a}_n + \boldsymbol{\phi}_n \boldsymbol{\rho}, n \in \mathbb{N}^* \tag{4}$$

where the vectors $\mathbf{a}_n$*, called the nth order vector invariant, and the* nth *order acce* *lerations tensors* $\boldsymbol{\phi}_n$ *satisfy the following relationships*:

$$\begin{cases} \mathbf{a}_{n+1} = \dot{\mathbf{a}}_n + \boldsymbol{\phi}_n v \\ \mathbf{a}_1 = v \end{cases}, n \in \mathbb{N}^* \tag{5}$$

$$\begin{cases} \boldsymbol{\phi}_{n+1} = \dot{\boldsymbol{\phi}}_n + \boldsymbol{\phi}_n \tilde{\boldsymbol{\omega}} \\ \boldsymbol{\phi}_1 = \tilde{\boldsymbol{\omega}} \end{cases}, n \in \mathbb{N}^*. \tag{6}$$

Let $\mathcal{A}$ be the noncommutative unitary ring of the Euclidean tensors, and let $\mathcal{A}[X]$ be the set of the polynomials with coefficients in the noncommutative ring $\mathcal{A}$. A generic element of $\mathcal{A}[X]$ has the form

$$\mathbf{P}(X) = \mathbf{A}_m X^m + \cdots + \mathbf{A}_1 X + \mathbf{A}_0, \mathbf{A}_k \in \mathcal{A}, k = \overline{0, m} \tag{7}$$

Also, we will consider the following theorem:

Theorem 2 ([14]) *There is a unique polynomial with its' coefficients in the noncommutative ring of Euclidean tensors such that the vector, respectively, the tensor invariants of the* n^{th} *order accelerations may be written as*

$$\begin{aligned} \mathbf{a}_n &= \mathbf{P}_n(\mathbf{D})\mathbf{v} \\ \boldsymbol{\Phi}_n &= \mathbf{P}_n(\mathbf{D}) \tilde{\boldsymbol{\omega}} \end{aligned}, n \in \mathbb{N}^* \tag{8}$$

where $\mathbf{P}_n$ *fulfills the relationship of recurrence, with* $\mathbf{D}= \frac{d}{dt}$ *the time derivative operator*

$$\begin{cases} \mathbf{P}_{n+1} = \mathbf{D}\mathbf{P}_n + \mathbf{P}_n\left(\tilde{\boldsymbol{\omega}}\right), n \in \mathbb{N}^* \\ \mathbf{P}_1 = I \end{cases} \tag{9}$$

It follows that

$$\begin{aligned} \mathbf{P}_1 &= \mathbf{I} \\ \mathbf{P}_2 &= \mathbf{D} + \tilde{\boldsymbol{\omega}} \\ \mathbf{P}_3 &= \mathbf{D}^2 + \tilde{\boldsymbol{\omega}}\,\mathbf{D} + 2\dot{\tilde{\boldsymbol{\omega}}} + \tilde{\boldsymbol{\omega}}^2 \\ \mathbf{P}_4 &= \mathbf{D}^3 + \tilde{\boldsymbol{\omega}}\,\mathbf{D}^2 + \left(3\dot{\tilde{\boldsymbol{\omega}}} + \tilde{\boldsymbol{\omega}}^2\right)\mathbf{D} + 3\ddot{\tilde{\boldsymbol{\omega}}} + 2\,\tilde{\boldsymbol{\omega}}\,\dot{\tilde{\boldsymbol{\omega}}} + 3\dot{\tilde{\boldsymbol{\omega}}}\,\tilde{\boldsymbol{\omega}} + \tilde{\boldsymbol{\omega}}^3 \end{aligned} \tag{10}$$

Unfortunately, the free term of the polynomial $\mathbf{P}_4$ presented in reference [14] is wrong. The correct result is that of Eq. (10).

2 Dual Lie Algebra and Higher-Order Kinematics

The parametric Eq. (2) describe the rigid body motion. The orthogonal dual tensor that parameterizes the rigid body motion is defined in [1, 4, 12]:

$$\underline{\mathbf{R}} = \left(\mathbf{I} + \varepsilon\,\tilde{\boldsymbol{\rho}}\right)\mathbf{R} \tag{11}$$

In the relation (11), the skew-symmetric tensor associated to the vector $\boldsymbol{\rho}$ is denoted by $\tilde{\boldsymbol{\rho}}$, and ε is a dual quantity, $\varepsilon^2 = 0$, $\varepsilon \neq 0$ [1, 4, 12]. We will denote with $\underline{\mathbf{V}}_3$ the set of dual vectors. The dual tensors $\underline{\mathbf{R}}$ belong to a Lie group of orthogonal dual denoted by $\underline{\mathbb{SO}}_3 = \{\underline{\mathbf{R}} \in \mathbf{L}(\underline{\mathbf{V}}_3, \underline{\mathbf{V}}_3) | \underline{\mathbf{R}}\underline{\mathbf{R}}^{\mathrm{T}} = \underline{\mathbf{I}}, \det\underline{\mathbf{R}} = 1\}$. The Lie group $\underline{\mathbb{SO}}_3$ is isomorphic to the Lie group of the rigid body displacement $S\mathbb{E}_3$ [4]. The Lie algebra of the Lie group $\underline{\mathbb{SO}}_3$ is isomorphic to the dual vector algebra $\underline{\mathbf{V}}_3$, having as internal operation the cross product for dual vectors. It can be demonstrated that there exists a unit dual vector $\underline{\mathbf{u}}$ and a dual angle $\underline{\alpha}$ such that: $\underline{\mathbf{R}} = \underline{\mathbf{I}} + \sin\underline{\alpha}\,\tilde{\underline{\mathbf{u}}} + (1 - \cos\underline{\alpha})\tilde{\underline{\mathbf{u}}}^2 = \exp(\underline{\alpha}\,\tilde{\underline{\mathbf{u}}})$.

The rigid body motion given by the Eq. (2) is a parameterized curve: $\underline{\mathbf{R}} = \underline{\mathbf{R}}(t)$ in the $\underline{\mathbb{SO}}_3$ Lie group. The dual angular velocity for the rigid body motion is given by [4]

$$\underline{\boldsymbol{\omega}} = \mathrm{vect}\dot{\underline{\mathbf{R}}}\underline{\mathbf{R}}^{\mathrm{T}} \tag{12}$$

It can be demonstrated by the following relationship:

$$\underline{\boldsymbol{\omega}} = \boldsymbol{\omega} + \varepsilon \mathbf{v} \tag{13}$$

where $(\boldsymbol{\omega}, \mathbf{v})$ is the spatial twist of rigid body motion. The dual angular velocity denoted by $\underline{\boldsymbol{\omega}}$ is called "dual twist".

By taking the $(n-1)$-th derivative with respect to time of the equation $\mathbf{v} = \dot{\boldsymbol{\rho}} - \boldsymbol{\omega} \times \boldsymbol{\rho}$, we will obtain

$$\mathbf{v}^{(n-1)} = \boldsymbol{\rho}^{(n)} - \sum_{k=0}^{n-1} C_{n-1}^{k} \tilde{\boldsymbol{\omega}}^{(n-1-k)} \boldsymbol{\rho}^{(k)} \tag{14}$$

For $\boldsymbol{\rho} = 0$, from Eq. (14), it follows:

$$\mathbf{v}^{(n-1)} = \mathbf{a}_n - \sum_{k=1}^{n-1} C_{n-1}^{k} \tilde{\boldsymbol{\omega}}^{(n-1-k)} \mathbf{a}_k \tag{15}$$

with $\mathbf{a}_k$, $k = \overline{1, n}$, the kth order vector invariant of the rigid body motion.

By taking the $(n-1)^{\mathrm{th}}$ order derivative of the equation $\dot{\mathbf{R}} = \tilde{\boldsymbol{\omega}}\,\mathbf{R}$, considering the fact that the nth acceleration tensor is $\boldsymbol{\phi}_n = \mathbf{R}^{(n)}\mathbf{R}^T$, it follows:

$$\tilde{\boldsymbol{\omega}}^{(n-1)} = \boldsymbol{\phi}_n - \sum_{k=1}^{n-1} C_{n-1}^{k} \tilde{\boldsymbol{\omega}}^{(n-1-k)} \boldsymbol{\phi}_k \tag{16}$$

The Eqs. (15), (16), and (8) can be written in the compact matrix form for the case of the velocities, accelerations, jerks and jounces as

$$\begin{bmatrix} \mathbf{v} \\ \dot{\mathbf{v}} \\ \ddot{\mathbf{v}} \\ \dddot{\mathbf{v}} \end{bmatrix} = \begin{bmatrix} \mathbf{I} & \mathbf{0} & \mathbf{0} & \mathbf{0} \\ -\tilde{\boldsymbol{\omega}} & \mathbf{I} & \mathbf{0} & \mathbf{0} \\ -2\dot{\tilde{\boldsymbol{\omega}}} & -\tilde{\boldsymbol{\omega}} & \mathbf{I} & \mathbf{0} \\ -3\ddot{\tilde{\boldsymbol{\omega}}} & -3\dot{\tilde{\boldsymbol{\omega}}} & -\tilde{\boldsymbol{\omega}} & \mathbf{I} \end{bmatrix} \begin{bmatrix} \mathbf{a}_1 \\ \mathbf{a}_2 \\ \mathbf{a}_3 \\ \mathbf{a}_4 \end{bmatrix} \tag{17}$$

$$\begin{bmatrix} \tilde{\underline{\omega}} \\ \dot{\tilde{\underline{\omega}}} \\ \ddot{\tilde{\underline{\omega}}} \\ \dddot{\tilde{\underline{\omega}}} \end{bmatrix} = \begin{bmatrix} \mathbf{I} & \mathbf{0} & \mathbf{0} & \mathbf{0} \\ -\tilde{\underline{\omega}} & \mathbf{I} & \mathbf{0} & \mathbf{0} \\ -2\dot{\tilde{\underline{\omega}}} & -\tilde{\underline{\omega}} & \mathbf{I} & \mathbf{0} \\ -3\ddot{\tilde{\underline{\omega}}} & -3\dot{\tilde{\underline{\omega}}} & -\tilde{\underline{\omega}} & \mathbf{I} \end{bmatrix} \begin{bmatrix} \boldsymbol{\phi}_1 \\ \boldsymbol{\phi}_2 \\ \boldsymbol{\phi}_2 \\ \boldsymbol{\phi}_3 \end{bmatrix}$$

$$\begin{bmatrix} \mathbf{a}_1 \\ \mathbf{a}_2 \\ \mathbf{a}_3 \\ \mathbf{a}_4 \end{bmatrix} = \begin{bmatrix} \mathbf{I} & \mathbf{0} & \mathbf{0} & \mathbf{0} \\ \tilde{\boldsymbol{\omega}} & \mathbf{I} & \mathbf{0} & \mathbf{0} \\ 2\dot{\tilde{\boldsymbol{\omega}}} + \tilde{\boldsymbol{\omega}}^2 & \tilde{\boldsymbol{\omega}} & \mathbf{I} & \mathbf{0} \\ 3\ddot{\tilde{\boldsymbol{\omega}}} + 2\tilde{\boldsymbol{\omega}}\dot{\tilde{\boldsymbol{\omega}}} + 3\dot{\tilde{\boldsymbol{\omega}}}\tilde{\boldsymbol{\omega}} + \tilde{\boldsymbol{\omega}}^3 & 3\dot{\tilde{\boldsymbol{\omega}}} + \tilde{\boldsymbol{\omega}}^2 & \tilde{\boldsymbol{\omega}} & \mathbf{I} \end{bmatrix} \begin{bmatrix} \mathbf{v} \\ \dot{\mathbf{v}} \\ \ddot{\mathbf{v}} \\ \dddot{\mathbf{v}} \end{bmatrix} \tag{18}$$

:

$$\begin{bmatrix} \boldsymbol{\phi}_1 \\ \boldsymbol{\phi}_2 \\ \boldsymbol{\phi}_2 \\ \boldsymbol{\phi}_3 \end{bmatrix} = \begin{bmatrix} \mathbf{I} & \mathbf{0} & \mathbf{0} & \mathbf{0} \\ \tilde{\boldsymbol{\omega}} & \mathbf{I} & \mathbf{0} & \mathbf{0} \\ 2\dot{\tilde{\boldsymbol{\omega}}} + \tilde{\boldsymbol{\omega}}^2 & \tilde{\boldsymbol{\omega}} & \mathbf{I} & \mathbf{0} \\ 3\ddot{\tilde{\boldsymbol{\omega}}} + 2\tilde{\boldsymbol{\omega}}\dot{\tilde{\boldsymbol{\omega}}} + 3\dot{\tilde{\boldsymbol{\omega}}}\tilde{\boldsymbol{\omega}} + \tilde{\boldsymbol{\omega}}^3 & 3\dot{\tilde{\boldsymbol{\omega}}} + \tilde{\boldsymbol{\omega}}^2 & \tilde{\boldsymbol{\omega}} & \mathbf{I} \end{bmatrix} \begin{bmatrix} \tilde{\underline{\omega}} \\ \dot{\tilde{\underline{\omega}}} \\ \ddot{\tilde{\underline{\omega}}} \\ \dddot{\tilde{\underline{\omega}}} \end{bmatrix}$$

The following theorem results from the previous results:

Theorem 3. *The higher-order time derivative of a dual twist uniquely determined the structure of the higher-order accelerations field of a rigid body motion.*

3 Higher-Order Kinematics of Multibody System Analysis Using Hypercomplex Algebras

Let $C_k, k = \overline{0, m}$ a spatial kinematic chain of the m + 1 bodies. The relative motion of the rigid body C_k with respect to the reference frame attached to C_{k-1} is described by the orthogonal dual tensor ${}^{k-1}\underline{\mathbf{R}}_k \in \underline{\mathbf{SO}}_3^{\mathbb{R}}$. The relative motion properties of the terminal body C_m with respect to the reference frame attached to C_0 are described by the orthogonal dual tensor (Fig. 1).

$$\underline{\mathbf{R}} = {}^0\underline{\mathbf{R}}_1 {}^1\underline{\mathbf{R}}_2 \ldots {}^{m-1}\underline{\mathbf{R}}_m \tag{19}$$

Instantaneous dual angular velocity (dual twist) of the rigid body C_m, with respect to the rigid body C_0, will be given by the equation

$${}_0\underline{\boldsymbol{\omega}}_m = \text{vect}\dot{\underline{\mathbf{R}}}\,\underline{\mathbf{R}}^T \tag{20}$$

Using the notation

$$\underline{\boldsymbol{\omega}}_k = {}^0\underline{\mathbf{R}}_1 {}^1\underline{\mathbf{R}}_2 \ldots {}^{k-2}\underline{\mathbf{R}}_{k-1}\underline{\boldsymbol{\Omega}}_k,$$

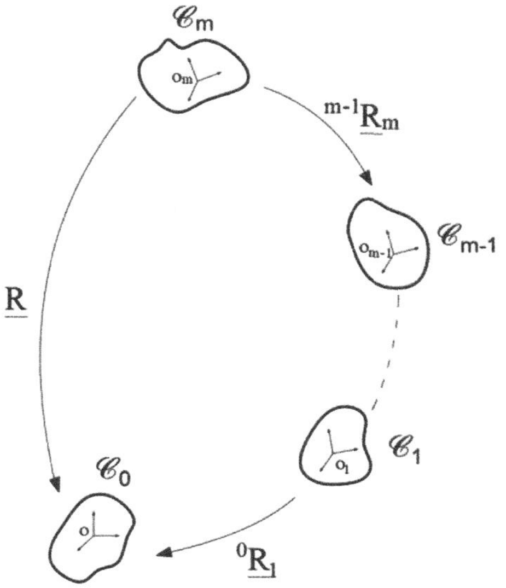

Fig. 1. Orthogonal dual tensors of spatial kinematic chain of rigid bodies

where $\underline{\mathbf{\Omega}}_k = \text{vect}\left({}^{k-1}\underline{\dot{\mathbf{R}}}_k {}^{k-1}\underline{\mathbf{R}}_k^T\right)$ the Eqs. (19) and (20) becomes

$$_0\underline{\boldsymbol{\omega}}_m = \underline{\boldsymbol{\omega}}_1 + \underline{\boldsymbol{\omega}}_2 + \cdots + \underline{\boldsymbol{\omega}}_m \quad (21)$$

In Eq. (21), $\underline{\boldsymbol{\omega}}_k$ is the dual twist of the relative motion of the body C_k with respect to the body C_{k-1}, observed from the reference frame attached to rigid body C_0.

Applying the multidual differential transformation defined in [15] to the relation given by Eq. (21), we determine the time derivative of the kth order of a dual twist $_0\underline{\boldsymbol{\omega}}_m$, $k = \overline{0, n-1}$.

Thus

$$_0\widehat{\underline{\boldsymbol{\omega}}}_m = \widehat{\underline{\boldsymbol{\omega}}}_1 + \widehat{\underline{\boldsymbol{\omega}}}_2 + \ldots \widehat{\underline{\boldsymbol{\omega}}}_m \quad (22)$$

where $\widehat{f}$ represents the multidual differential transformation of time-function f [15]. The multidual vector $_0\widehat{\underline{\boldsymbol{\omega}}}_m$ is named multidual twist of the terminal body C_m. We will obtain

$$_0\underline{\boldsymbol{\omega}}_m{}^{(k)} = k!\frac{d}{d\varepsilon^k}{}_0\widehat{\underline{\boldsymbol{\omega}}}_m, k = \overline{1, n} \quad (23)$$

Multidual twist of the rigid body contains all the information concerning the properties of the vector field of higher-order accelerations.

4 General 2C Manipulator and Higher-Order Kinematics

Either the particular case of the general 2C manipulator with four degrees of freedom. We will apply the available results obtained previously to determine the velocity, acceleration, jerk and jounce fields. In this case, the relative motions of three bodies C_0, C_1, andC_1, C_2 are known. The velocity, acceleration, jerk and jounce fields of the terminal body C_2 must be determined. The spatial motion of the C_2 terminal body with respect to the rigid body C_0 has been described by dual orthogonal tensor as follows:

$$^0\underline{\mathbf{R}}_2 = {}^0\underline{\mathbf{R}}_1 {}^1\underline{\mathbf{R}}_2 \tag{24}$$

where

$$^0\underline{\mathbf{R}}_1 = \exp[\underline{\varphi}_1(t)_0\underline{\tilde{\mathbf{u}}}_1] \tag{25}$$

$$^1\underline{\mathbf{R}}_2 = \exp[\underline{\varphi}_2(t)_1\underline{\tilde{\mathbf{u}}}_2] \tag{26}$$

described the relative motion of C_0, C_1, andC_1, C_2.

In Eqs. (25) and (26), the dual angles $\underline{\varphi}_1(t)$ and $\underline{\varphi}_2(t)$ are four times differentiable functions, and unit dual vectors $_0\underline{\mathbf{u}}_1$ and $_1\underline{\mathbf{u}}_2$ being constant. We will denote

$$_0\underline{\mathbf{u}}_1 = \cdot\underline{\mathbf{u}}_1 \tag{27}$$

$$_0\underline{\mathbf{u}}_2 = \left(I + sin\underline{\varphi}_1(t)_0\tilde{\underline{\mathbf{u}}}_1 + \left(1 - cos\underline{\varphi}_1(t)\right)_0\tilde{\underline{\mathbf{u}}}_1^2\right)_1\underline{\mathbf{u}}_2 = \cdot\underline{\mathbf{u}}_2 \tag{28}$$

$$\underline{\omega}_1 = \dot{\varphi}_1(t) + \varepsilon\dot{d}_1(t) \tag{29}$$

$$\underline{\omega}_2 = \dot{\varphi}_2(t) + \varepsilon\dot{d}_2(t) \tag{30}$$

Taking into consideration the observations from Sect. 4 and the results of [15], the vector field of the velocity, the acceleration, the jerk, the jounce is uniquely determined by the dual vectors $\underline{\boldsymbol{\omega}}, \dot{\underline{\boldsymbol{\omega}}}, \ddot{\underline{\boldsymbol{\omega}}}, \dddot{\underline{\boldsymbol{\omega}}}$. Also, from the previous results, after some algebra, we will have

$$\underline{\boldsymbol{\omega}} = \underline{\omega}_1\underline{\mathbf{u}}_1 + \underline{\omega}_2\underline{\mathbf{u}}_2 \tag{31}$$

The multidual twist of rigid body C_2 is $\widehat{\underline{\boldsymbol{\omega}}} = \widehat{\underline{\omega}}_1\underline{\mathbf{u}}_1 + \widehat{\underline{\omega}}_2\widehat{\underline{\mathbf{u}}}_2$. From Eq. 23, one obtains

$$\dot{\underline{\boldsymbol{\omega}}} = \dot{\underline{\omega}}_1\underline{\mathbf{u}}_1 + \dot{\underline{\omega}}_2\underline{\mathbf{u}}_2 + \underline{\omega}_1\underline{\omega}_2\underline{\mathbf{u}}_1 \times \underline{\mathbf{u}}_2 \tag{32}$$

$$\ddot{\underline{\boldsymbol{\omega}}} = \ddot{\underline{\omega}}_1\underline{\mathbf{u}}_1 + \ddot{\underline{\omega}}_2\underline{\mathbf{u}}_2 + \left(2\underline{\omega}_1\dot{\underline{\omega}}_2 + \underline{\omega}_1^2\underline{\omega}_2 + \dot{\underline{\omega}}_1\underline{\omega}_2\right)\underline{\mathbf{u}}_1 \times \underline{\mathbf{u}}_2 \tag{33}$$

$$\begin{aligned}\dddot{\underline{\boldsymbol{\omega}}} &= \left[\dddot{\underline{\omega}}_1 + 3\left(\underline{\omega}_1^2\dot{\underline{\omega}}_2 - \dot{\underline{\omega}}_1\underline{\omega}_1\underline{\omega}_2\right)\underline{\mathbf{u}}_1 \cdot \underline{\mathbf{u}}_2\right]\underline{\mathbf{u}}_1 + \left[\dddot{\underline{\omega}}_2 - 3\left(\underline{\omega}_1^2\dot{\underline{\omega}}_2 - \dot{\underline{\omega}}_1\underline{\omega}_1\underline{\omega}_2\right)\right]\underline{\mathbf{u}}_2 \\ &+ \left(3\underline{\omega}_1\ddot{\underline{\omega}}_2 + 3\underline{\omega}_1^2\dot{\underline{\omega}}_2 + 3\dot{\underline{\omega}}_1\dot{\underline{\omega}}_2 - \underline{\omega}_1^3\underline{\omega}_2\right)\underline{\mathbf{u}}_1 \times \underline{\mathbf{u}}_2\end{aligned} \tag{34}$$

The results for six degrees of freedom general 3C manipulator can be obtained in a similar way. This is the subject of future work.

5 Conclusions

It is proved that the multidual twist of the rigid body contains all the information concerning the properties of the vector field of higher-order accelerations. Using the multidual algebraic calculus, the equations that determine the vector field of higher-order accelerations for spatial kinematic chains are given. The results that have been obtained are interesting in the fields of computational kinematics, jerk and jounce analysis of robotic systems, sensitivity and singularity analysis of multibody kinematics.

References

1. Angeles, J.: The Application of dual algebra to kinematic analysis. Comput. Methods Mech. Syst. **161**, 3–32 (1998)
2. Ball, R.S.: The Theory of Screws. Cambridge University Press (1900)
3. Bokelberg, E.H., Hunt, K.H., Ridley, P.R.: Spatial motion I points of inflection and the differential geometry of screws. Mech. Mach. Theory **27**(1), 1–15 (1992)
4. Condurache, D., Burlacu, A.: Orthogonal dual tensor method for solving the AX= XB sensor calibration problem. Mech. Mach. Theory **104**(October), 382–404 (2016)
5. Condurache, D., Matcovschi, M.H.: Computation of angular velocity and acceleration tensors by direct measurements. Acta Mech. **153**(3–4), 147–167 (2002)
6. Hunt, K.H.: Kinematic Geometry of Mechanisms. Oxford University Press (1978)
7. Karger, A.: Singularity analysis of serial robot-manipulators. ASME J. Mech. Des. **118**(4), 520–525 (1996)
8. Lerbet, J.: Analytic geometry and singularities of mechanisms, ZAMM. Z. Angew. Math. Mech. **78**(10b), 687–694 (1999)
9. Müller, A.: Higher derivatives of the kinematic mapping and some applications. Mech. Mach. Theory **76**, 70–85 (2014)
10. Phillips, J.: Freedom in Machinery. Introducing Screw Theory, vol. 1. Cambridge University Press (1984).
11. Rico, J.M., Gallardo, J., Duffy, J.: Screw theory and higher-order kinematic analysis of open serial and closed chains. Mech. Mach. Theory **34**(4), 559–586 (1999)
12. Samuel, A.E., McAree, P.R., Hunt, K.H.: Unifying screw geometry and matrix transformations. Int. J. Robot. Res. **10**(5), 454–472 (1991)
13. Veldkamp, G.R.: Canonical systems and instantaneous invariants in spatial kinematics. J. Mech. **3**(3), 329–388 (1967)
14. Condurache, D.: Higher-Order Relative Kinematics of Rigid Body Motions: A Dual Lie Algebra Approach, Advances in Robot Kinematics. Springer Proceedings in Advanced Robotics, vol. 8, pp 83–91. Springer, Cham (2018)
15. Condurache, D.: Multidual Algebra and Higher-Order Kinematics, New Trends in Mechanism and Machine. Mechanisms and Machine Science, vol. 89, pp 48–55. Springer, Cham (2020)

Preliminary Results on a New Semi-Analytical Technique for the Study of Vibro-Impact Problems

Nicolae Herisanu(✉) and Vasile Marinca

University Politehnica Timisoara, Timisoara, Romania
nicolae.herisanu@upt.ro

Abstract. This paper presents some preliminary results emphasizing the applicability and usefulness of a new approach to study the dynamical behavior of a vibro-impact oscillator. A technique combining an analytical and a numerical approach is proposed involving the Optimal Auxiliary Functions Method. The proposed technique is applied to investigate a two-degree-of-freedom vibro-impact oscillator subject to a harmonic perturbing force. The efficiency of the proposed approach is proved and the analytical solutions were used in the analysis of the kinetic energy loss and contact force.

Keywords: Vibro-impact · Optimal auxiliary functions method

1 Introduction

Various engineering applications make use of vibro-impact devices designed to use vibration and impact effects to achieve various goals [1, 2]. Among them, the most recent applications are developed in the field of energy harvesting systems [3–5].

Besides numerical and experimental approaches, analytical approaches received great attention from scientists in the last decades. The method of Kryloff and Bogoliubov is applied in [6] to analyze the nonlinear equation of a single-degree-of-freedom oscillator with an impact damper. Some mechanical models of contact vibrations are proposed and discussed in [7] with focus on controlling multibody vibro-impact dynamics. Analytical and numerical solutions are obtained in [8] for a vibro-impact system consisting of a crank-slider mechanism and one oscillator attached to it. The standard Poincare–Bendixson theory, Lyapunov second method, and Zhuravlev transformations are used in [9] to analytically investigate a mechanical system with one-degree-of-freedom.

In the present paper, the behavior of a two-degree-of-freedom vibro-impact system subject to a harmonic perturbing force is analyzed by means of a new approach involving an analytical technique, namely the Optimal Auxiliary Functions Method, which provide accurate explicit analytical solutions to complex nonlinear systems. Non-periodic motions are analyzed and the contact force, as well as the kinetic energy loss, are investigated for the considered system. Comparisons with numerical integration results are developed in order to validate the analytical results which proved to be highly accurate.

N. Herisanu and V. Marinca (eds.), *Acoustics and Vibration of Mechanical Structures – AVMS-2021*, Springer Proceedings in Physics 274,
https://doi.org/10.1007/978-3-030-96787-1_6

2 Dynamical Model

The considered 2DOF vibro-impact system is presented in Fig. 1.

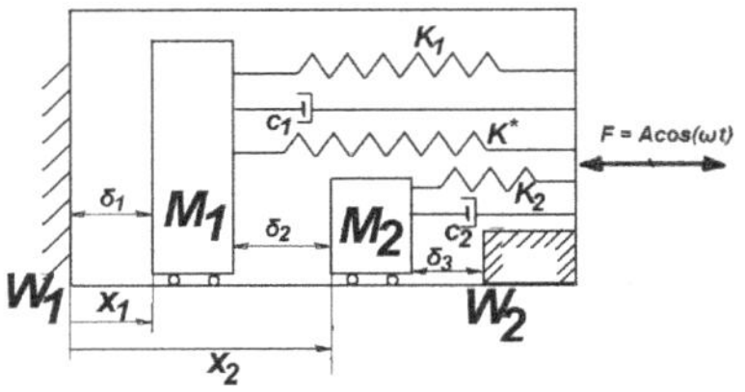

Fig. 1. 2DOF model of the vibro-impact system

The system under study consists of two oscillators of mass M_1 and M_2 moving in the horizontal plane, separated by three gaps: δ_1 between the wall W_1 and the mass M_1, δ_2 between the masses M_1 and M_2, and δ_3 between the mass M_2 and the wall W_2. The first oscillator is connected to a linear spring K_1, a damper c_1, and a nonlinear spring K^*, while the second oscillator is connected to a linear spring K_2 and a damper c_2. Here x_1 and x_2 denote the displacements between the two oscillators and the left vertical wall W_1. The whole system is excited by a harmonical force $F = A\cos\omega t$ in the horizontal direction.

The governing equations for the above-described vibro-impact system are

$$M_1 x_1'' + c_1(x_1' - F') + K_1(x_1 - F) + K^*(x_1 - F)^3 = 0 \tag{1}$$

$$M_2 x_2'' + c_2(x_2' - F') + K_2(x_2 - F) = 0 \tag{2}$$

where prime denotes derivative with respect to time.

Taking into account the transformations

$$\begin{aligned} &\omega t = \Omega\tau\ ;\ \Omega_1 = (K_1/M_1)^{1/2}\ ;\ \Omega_2 = (K_2/M_2)^{1/2};\ \tau = \Omega_1 t \\ &x_1 = \delta_1 X\ ;\ x_2 = \delta_1 Y\ ;\ \alpha_1 = c_1/(2M_1\Omega_1) \\ &\alpha_2 = c_2/(2M_2\Omega_1);\ \beta = K^*\delta_1^2/K_1\ ;\ a = A/\delta_1\ ;\ k = \tfrac{\Omega_2}{\Omega_1} \end{aligned} \tag{3}$$

the system (1)–(2) becomes

$$\begin{aligned} &\ddot{X} + 2\alpha_1\dot{X} + \left(1 + \frac{3}{2}a^2\right)X + \beta X^3 = -2\alpha_1 a\Omega\sin(\Omega\tau) + (a + 3a\beta X^2 + \\ &+\frac{3}{4}\beta a^3)\cos(\Omega\tau) - \frac{3}{2}a^2\beta X\cos(2\Omega\tau) + \frac{\beta}{4}a^3\cos(3\Omega\tau) \end{aligned} \tag{4}$$

$$\ddot{Y} + 2\alpha_2 k\dot{Y} + k^2 Y = ak^2\cos(\Omega\tau) - 2\alpha_2 ka\Omega\sin(\Omega\tau) \tag{5}$$

where dot denotes derivative with respect to the new variable τ.

The initial conditions are

$$X(0) = 1\,,\ \dot{X}(0) = V_1\,,\ Y(0) = 1 + \frac{\delta_2}{\delta_1}\,,\ \dot{Y}(0) = V_2 \tag{6}$$

In order to investigate the above-described system, the Optimal Auxiliary Functions Method will be used to obtain explicit analytical solutions.

3 Application of OAFM to Analyze the Vibro-Impact System

According to OAFM [10–15], in order to solve the system (4)–(5), we will find solutions as

$$\overline{X}(\tau) = X_0(\tau) + X_1(\tau, C_i)\,,\ i = 1, 2, ..., p \tag{7}$$

$$\overline{Y}(\tau) = Y_0(\tau) + Y_1(\tau, C_i + C_j)\,,\ j = 1, 2, ..., q \tag{8}$$

containing the initial approximation and the approximation of the first order.

The linear operators corresponding to the governing equations are

$$L(X) = \ddot{X} + 2\alpha_1\dot{X} + \left(1 + \frac{3}{2}a^2\right)X \tag{9}$$

$$L(Y) = \ddot{Y} + 2\alpha_2 k\dot{Y} + k^2 Y \tag{10}$$

and consequently, the linear equations which will allow finding the initial approximations will be

$$\ddot{X}_0 + 2\alpha_1\dot{X}_0 + \left(1 + \frac{3}{2}a^2\right)X_0 = 0,\ X_0(0) = 1\,,\ \dot{X}_0(0) = V_1 \tag{11}$$

$$\ddot{Y}_0 + 2\alpha_2 k\dot{Y}_0 + k^2 Y_0 = 0,\ Y_0(0) = 1 + \tfrac{\delta_2}{\delta_1}\,,\ \dot{Y}_0(0) = V_2 \tag{12}$$

with the solutions

$$X_0(\tau) = e^{-\alpha_1\tau}\left(C_1\cos\sqrt{\overline{\Omega}^2 - \alpha_1^2}\tau + C_2\sin\sqrt{\overline{\Omega}^2 - \alpha_1^2}\right),\ \overline{\Omega}^2 = 1 + \frac{3}{2}a^2 \tag{13}$$

$$Y_0(\tau) = e^{-\alpha_2 k\tau}\left[C_3\cos\sqrt{k^2 - k\alpha_2}\tau + C_4\sin\sqrt{k^2 - k\alpha_2}\tau\right] \tag{14}$$

The nonlinear operators corresponding to (4) and (5) are given by

$$\begin{aligned} N(X) &= \beta X^3 + 2\alpha_1 a\Omega\sin(\Omega\tau) - (a + 3a\beta X^2 + \frac{3}{4}\beta a^3)\cos(\Omega\tau) + \\ &+ \frac{3}{2}a^2\beta X\cos(2\Omega\tau) - \frac{\beta}{4}a^3\cos(3\Omega\tau) \end{aligned} \tag{15}$$

$$N(Y) = ak^2\cos(\Omega\tau) - 2\alpha_2 ka\Omega\sin(\Omega\tau) \tag{16}$$

so that the first approximation may be chosen as

$$X_1(\tau) = Q_1[\cos(3\Omega\tau) - \cos(\Omega\tau)] + Q_2[\sin(3\Omega\tau) - 3\sin(\Omega\tau)] \tag{17}$$

$$Y_1(\tau) = Q_3\cos(\Omega\tau) + Q_4\sin(\Omega\tau) \tag{18}$$

and therefore, the final solutions will be of the form

$$\begin{aligned}\overline{X}(\tau) = e^{-\alpha_1\tau}\left(\cos\sqrt{\overline{\Omega}^2 - \alpha_1^2}\tau + \frac{\alpha_1 + V_1}{\sqrt{\overline{\Omega}^2 - \alpha_1^2}}\sin\sqrt{\overline{\Omega}^2 - \alpha_1^2}\right) + \\ +Q_1[\cos(3\Omega\tau) - \cos(\Omega\tau)] + Q_2[\sin(3\Omega\tau) - 3\sin(\Omega\tau)]\end{aligned} \tag{19}$$

$$Y(\tau) = e^{-\alpha_2 k\tau}\left[C_3\cos\sqrt{k^2 - k\alpha_2}\tau + C_4\sin\sqrt{k^2 - k\alpha_2}\tau\right] + Q_3\cos(\Omega\tau) + Q_4\sin(\Omega\tau) \tag{20}$$

where the parameters Q_1 and Q_2 may be determined by Galerkin method and Q_3, Q_4 may be determined by identification of coefficients

$$\begin{aligned}&Q_1 = -\frac{9a}{8 + 12a^2},\ Q_2 = \frac{(36a^3 - 3a)\Omega\alpha_1}{8(2 + 3a^2)^2},\ C_3 = 1 + \frac{\delta_2}{\delta_1} - \frac{ak^2(k^2 - \Omega^2) + 2a\alpha_2^2 k^2\Omega^2}{(\Omega^2 - k^2)^2 + 4\alpha_2^2 k^2\Omega^2}.\\ &C_4 = \frac{V_2}{\sqrt{k^2 - k\alpha_2}} - \frac{a\alpha_2 k\Omega^4}{\sqrt{k^2 - k\alpha_2}\left((\Omega^2 - k^2)^2 + 4\alpha_2^2 k^2\Omega^2\right)} + \\ &+\alpha_2 k\left(1 + \frac{\delta_2}{\delta_1} - \frac{ak^2(k^2 - \Omega^2) + 2a\alpha_2^2 k^2\Omega^2}{(\Omega^2 - k^2)^2 + 4\alpha_2^2 k^2\Omega^2}\right)\\ &Q_3 = \frac{(k^2 - \Omega^2)ak^2 + 4\alpha_2^2 ak^2\Omega^2}{(\Omega^2 - k^2) + 4\alpha_2^2 k^2\Omega^2},\ Q_4 = \frac{2\alpha_2 ak\Omega^3}{(\Omega^2 - k^2) + 4\alpha_2^2 k^2\Omega^2}\end{aligned} \tag{21}$$

Finally, the explicit analytical solutions of the considered system will be

$$\begin{aligned}\overline{X}(\tau) = e^{-\alpha_1\tau}\left(\cos\sqrt{\overline{\Omega}^2 - \alpha_1^2}\tau + \frac{\alpha_1 + V_1}{\sqrt{\overline{\Omega}^2 - \alpha_1^2}}\sin\sqrt{\overline{\Omega}^2 - \alpha_1^2}\tau\right) + \\ +\frac{9a}{8 + 12a^2}[\cos(\Omega\tau) - \cos(3\Omega\tau)] + \frac{(36a^3 - 3a)\Omega\alpha_1}{8(2 + 3a^2)^2}[\sin(3\Omega\tau) - 3\sin(\Omega\tau)]\end{aligned} \tag{22}$$

$$\begin{aligned}Y(\tau) = e^{-\alpha_2 k\tau}[C_3\cos\sqrt{k^2 - k\alpha_2}\tau + C_4\sin\sqrt{k^2 - k\alpha_2}\tau\Big] + \\ +\frac{ak^2(k^2 - \Omega^2) + 2a\alpha_2^2 k^2\Omega^2}{(k^2 - \Omega^2)^2 + 4\alpha_2^2 k^2\Omega^2}\cos\Omega\tau + \frac{a\alpha_2 k\Omega^3}{\left((\Omega^2 - k^2) + 4\alpha_2^2 k^2\Omega^2\right)}\sin\Omega\tau\end{aligned} \tag{23}$$

The accuracy of the obtained results is proved by graphical comparisons between analytical and numerical solutions for a set of physical parameters chosen as $M_1 = 0.072$, $M_2 = 0.086$, $K_1 = 110$, $K_2 = 141$, $a = 3$ (Figs. 2 and 3).

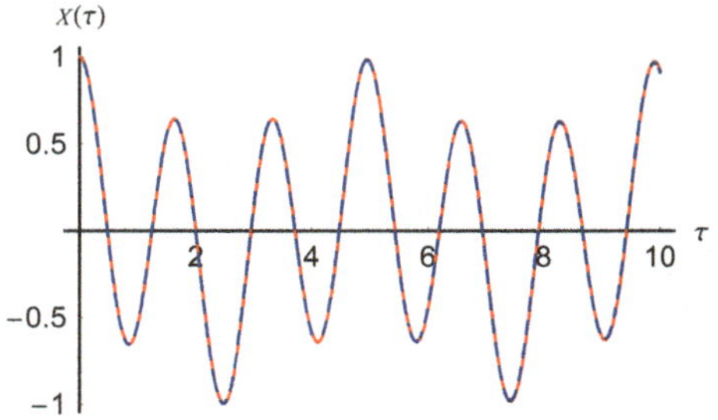

Fig. 2. Comparison between the approximate analytical solution (22) and numerical integration results for $\omega = 49.61$ and $\delta_1 = \delta_2 = 0.02$ ——numerical - - - -analytical solution

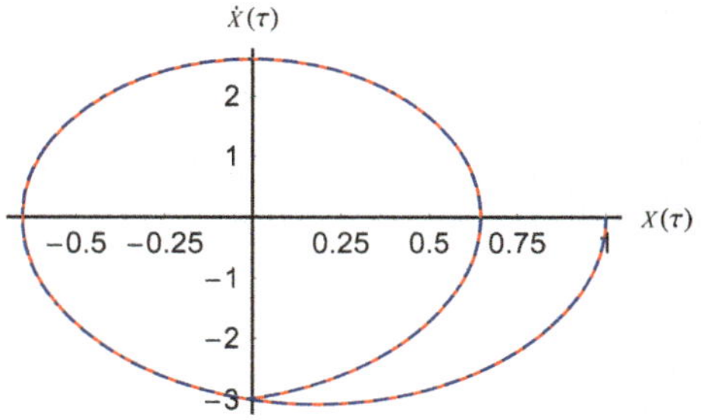

Fig. 3. Comparison of the phase portraits for analytical solution (22) and numerical integration results for $\omega = 49.61$ and $\delta_1 = \delta_2 = 0.02$ n ——umerical - - - -analytical solution

4 Analysis of Non-Periodic Motion

Depending on the possible motions of the masses M_1 and M_2, there are more identifiable scenarios of non-periodic motions.

In the first possible case, the oscillator of mass M_1 is moving from the static equilibrium position to the position of impact with the wall W_1. If the moment of this impact is τ_1, then this case takes place if $X(\tau) \leq 1$, $X(\tau_1) = 0$ and $\dot{X}(\tau_1) \leq 0$. In particular, if $\dot{X}(\tau_1) = 0$, then the oscillator M_1 stops at the wall W_1. The velocity of the mass M_1 after the impact will be $\dot{X}(\tau_1^+) = -R_2\dot{X}(\tau_1^-)$, where the symbols " + " and "–" indicate the moments after and before the impact, respectively, and R_2 is the coefficient of restitution. For the oscillator of mass M_2, there are possible the following situations (types of movement):

(a) The oscillator M_2 is moving between the oscillator of mass M_1 and the wall W_2 without impact and without sticking motion. The condition is equivalent to the existence of the inequalities $X(\tau) < Y(\tau) < 1 + \delta_2/\delta_1 + \delta_3/\delta_1$ and $\tau < \tau_1$. It follows that there exists a moment of time τ_1^* when M_2 stops, $\dot{Y}(\tau_1^*) = 0$.

(b) The oscillator M_2 impacts the oscillator M_1 before the collision of the mass M_1 with the wall W_1. In this case, there exists a moment $\tau_2 < \tau$ so that $X(\tau_2) = Y(\tau_2)$, $\dot{X}(\tau_2) < 0$, $\dot{Y}(\tau_2) < 0$, $\dot{X}(\tau_2) \neq \dot{Y}(\tau_2)$. After this collision, the velocities can be written as

$$\begin{aligned} \dot{X}(\tau_2^+) &= \frac{M_1 - RM_2}{M_1 + M_2}\dot{X}(\tau_2^-) + \frac{M_2(1+R)}{M_1 + M_2}\dot{Y}(\tau_2^-) \\ \dot{Y}(\tau_2^+) &= \frac{M_1(1+R)}{M_1 + M_2}\dot{X}(\tau_2^-) + \frac{M_2 - RM_1}{M_1 + M_2}\dot{Y}(\tau_2^-);\ \dot{X}(\tau_2^-) = \dot{X}(\tau_2)\ ;\ \dot{Y}(\tau_2^+) = \dot{Y}(\tau_2) \end{aligned} \tag{24}$$

(c) The oscillator of mass M_1 is sticking to M_2 (sticking motion). The condition is written as $X(\tau_2) = Y(\tau_2)$ and $\dot{X}(\tau_2) = \dot{Y}(\tau_2)$. The contact force corresponding to masses M_1 and M_2 is obtained from Eqs. (4) and (5) as

$$\begin{aligned} F_{12}(\tau) &= 2a\Omega(k\alpha_2 - \Omega\alpha_1)\sin\Omega\tau + (a - ak^2 + 3a\beta X^2 + \frac{3}{4}\beta a^3)\cos\Omega\tau - \\ &-\frac{3}{2}a^2\beta X\cos 2\Omega\tau + \frac{\beta}{4}a^3\cos 3\Omega\tau + 2\alpha_2\dot{Y} - 2\alpha_1\dot{X} + k^2Y - \left(1 + \frac{3}{2}a^2\right)X - \beta X^3 \end{aligned} \tag{25}$$

The sticking takes place only if $F_{12} > 0$.

(d) The oscillator of mass M_2 impacts the wall W_2 at the moment $\tau = \tau_3$. The conditions are $Y(\tau_3) = 1 + \delta_2/\delta_1 + \delta_3/\delta_1$ and $\dot{Y}(\tau_3) \geq 0$. If $\dot{Y}(\tau_3) = 0$, then the oscillator M_2 stops without impacting the wall W_2.

It is to remark that after the system has one of the movements described above, the movement of each oscillator should be studied by considering the conditions established for the corresponding case. For example, after the first case takes place, in the condition $\dot{X}(\tau_1) < 0$, and $\dot{X}(\tau_1^+) = -R\dot{X}(\tau_1^-)$, and $\dot{Y}(\tau_1^*) = 0$, then the approximate solutions (22) and (23) will be written as

$$\begin{aligned} \overline{X}(\tau) &= e^{-\alpha_1(\tau-\tau_1)}\frac{R_2\dot{X}(\tau_1^-)}{\sqrt{\overline{\Omega}^2 - \alpha_1^2}}\sin\sqrt{\overline{\Omega}^2 - \alpha_1^2}(\tau - \tau_1) + \frac{9a}{8 + 12a^2}[\cos\Omega(\tau - \tau_1) - \\ &- \cos 3\Omega(\tau - \tau_1)] + \frac{(36a^3 - 3a)\Omega\alpha_1}{8(2 + 3a^2)^2}[\sin 3\Omega(\tau - \tau_1) - 3\sin\Omega(\tau - \tau_1)],\ \tau \geq \tau_1 \end{aligned} \tag{26}$$

$$\begin{aligned} Y(\tau) &= e^{-\alpha_2 k(\tau-\tau_1^*)}\Big[(\gamma - Q_3)\cos\sqrt{k^2 - k\alpha_2}(\tau - \tau_1^*) + \\ &+ \frac{\alpha_2 k(\gamma - Q_3) - \Omega Q_4}{\sqrt{k^2 - k\alpha_2}}\sin\sqrt{k^2 - k\alpha_2}(\tau - \tau_1^*)\Big] \\ &+ Q_3\cos\Omega(\tau - \tau_1^*) + Q_4\sin\Omega(\tau - \tau_1^*) \end{aligned} \tag{27}$$

where τ_1^* is given by the equation $\dot{Y}(\tau_1^*) = 0$, $\gamma = Y(\tau_1^*)$, and $Y(\tau_1^*)$ is given by Eq. (23).

Concerning the kinetic energy loss in the analyzed case, it is known that this is given by the difference between the kinetic energy at the beginning and at the end of impact

$$\Delta T = \frac{1}{2}M_1\dot{X}^2(\tau^-) + \frac{1}{2}M_2\dot{Y}^2(\tau^-) - \frac{1}{2}M_1\dot{X}^2(\tau^+) - \frac{1}{2}M_2\dot{Y}^2(\tau^+) \tag{28}$$

and consequently, in our case, the kinetic energy loss becomes

$$\Delta T = \frac{M_1 M_2}{2(M_1 + M_2)}\left[\dot{X}(\tau) - \dot{Y}(\tau)\right]^2(1 - R^2) \tag{29}$$

The contact force (25) and the kinetic energy loss (29) are graphically presented in Figs. 4 and 5.

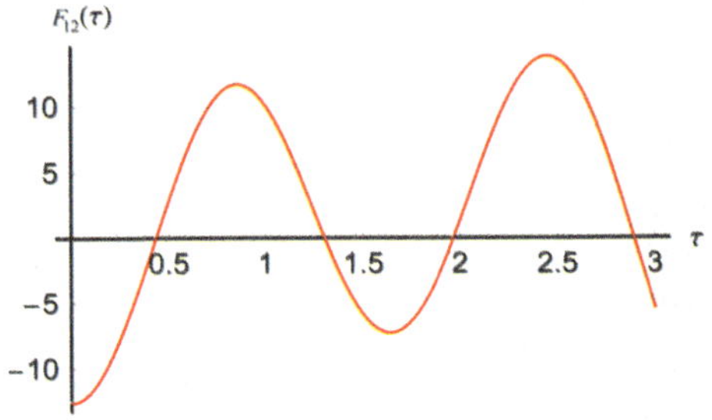

Fig. 4. The contact force

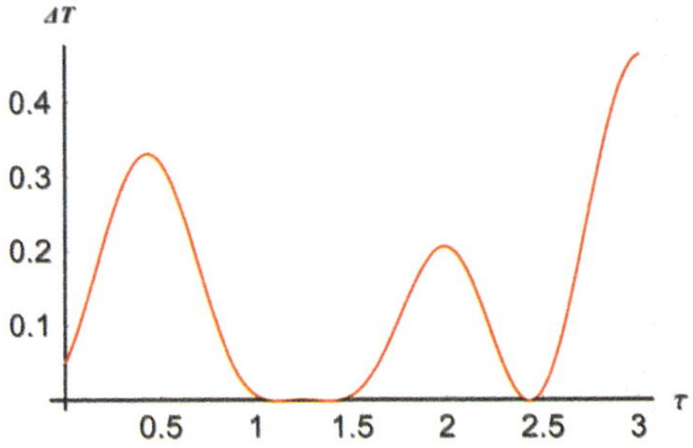

Fig. 5. The kinetic energy loss

5 Conclusions

A novel analytical approach is proposed in this work to investigate the behavior of a two-degree-of-freedom vibro-impact oscillator subject to a harmonic perturbing force. More possible motions were identified and analyzed. The Optimal Auxiliary Functions Method (OAFM) is employed in this analysis, where a new analytical tool provides highly accurate explicit analytical solutions. The contact force and kinetic energy loss are analyzed based on the obtained solutions. Further investigation should be developed to study the stability of periodic motions of the system based on the results obtained in this work.

References

1. Babitsky, V.I.: Theory of Vibro-Impact Systems and Applications. Springer, Berlin, Heidelberg, Germany (1988)
2. Awrejcewicz, J., Tomczak, K.: Analytical stability improvement of the periodic vibro-impact processes. In: Proceedings of the ASME Design Engineering Technical Conference, 2587–2594 Las Vegas, Nevada, USA (1999)
3. Vijayan, K., Friswell, M.I., Haddad Khodaparast, H., Adhikari, S.: Non-linear energy harvesting from coupled impacting beams. Int. J. Mech. Sci. **96–97**, 101–109 (2015)
4. Lai, Z.H., Wang, J.L., Zhang, C.L., Zhang, G.Q., Yurchenko, D.: Harvest wind energy from a vibro-impact DEG embedded into a bluff body. Energy Convers Manag. **199**, 111993 (2019)
5. Fu, Y., Ouyang, H., Davis, R.B.: Nonlinear dynamics and triboelectric energy harvesting from a three-degree-of-freedom vibro-impact oscillator. Nonlinear Dyn. **92**, 1985–2004 (2018)
6. Cronin, D.I., Van, N.K.: Substitute for the impact damper. J. Eng. Industry 1295–1300 (1975)
7. Awrejcewicz, J., Tomczak, K., Lamarque, C.H.: Controlling systems with impacts. Int. J. Bifurcation Chaos **9**, 547–553 (1999)
8. Zukovic, M., Hajradinovic, D., Kovacic, I.: On the dynamics of vibro-impact systems with ideal and non-ideal excitation. Meccanica **56**, 439–460 (2021)
9. Ivanov, A.P.: Analytical methods in the theory of vibro-impact systems. J. Appl. Mech. **57**, 221–236 (1993)
10. Marinca, V., Herisanu, N.: The nonlinear thermomechanical vibration of a functionally graded beam on Winkler-Pasternak foundation. MATEC Web Conf. **148**, 13004 (2018)
11. Herisanu, N., Marinca, V.: An effective analytical approach to nonlinear free vibration of elastically actuated microtubes. Meccanica **56**, 813–823 (2021)
12. Marinca, V., Herisanu, N.: Optimal auxiliary functions method for a pendulum wrapping on two cylinders. Mathematics **8**, 1364 (2020)
13. Marinca, V., Herisanu, N.: Construction of analytic solutions to axisymmetric flow and heat transfer on a moving cylinder. Symmetry **12**, 1335 (2020)
14. Herisanu, N., Marinca, V.: An efficient analytical approach to investigate the dynamics of a misaligned multirotor system. Mathematics **8**, 1083 (2020)
15. Herisanu, N., Marinca, V., Madescu, G.: Application of the optimal auxiliary functions method to a permanent magnet synchronous generator. Int. J. Nonlinear Sci. Numer. Simul. **20**, 399–406 (2019)

Subharmonic Generation in GeSbSe Chalcogenide Cantor-Like Glasses

V. Chiroiu[1(✉)], N. Nedelcu[1], L. Munteanu[1], A. Stan[1,2], and N. Stan[1]

[1] Romanian Academy, Institute of Solid Mechanics, Bucharest, Romania
[2] University of Pitesti, Piteşti, Romania

Abstract. We report in this paper the subharmonic generation of waves in GeSbSe chalcogenide glasses architectured as Cantor-like structures. The subharmonic waves generation is explained as a superposition of waves in both phonon and fracton vibration regimes.

1 Introduction

Chalcogenides contain one or more chalcogenide components (Sulfur, Selenium, Germanium, and Tellurium, but excluding Oxygen). These components can be covalently bonded and can be classified as covalent networking solids. The chalcogenides are based on Sulfur, As-S or Ge–S. The glass-forming abilities decrease with increasing of the molar weight of the constituent elements [1–3].

Glassy systems, namely $Ge_xSb_{40\text{-}x}Se_{60}$ alloys with composition x = 12, 25, and 30 at. % were synthesized from elements with 5 N purity (Ge, Sb, Se) by the conventional melt-quenching method.

The mixture of the elements with proper weight percent was put in a quartz ampoule which was evacuated down to 10^{-3} Pa [4].

The ampoules were loaded in a rotary furnace and were heated up to 950 °C. For obtaining homogeneous melting, the glassy mixture was kept at this temperature, for 12 h continuously, rotating the furnace. After ending the process, the ampoules were pulled out, and the melts were rapidly cooled down in ice water. Parts of the synthesized glasses were served as parent material for deposition on quartz substrate using a resistive crucible of 0.1 mm molybdenum sheet from having the shape and size of the 0482070 from the UMICORE catalog. Degassing took place at a pressure of 10–1 Pa for 12 min in the atmosphere of N2, and during all that time, the dome, with has the quartz substrates is rotated (about 10 rot/min) [5].

This process was necessary to remove the impurities on the walls from the vacuum deposition installation and on the bulk. At 4 $\text{Pa}\times 10^{-3}$ pressure, the heating resistance is coupled, ensuring the temperature to be 300 °C at the evaporation chamber, while the rotating dome is at the speed rotation of 10 rpm. Powder material was evaporated at an electric current of at least 400 mA.

N. Herisanu and V. Marinca (eds.), *Acoustics and Vibration of Mechanical Structures – AVMS-2021*, Springer Proceedings in Physics 274,
https://doi.org/10.1007/978-3-030-96787-1_7

The monitoring of the deposition control was performed for approximately 6 min with a deposition rate, of 3 nm/sec, and showed 1300 nm thickness. After evaporation, the samples were measured by Lambda 950 Spectrophotometer [6].

Using the transmission spectra, we can calculate the optical and dielectric properties by the Swanepoel method, and also the bandgap [7, 8].

The nonlinear elastic response of materials has been observed, recently, as a macroscopic interaction of waves with localized inhomogeneities into the material such as grains and cracks [9–11]. Besides the harmonic and subharmonic generation of waves, the material exhibits also hysteresis [12], self-modulation, and resonance of the contact vibrations [13].

Nonlinear effects have also been observed at a nanoscale level [14] due to the variation of a structural characteristic distance to a simple harmonic excitation at one of the normal modes [14–17]. In such a case, the nonlinearity is distributed all over the structure.

Experiments on 1D artificial piezoelectric plates with Cantor-like structure put into evidence an extremely low thresholds in the subharmonic generation of ultrasonic waves as compared to the corresponding homogeneous and periodic plates [18–20]. The subharmonic waves generation is a complex problem that may lead the structure into chaotic dynamics [21].

In this paper, the subharmonic waves generation in chalcogenide structures with Cantor structure is analyzed. The subharmonic generation of waves in chalcogenides takes place at a moment of time after the driving force exceeds a threshold value.

2 Generation of Subharmonic Waves

The GeSbSe chalcogenide beam with Cantor structure is presented in Fig. 1. The Cantor structure assumes a triadic sequence up to the fourth generation with 31 elements. The Cantor structure follows the rule of Henry John Stephen Smith [18] and Georg Cantor [19] of iteratively deleting the middle third segment from a set of line segments. This structure suggests fractality and self-similarity because two copies of it are equal if each copy is reduced by a factor of 3 and translated.

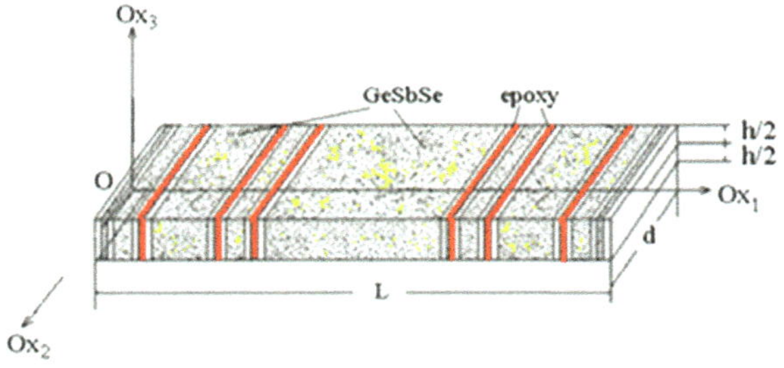

Fig. 1. The chalcogenide beam with Cantor structure.

The length of the beam is L, the width of the smallest layer is $L/81$, and the thickness of the plate is h. The width of the plate is d.

The aim of this paper is to determine the displacement field $u(x,t)$, $x=(x_1,x_2,x_3)$ and the Cauchy stress distribution σ_{ij}, $i,j=1,2,3$, as functions of position in the reference configuration and time. The reference coordinate system $Ox_1x_2x_3$ has the origin at the left end of the beam, in the middle plane of the beam. The axis Ox_1 is in the plane normal to the layers, and the axis Ox_3 is out of plane normal to the beam.

In the following, index 1 refers to the chalcogenide and index 2 to the epoxy.

The beam is characterized by the mass density ρ_1 and ρ_2 of the chalcogenide and the epoxy, the specific internal energies E_i and E_2 and the stress response functions σ_{ij}^1 and σ_{ij}^2, $i,j=1,2,3$, respectively.

The deformation of the beam is expressed by the usual deformation gradient.

$$F_{ij}=\delta_{ij}+\frac{\partial u_1}{\partial x_j}. \tag{1}$$

The right and left Cauchy-Green deformation tensors are

$$C_{ij}=F_{ki}F_{kj},\ B_{ij}=F_{ik}F_{jk}, \tag{2}$$

and the Lagrange strain tensor is

$$E_{ij}=\frac{1}{2}\left(F_{ki}F_{kj}-\delta_{ij}\right). \tag{3}$$

In the constitutive laws, the invariants of B are usually used

$$I_1=\mathrm{tr}(B)=B_{kk}, \tag{4}$$

The Cauchy stress expresses the force per unit deformed area and is given by

$$n_i\sigma_{ij}=\lim_{dA\to 0}\frac{\mathrm{d}P_j^{(n)}}{\mathrm{d}A}. \tag{5}$$

The constitutive laws for elastic materials are given by

$$\sigma_{ij}-\frac{1}{J}F_{ik}\frac{\partial W}{\partial F_{jk}}. \tag{6}$$

where W is the Helmholtz free energy.

The GeSbSe chalcogenides exhibit important fluctuations of the atomic distances in material as suggested in Fig. 2a. The atomic structure of GeSbSe is presented in Fig. 2b [3]. The peaks in interatomic distances for Ge-Sb, Ge–Ge, Sb-Sb, and Se-Se atoms appear to be 3.80 ± 0.05 Å, 3.85 ± 0.05 Å, 3.80 ± 0.05 Å, and 3.80 ± 0.05 Å, respectively [14].

These fluctuations owe to the small length of action forces from the amorphous material. The response of glasses to external forces within a length and time scale which exceeds the domain of applicability of a classical theory depends on the ratio $(\lambda/l,\ \tau/\tau_0)$ or $(\lambda/l,\omega_0/\omega)$, where λ is a characteristic length of the glass (atomic distance, granular

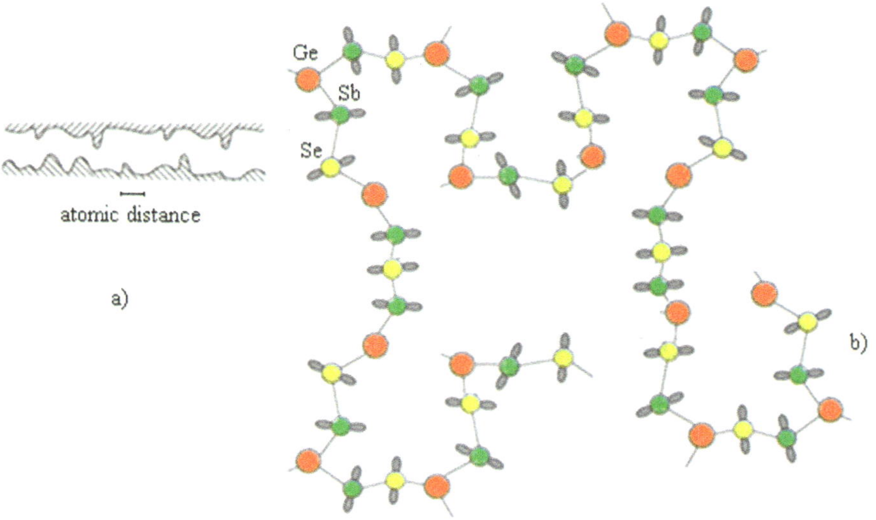

Fig. 2. **a** Fluctuations of the atomic distances and **b** the atomic structure of GeSbSe.

distance, etc.), l is the external characteristic length associated with the external forces (waves, distances over which load distribution change sharply, geometrical and surface discontinuities), τ the time scale (or frequency ω) which is the transmission time of a signal (or a frequency), and τ_0 is the external characteristic time or frequency) associated with the external forces [22, 23]. For $\lambda/l << 1$ and $\tau/\tau_0 << 1$, the external forces excite large numbers of subbodies simultaneously, so that subbodies interact and the result is a statistical average of the individual responses. For $\lambda/l = 1$ and $\tau/\tau_0 = 1$, the individual fields of subbodies (intermolecular and atomic forces) are important.

Forces with a short domain of action that characterize the structure of chalcogenides are identified by the distribution functions of the pairs of Ge-Se and Sb-Se atoms that can be performed using a 3D simulation in a configuration with a given number of atoms. A simulation expectation was realized in [5] for ternary films of $Ge_{15}Sb_{25}Se_{60}$ and $Ge_{35}Sb_5Se_{60}$ with a configuration of 200 atoms. The results are obtained from X-ray and neutron diffraction correlated with the Reverse Monte Carlo simulations in binary and ternary films indicate that Sb atoms incorporated in $Ge_{40}Se_{60}$ are covalently bonded to Se and form trigonal Sb-Se units.

The results are shown in Fig. 3. The distribution functions of the pairs of atoms are evidence of the presence of short-distance bonds in the amorphous material. By increasing the Ge concentration and respectively decreasing the Sb concentration, the interatomic distances in the Ge-Se bonds increase from 2.37 Å to 2.40 Å (Figs. 4, 5, 6, 7, and 8).

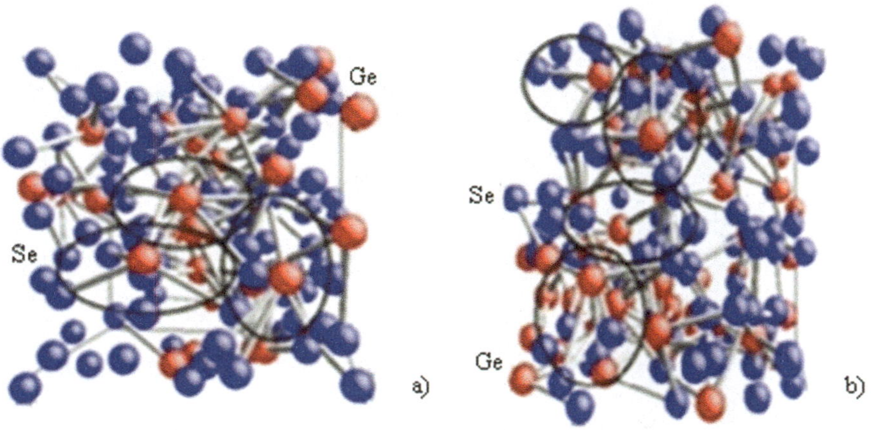

Fig. 3. Ge-Se pair distribution functions and their 3D bonds in $Ge_{15}Sb_{25}Se_{60}$ (**a**) and $Ge_{35}Sb_{5}Se_{60}$ (**b**).

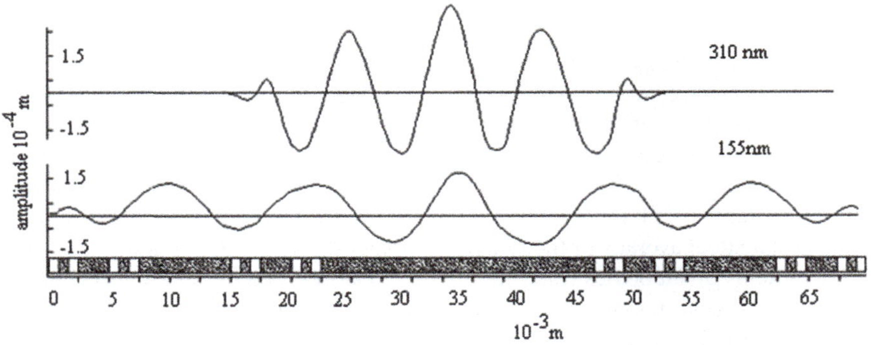

Fig. 4. Amplitudes of the surface displacement for 310 nm and the subharmonic 155 nm.

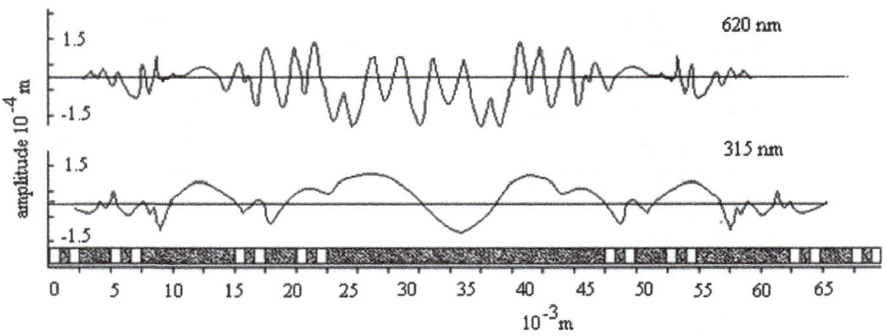

Fig. 5. Amplitudes of the surface displacement for 620 nm and the subharmonic 315 nm.

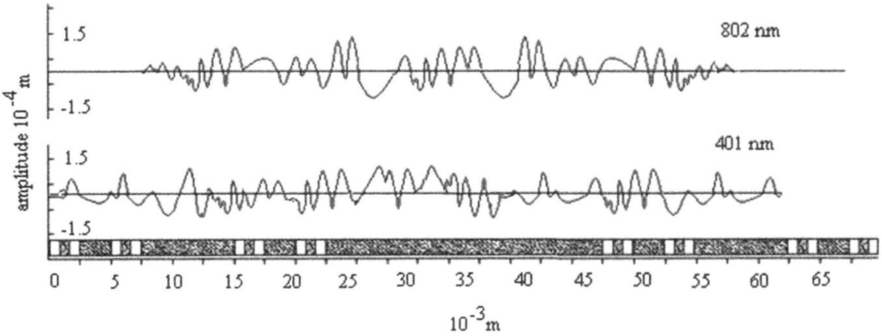

Fig. 6. Amplitudes of the surface displacement for 802 nm and the subharmonic 401 nm.

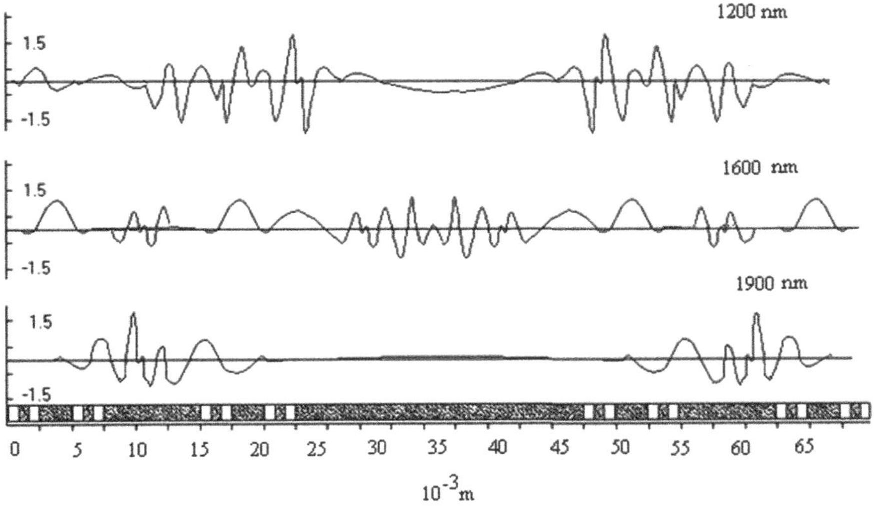

Fig. 7. The amplitudes of the surface displacement for localized vibration modes at 1200, 1600, and 1900 nm.

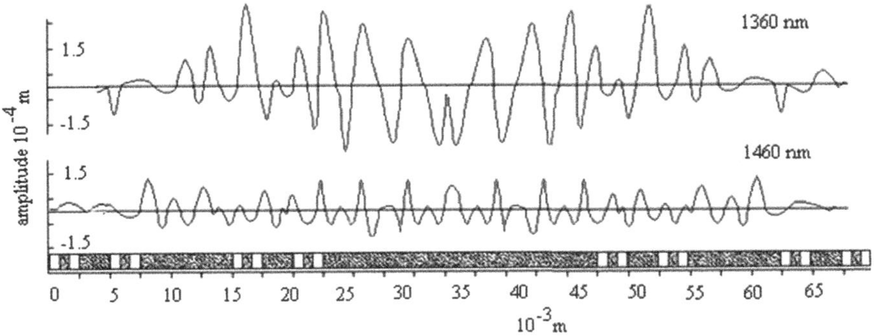

Fig. 8. The amplitudes of the surface displacement for extended vibration modes for 1360 nm and 1460 nm.

3 Conclusions

Two kinds of vibrations are found: a localized mode or the fracton regime and an extended vibration or the phonon regime. The fraction vibrations are localized on a few elements of the plate, while the phonon vibrations extend to the whole chalcogenide plate. In the case of a periodical plate, the dispersion prevents the matching of frequencies of the fundamental and appropriate subharmonic vibration modes. For the homogeneous plate, the mismatch interval is due to the symmetry of fundamental vibration modes. Only symmetric odd n can induce a subharmonic, but the normal amplitude never coincides with a plate vibration mode. For the chalcogenide Cantor plate, we have obtained the same qualitative results as Craciun and Alippi. Given a normal vibration mode, the value of the expected threshold which describes the ability to generate subharmonics is determined by the existence of a normal mode with small frequency mismatch between the fundamental and subharmonic displacements.

Acknowledgements. This work was supported by a grant from the Romanian Ministry of Research and Innovation, project PN-IIIP2-2.1-PED-2019-0085 CONTRACT 447PED/2020 (Acronim POSEIDON).

References

1. Nedelcu (Dulgheru), N., Chiroiu, V., Rugină, C., Munteanu, L., Ioan, R., Girip, I., Dragne, C.: Preparation of GeSbSe thin films by conventional melt-quenching method and studying their characteristics. Results in Physics, vol 16, Martie 2020, paper 102856 (2020)
2. Dulgheru (Nedelcu), N., Gartner, M., Anastasescu, M., Stoica, M., Nicolescu, M., Stroescu, H., Atkinson, I., Bratan, V., Stanculescu, I., Szekeres, A., Terziyska, P., Fabian, M.: Influence of compositional variation on the optical and morphological properties of GeSbSe films for optoelectronics application. Infrared Phys. Technol. **93**, 260–270 (2018)
3. Nedelcu, N., Chiroiu, V., Munteanu, L., Girip, I.: Optical nonlinearity in the GeSbSe chalcogenide glasses. Mater. Res. Expr. **7**(6), 1–15 (2020)
4. Nedelcu, N., Chiroiu, V., Munteanu, L., Girip, I.: Characterization of GeSbSe thin films synthesized by the conventional melt-quenching method. Spectroscopy - IR Spectroscopy for today's Spectroscopists, g 22–33, vol 35, S3, 22–33 (2020)
5. Dulgheru (Nedelcu), N.: Correlation of optical and morphostructural properties in chalcogenide compounds with applications in optoelectronics, PhD thesis, Romanian Academy (2019)
6. Dulgheru, N., Stoica, M., Calderon-Moreno, J.M., Anastasescu, M., Nicolescu, M., Stroescu, H., Atkinson, I., Stanculescu, I., Szekeres, A., Gartner, M.: Optical, morphological and durability studies of quaternary chalcogenide Ge-Sb(As)-(S, Te) films. Mater. Res. Bull. **106**, 234–242 (2018)
7. Nedelcu, N., Chiroiu, V., Munteanu, L., Rugină, C.: Uncertainties of transmittance and absorbance measurements of the chalcogenide thin films. Appl. Phys. A Mater. Sci. Process. 127–147, Article number:147 (2021)
8. Fabian, M., Dulgheru, N., Antonova, K., Szekeres, A., Gartner, M.: Investigation of the atomic structure of Ge-Sb-Se chalcogenide glasses. Adv. Condens. Matter Phys. Article 7158079 (2018)

9. Guyer, R.A., Johnson, P.A.: Nonlinear mesoscopic elasticity: evidence for a new class of materials. Phys. Today **52**, 30–35 (1999)
10. Delsanto, P.P. (ed.): The Universality of Nonclassical Nonlinearity, with Applications to NDE and Ultrasonics. Springer, New York (2006)
11. Hirsekorn, S., Delsanto, P.P.: On the universality of nonclassical nonlinear phenomena and their classification. Appl. Phys. Lett. **84**, 1413 (2004)
12. Moussatov, A., Gusev, V., Castagnede, B.: Self induced hysteresis for nonlinear acoustic waves in ´ cracked material. Phys. Rev. Lett. **90**, 124301-1–124301-4 (2003)
13. Solodov, I., Wackerl, J., Pfleiderer, K., Busse, G.: Nonlinear self-modulation and subharmonic acoustic spettroscopy for damage detection and location. Appl. Phys. Lett. **84**, 5386–5388 (2004)
14. Burnham, N.A., Kulik, A.J., Gremaud, G., Briggs, G.A.D.: Nanosubharmonics: The dynamics of small nonlinear contacts. Phys. Rev. Lett. **74**, 5092–5095 (1995)
15. Solodov, Y., Korshak, B.A.: Instability, chaos and "memory" in acoustic-wave–crack interaction. Phys. Rev. Lett. **88**, 014303-1–014303-3 (2002)
16. Nayfeh, A.H., Balachandran, B.: Modal interaction in dynamical and structural systems. ASME Appl. Mech. Rev. **42**, 175–202 (1989)
17. Nayfeh, A.H., Nayfeh, S.A., Anderson, T.A., Balachandran, B.: Transfer of energy from high frequency to low-frequency modes. In: Thompson, J.M., Bishop, S.R. (eds.) Nonlinearity and Chaos in Engineering Dynamics, pp. 39–58. New York, Wiley (1994)
18. Alippi, A., Shkerdin, G., Bettucci, A., Craciun, F., Molinari, E., Petri, A.: Low-threshold subharmonic generation in composites structures with Cantor–like code. Phys. Rev. Lett. **69**, 3318–3321 (1992)
19. Alippi, A., Angelici, M., Bettucci, A., Germano, M.: Surface mapping of nonresonant harmonic and subharmonic generated modes in finite piezoelectric structures. In Proceedings of of 16 th International Symphosium on Nonlinear Acoustics, Moscow, Vol. II, pp. 759–762 (2002)
20. Alippi, A., Albino, M., Angelici, M., Bettucci, A., Germano, M.: Space distribution of harmonic mode vibration amplitudes in nonlinear finite piezoelectric transducer. Ultrasonics **43**, 1–3 (2004)
21. Hayashi, C.: Nonlinear Oscillations in Physical Systems, pp. 86–97. Princeton University Press, Princeton, NJ (1985)
22. Eringen, A.C.: Nonlocal Continuum Field Theories. Springer Verlag (2002)
23. Eringen, A.C.: Microcontinuum Field Theories II Fluent Media. 1st Edition. Springer (2001)

Vibrations of a Rigid Hanged by an Elastic Kinematic Chain and a Rigid Kinematic Chain

Alin-Florentin Stan[1,2], Nicolae Pandrea[1], and Nicolae-Doru Stănescu[1(✉)]

[1] University of Pitesti, 110040 Pitesti, Romania
[2] IMSAR, 010141 Bucharest, Romania

Abstract. This paper presents a structure consisting of two simple kinematic chains for a hanged rigid. The analysis is performed in two cases: in case 1, both kinematic chains are elastic ones, and in case 2, one kinematic chain is elastic, while the other one is rigid. In both cases, we determine the equations of the free and forced vibrations. A numerical example highlights the theory.

Keywords: Screw coordinates · Elastic kinematic chain · Vibration

1 Introduction

The problems of elastic kinematic chains may be divided into two categories: problems concerning the mechanism in which one or more components are elastic ones, or problems dealing with platforms hanged by elastic kinematic chains. In the first situation, researchers study the case of two flexible elements linked one to another by different kinematic linkages, develop some numerical algorithms, and then apply the theory to a four-bar mechanism [1]. In the case of a simple mechanism, like the crank-shaft one, they [2] start from the second-order Lagrange equations and use a multibody approach. The analysis of the rigidity of kinematic chains in a particular case of manipulator for a platform is made in [3], while the case of a platform with elastic elements in kinematic chains is discussed in [4]. For a Stewart platform, one may consider that linkages are flexible, and use the Kane equations and principle of virtual work in order to perform the kinematic analysis and, by linearization of the moving equations, one may deduce the small oscillations of a satellite [5]. For the case of a double platform with flexible kinematic linkages, in Ref. [6] the authors performed the static analysis and proved that the rigidity of the manipulator depends on its position. The study of the mobile platforms may be realized using different algorithms [7], or writing special programs [8]. Pandrea [9] underlay the study of the mechanism and kinematic chains in plückerian coordinates and developed the corresponding linear elastic calculation in the screw coordinates.

2 Mechanical System

One considers the rigid AB (Fig. 1) hanged by the kinematic chains AC_1D_1 and AC_2D_2, each kinematic chain consisting of two elastic bars AC_1, C_1D_1, and BC_2, C_2D_2, respectively. The kinematic linkages at the points C_1 and C_2 are cylindrical ones. We attach the

N. Herisanu and V. Marinca (eds.), *Acoustics and Vibration of Mechanical Structures – AVMS-2021*, Springer Proceedings in Physics 274,
https://doi.org/10.1007/978-3-030-96787-1_8

reference system $Cxyz$ to the rigid AB and the reference frames $C_1x_1y_1z_1$ and $C_2x_2y_2z_2$ as in Fig. 1. One knows the dimensions l and a, and the mechanical and geometric properties of each bar. The rigid AB is acted by the forces **P**, which are parallel to the Cx-axis. The requirements are the study of the free and forced vibrations of the rigid AB in two cases: each kinematic chain is elastic, and the BC_2D_2 kinematic chain is rigid.

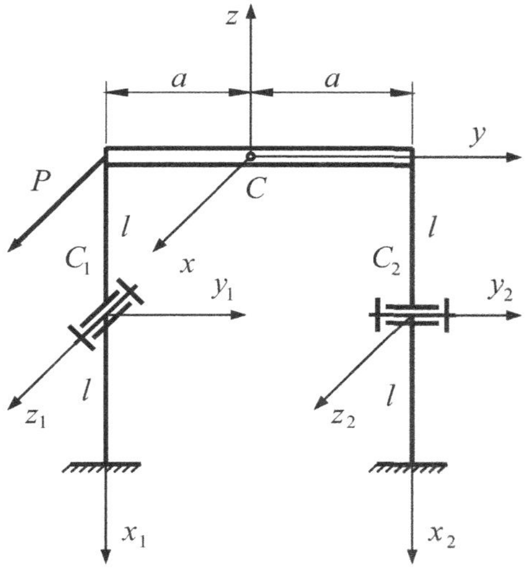

Fig. 1. Mechanical model.

3 Rigidity Matrix

We calculate:

- rotational matrices of the systems $C_ix_iy_iz_i$ with respect to the reference system $Cxyz$

$$[\mathbf{R}_1] = [\mathbf{R}_2] = \begin{bmatrix} 0 & 0 & 1 \\ 0 & 1 & 0 \\ -1 & 0 & 0 \end{bmatrix};$$

- the matrices $[\mathbf{G}_{Ci}] = \begin{bmatrix} 0 & -Z_{C_i} & Y_{C_i} \\ Z_{C_i} & 0 & -X_{C_i} \\ -Y_{C_i} & X_{C_i} & 0 \end{bmatrix}$, where X_{C_i}, Y_{C_i}, Z_{C_i} are the coordinates of the point C_i, $i = \overline{1,2}$, with respect to the reference frame $Cxyz$;
- the matrices $[\mathbf{T}_{Ci}] = \begin{bmatrix} [\mathbf{R}_i] & [\mathbf{0}] \\ [\mathbf{G}_{Ci}] & [\mathbf{R}_i] \end{bmatrix}$, $[\mathbf{T}_{Ci}]^{-1} = \begin{bmatrix} [\mathbf{R}_i]^{\mathrm{T}} & [\mathbf{0}] \\ [\mathbf{R}_i]^{\mathrm{T}}[\mathbf{G}_{Ci}]^{\mathrm{T}} & [\mathbf{R}_i]^{\mathrm{T}} \end{bmatrix}$, $i = \overline{1,2}$;
- the column matrices of the linkages at C_1 and C_2 relative to the $C_1x_1y_1z_1$ and $C_2x_2y_2z_2$ systems, $\{\mathbf{u}_{C_1}\} = \begin{bmatrix} 0 & 0 & 1 & 0 & 0 & 0 \end{bmatrix}^{\mathrm{T}}$, and $\{\mathbf{u}_{C_2}\} = \begin{bmatrix} 0 & 1 & 0 & 0 & 0 & 0 \end{bmatrix}^{\mathrm{T}}$, respectively;
- the column matrices of the plückerian coordinates of the linkages at points C_i, $i = \overline{1,2}$, relative to the $Cxyz$ reference system $\{\mathbf{U}_{C_i}\} = [\mathbf{T}_{C_i}]\{\mathbf{u}_{C_i}\}$, $i = \overline{1,2}$;

- the rigidity matrices of each kinematic chain ($i = \overline{1,2}$), relative to the $C_i x_i y_i z_i$ reference system

$$[\mathbf{k}_i] = \begin{bmatrix} 0 & 0 & 0 & \frac{E_i A_i}{l} & 0 & 0 \\ 0 & 0 & \frac{6E_i I_{z_i}}{l^2} & 0 & \frac{12E_i I_{z_i}}{l^3} & 0 \\ 0 & \frac{6E_i I_{y_i}}{l^2} & 0 & 0 & 0 & \frac{12E_i I_{y_i}}{l^3} \\ \frac{G_i I_{x_i}}{l} & 0 & 0 & 0 & 0 & 0 \\ 0 & \frac{4E_i I_{y_i}}{l} & 0 & 0 & 0 & -\frac{6E_i I_{y_i}}{l^2} \\ 0 & 0 & \frac{4E_i I_{z_i}}{l} & 0 & \frac{6E_i I_{z_i}}{l^2} & 0 \end{bmatrix}; \tag{1}$$

- the rigidity matrices of each kinematic chain, $i = \overline{1,2}$, relative to the $Cxyz$ reference system $\{\mathbf{K}_i\} = \left[\mathbf{T}_{C_i}\right][\mathbf{k}_i]\left[\mathbf{T}_{C_i}\right]^{-1}$.

4 Case of Elastic Kinematic Chains

In this situation, the rigidity matrix of the system reads

$$[\mathbf{K}] = [\mathbf{K}_1] + [\mathbf{K}_2]. \tag{2}$$

Denoting the inertial matrix of the rigid AB by $[\mathbf{M}]$,

$$[\mathbf{M}] = \begin{bmatrix} 0 & 0 & 0 & m & 0 & 0 \\ 0 & 0 & 0 & 0 & m & 0 \\ 0 & 0 & 0 & 0 & 0 & m \\ J_x & 0 & 0 & 0 & 0 & 0 \\ 0 & J_y & 0 & 0 & 0 & 0 \\ 0 & 0 & J_z & 0 & 0 & 0 \end{bmatrix}, \tag{3}$$

the column matrix of the displacements of the point C by $\{\mathbf{\Delta}\}$,

$$\{\mathbf{\Delta}\} = \left[\theta_X \; \theta_Y \; \theta_Z \; \delta_X \; \delta_Y \; \delta_Z\right]^{\mathrm{T}}, \tag{4}$$

one obtains the matrix equation of the free vibration as

$$[\mathbf{M}]\{\ddot{\Delta}\} + [\mathbf{K}]\{\mathbf{\Delta}\} = \{\mathbf{0}\}. \tag{5}$$

If the expression of force $\mathbf{P}$ is $\mathbf{P} = \mathbf{P}_0 \cos(\omega t + \varphi)\mathbf{i}$, where $\mathbf{i}$ is the unit vector of the Cx direction, then the column matrix of the forces reads

$$\{\mathbf{F}\} = \left[P_0 \cos(\omega t + \varphi) \; 0 \; 0 \; 0 \; 0 \; aP_0 \cos(\omega t + \varphi)\right]^{\mathrm{T}},$$

while the matrix equation of the forced vibrations is

$$[\mathbf{M}]\{\ddot{\Delta}\} + [\mathbf{K}]\{\mathbf{\Delta}\} = \{\mathbf{F}\}. \tag{6}$$

5 Case of a Rigid Kinematic Chain

In this situation, we denote by $\{\boldsymbol{\xi}_{C_2}\}$ the matrix of possible displacement in the linkage at the point C_2.

The equation of the free vibrations takes the from

$$\{\mathbf{U}_{C_2}\}^{\mathrm{T}}[\boldsymbol{\eta}][\mathbf{M}]\{\mathbf{U}_{C_2}\}\{\ddot{\xi}_{C_2}\} + \{\mathbf{U}_{C_2}\}^{\mathrm{T}}[\boldsymbol{\eta}][\mathbf{K}_1]\{\mathbf{U}_{C_2}\}\{\boldsymbol{\xi}_{C_2}\} = \{\mathbf{0}\}. \tag{7}$$

Taking into account that $\{\boldsymbol{\xi}_{C_2}\} = \left[0\ \tau_Y\ 0\ 0\ 0\ 0\right]^{\mathrm{T}}$ and performing the calculations required by Eq. (7), one gets

$$\left(ml^2 + J_Y\right)\ddot{\tau}_Y + \frac{4E_1 I_{y_1}}{l}\tau_Y = 0, \tag{8}$$

that is, a periodical motion in the form $\tau_Y = \tau_{Y_0}\cos(\omega_0 t + \varphi_0)$, where

$$\omega_0 = \sqrt{\frac{4E_1 I_{y_1}}{(ml^2 + J_Y)l}}, \tag{9}$$

while τ_{Y_0} and φ_0 are constants obtained from the initial conditions.

In the case of the forced vibrations, we have the equation

$$\{\mathbf{U}_{C_2}\}^{\mathrm{T}}[\boldsymbol{\eta}][\mathbf{M}]\{\mathbf{U}_{C_2}\}\{\ddot{\xi}_{C_2}\} + \{\mathbf{U}_{C_2}\}^{\mathrm{T}}[\boldsymbol{\eta}][\mathbf{K}_1]\{\mathbf{U}_{C_2}\}\{\boldsymbol{\xi}_{C_2}\} = \{\mathbf{U}_{C_2}\}^{\mathrm{T}}[\boldsymbol{\eta}]\{\mathbf{F}\}, \tag{10}$$

where

$$\{\mathbf{F}\} = \left[P_0\cos(\omega t + \varphi)\ 0\ 0\ 0\ 0\ 2aP_0\cos(\omega t + \varphi)\right]^{\mathrm{T}}. \tag{11}$$

It results in the equation

$$\left(ml^2 + J_Y\right)\ddot{\tau}_Y + \frac{4E_1 I_{y_1}}{l}\tau_Y = P_0 l\cos(\omega t + \varphi). \tag{12}$$

The square matrix $[\boldsymbol{\eta}]$ is given by

$$[\boldsymbol{\eta}] = \begin{bmatrix} 0\,0\,0\,1\,0\,0 \\ 0\,0\,0\,0\,1\,0 \\ 0\,0\,0\,0\,0\,1 \\ 1\,0\,0\,0\,0\,0 \\ 0\,1\,0\,0\,0\,0 \\ 0\,0\,1\,0\,0\,0 \end{bmatrix}. \tag{13}$$

6 Numerical Example

In this section, we will consider six cases. The following values are common for all cases: $E_1 = E_2 = 2.1 \times 10^{11}\,\mathrm{N/m^2}$, $G_1 = G_2 = 8.1 \times 10^{10}\,\mathrm{N/m^2}$, $l = 0.2$ m, $a = 0.2$ m, $m = 50$ kg, $r = 0.1$ m (the radius of the rigid AB considered as a homogeneous bar), $\omega = 20\,\mathrm{rad/s}$, $\varphi = 0$ rad, $t_{\max} = 3$ s (the interval of simulations), the bars of the kinematic chains are homogeneous. For the first, the third, and the fifth case, one has $P = 0$ N (free vibration), while for the second, the fourth, and the sixth case $P = 1000$ N (force vibrations). The first and the second case are characterized by $d_1 = 0.005$ m (the diameter of the bars of the first kinematic chain), $d_2 = 0.005$ m (the diameter of the bars of the second kinematic chain), the force $P = 0$ N, while the initial conditions are: $\theta_X = 0.002$ rad, $\theta_Y = -0.001$ rad, $\theta_Z = 0.001$ rad, $\delta_X = 0.001$ m, $\delta_Y = 0.002$ m, $\delta_Z = -0.001$ m, $\dot{\theta}_X = \dot{\theta}_Y = \dot{\theta}_Z = 0\,\mathrm{rad/s}$, $\dot{\delta}_X = \dot{\delta}_Y = \dot{\delta}_Z = 0$ m/s. For the third and the fourth case, we consider $d_1 = 0.010$ m, and $d_2 = 0.005$ m, while for the fifth and the second case $d_1 = 0.005$ m, $d_2 = 0.005$ m. The initial conditions for the second, the third, and the fourth case are the same as in the first case, while for the fifth and the sixth case, the initial conditions are: $\tau_Y = 0.002$ rad, $\dot{\tau}_Y = 0\,\mathrm{rad/s}$. The results are captured in the next figures (Figs. 2, 3, 4, 5, 6, and 7). In all figures θ_X and δ_X were drawn in blue, θ_Y and δ_Y were drawn in green, while θ_Z and δ_Z were drawn in red. The reader may easily observe the periodicity of the free and forced vibrations in all cases. In the first four cases, the variable δ_Z presents a transitory period in which the amplitude decreases, after this period, the oscillations being periodical.

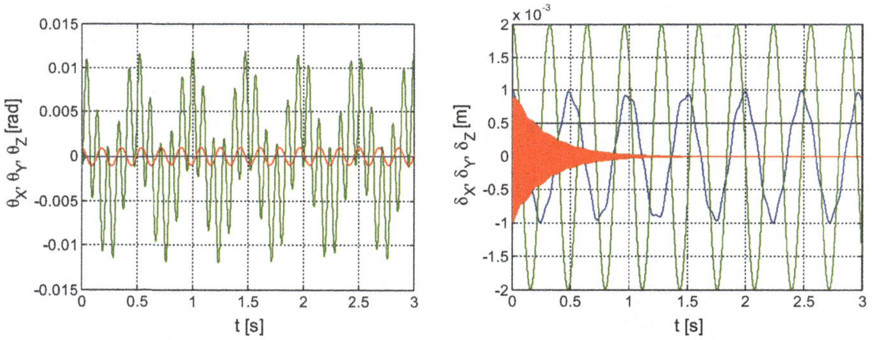

Fig. 2. Time histories in the first case.

In the fifth and the sixth case, the amplitudes of the parameters θ_X, θ_Z, δ_Y, and δ_Z are equal to zero.

It results that when the two kinematic chains are elastic, then the vibration of the rigid AB is a complex one, the motion having non-zero components. In the case when a kinematic chain is a rigid one, the dynamics of the rigid simplifies and consists only in a rotational motion about the Y-axis, and a translational one along the X axis. The same observation concerning the diminishing of the amplitude of vibrations may be also stated in the third and fourth case when the bars which compound the first kinematic chain have greater rigidities.

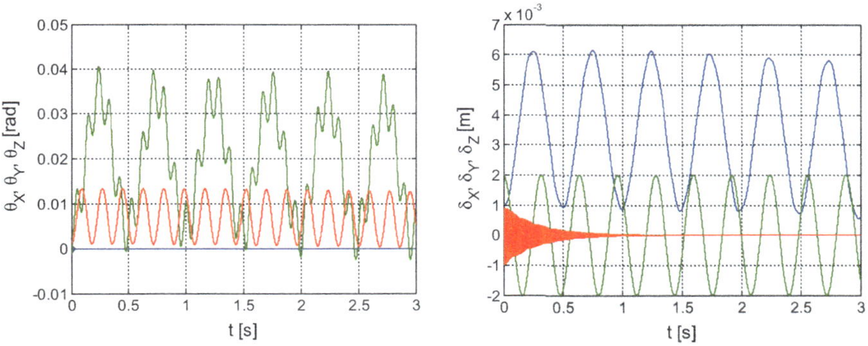

Fig. 3. Time histories in the second case.

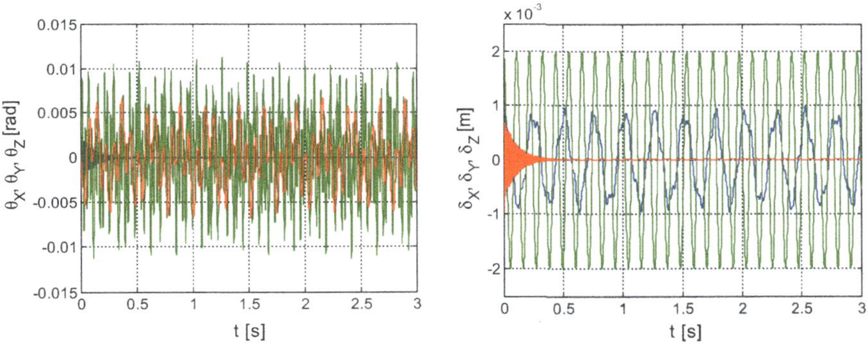

Fig. 4. Time histories in the third case.

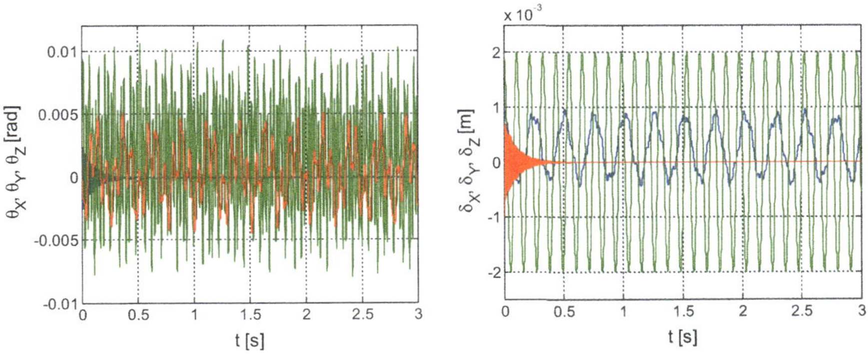

Fig. 5. Time histories in the fourth case.

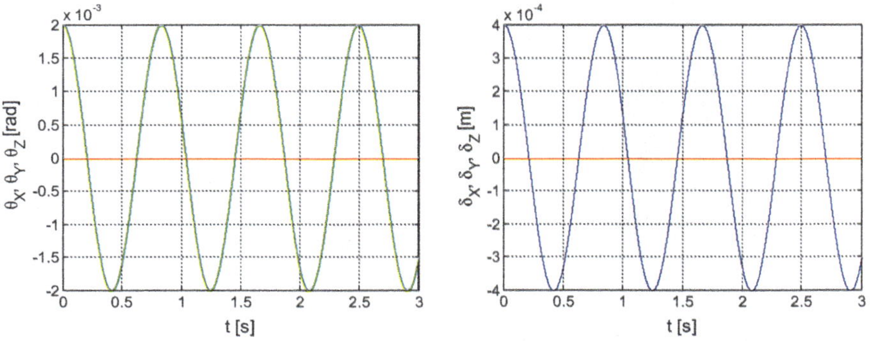

Fig. 6. Time histories in the fifth case.

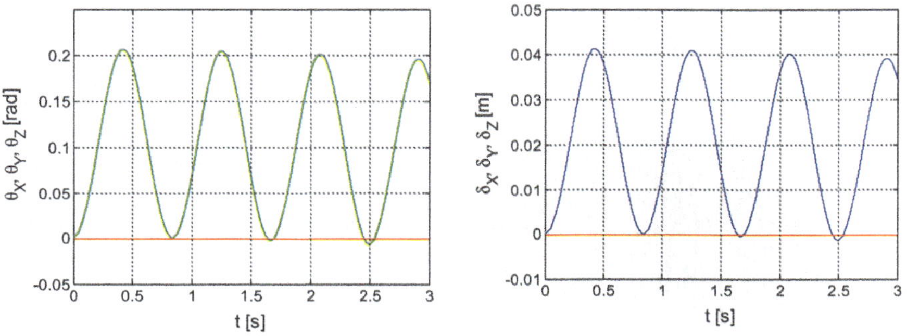

Fig. 7. Time histories in the sixth case.

7 Conclusions

This paper deals with the study of the dynamics of the rigid hanged by two elastic kinematic chains or hanged by an elastic kinematic chain and a rigid kinematic chain. In both situations, we determined the rigidity matrix of the system and the equations of the free and forced vibration.

The numerical results prove that in the case when both kinematic chains are elastic, the dynamics of the rigid simplifies, remaining periodic

In many cases, the platforms are acted by asymmetric loads resulting in different reactions in the legs; consequently, the legs may have different rigidities. In this way, it is possible that one leg has a rigidity much greater than the other ones. It results that the flexibility matrix that corresponds to the kinematic chain assigned to that leg may be neglected compared to the flexibility matrices of the rest of the kinematic chain and, consequently, the flexibility matrix of the system may be calculated only as a function of the flexibility matrices of the rest of the kinematic chains. The results obtained using this approach are different from those obtained using the approach proposed in this paper. Our model considers that the hanged rigid solid with one rigid kinematic chain has a constrained motion; hence the model may be generalized to systems with more general constraints and kinematic chains.

In our future studies, we will consider more complicated cases of the mechanical system with one rigid kinematic chain and several elastic kinematic chains with different acting forces, and we will discuss their dynamics. Greater attention will be paid to the platforms, especially the Stewart ones.

References

1. Hwang, Y.-L.: Dynamic recursive decoupling method for closed-loop flexible mechanical systems. Int. J. Non-Linear Mech. **41**, 1181–1190 (2006)
2. Marques, F., Roupa, I., Silva, M., T., Flores, P., Lankarani, H.M.: Examination and comparison of different methods to model closed loop kinematic chains using Lagrangian formulation with cut joint, clearance joint constraint and elastic joint approaches. Mech. Mach. Theory **160**, 104294 (2021)
3. Sun, T., Lian, B., Song, Y.: Stiffness analysis of a 2DoF over-constrained RPM with an articulated traveling platform. Mech. Mach. Theory **96**, 165–178 (2016)
4. Jia, Y.-H., Xu, S.-J., Hu, Q.: Dynamics of a spacecraft with large flexible appendage constrained by multi-strut passive damper. Acta. Mech. Sin. **29**(2), 294–308 (2013)
5. Jiao, J., Wu, Y., Yu, K., Zhao, R.: Dynamic modeling and experimental analyses of Stewrt platform with flexible hinges. J. Vib. Contr. **25**(1), 151–171 (2018)
6. Shan, X., Cheng, G.: Static analysis on a 2(3PUS+S) parallel manipulator with two moving platforms. J. Mech. Sci. Technol. **32**(8), 3869–3876 (2018)
7. Ma, S., Duan, W.: Dynamic coupled analysis of the floating platform using the asynchronous coupling algorithm. J. Mar. Sci. Appl. **13**, 85–91 (2014)
8. Inner, B., Kucuk, S.: A novel kinematic design, analysis and simulation tool for general Stewart platforms. Simul. Trans. Soc. Model. Simul. Int. **89(7)**, 876–897 (2013)
9. Pandrea, N.: Elements of Solids Mechanics in Plückerian Coordinates. The Publishing House of the Romanian Academy, Bucharest (2000)

Nonlinear Vibrations in a Milling Process

Elena-Larisa Predel and Nicolae-Doru Stănescu(✉)

University of Pitesti, 110040 Pitesti, Romania

Abstract. The present paper proposes a new more general model for the study of the stability of the milling process. The model starts from two previous ones with two and three degrees of freedom, respectively, and generalizes them to the most complex possible one, with four degrees of freedom. This model is expected to produce more complex dynamics than the previous models. The results of the simulations are also presented.

Keywords: Nonlinear vibrations · Milling · Simulation

1 Introduction

The problem of modeling the milling manufacturing process is an old and still actual one. The unequal tooth pitch leads to chatter vibration [1] and the modeling implies differential equations with multiple constant delays which are very difficult to be used in the determination of the stability zones. If the workpiece is a flexible one, then the manufacturing process leads to regenerative chatter vibration [2], which may be avoided using active control. Other researchers develop new models with multiple degrees of freedom in order to study the complex dynamics of milling [3]. The signals obtained in the manufacturing process are transformed and analyzed to determine if the process is a stable or an unstable one [4]. Testing of the stability of the milling process may be realized using virtual machining [5], or by experimental validation [6]. The study of the asymmetric structural elements and the influence of the nonlinear regeneration in the discontinuous milling process is presented in [7]. The machining behavior of some special milling cutters used in the manufacturing of some special alloys is investigated in reference [8]. The stability of the micro-milling operations is discussed in [9] using the phase portrait, Poincaré section, and instantaneous frequency. Chaotic dynamics in the milling process are studied using a two degree of freedom model, the investigation being conducted with the aid of the phase portrait, bifurcation diagram, and the largest Lyapunov exponent [10]. A new method for the prediction of chatter in milling is proposed in [11]. A new linear model for the study of the stability lobes in the milling process is described in [12], while in [13], the authors discuss the stability of a particular case of milling.

A general simple model for the study of the nonlinear vibrations in the cutting process was introduced by Grabec [14]. The model has only two degrees of freedom. This model was extended in [15] considering now that the vibrations of the technological

N. Herisanu and V. Marinca (eds.), *Acoustics and Vibration of Mechanical Structures – AVMS-2021*, Springer Proceedings in Physics 274,
https://doi.org/10.1007/978-3-030-96787-1_9

system are characterized by three degrees of freedom. Both models are nonlinear and may conduct chaotic vibrations. In our study, we will consider a more complex model with four degrees of motion which is assumed to present more complex dynamics than the previous ones.

2 Mechanical Model

We consider the milling of a circular axis of radius R_p with a cylindrical milling cutter of radius R_s (Fig. 1). We will denote by O_p the center of the workpiece, and by O_s the center of the tool, and let w_s be the depth of cut.

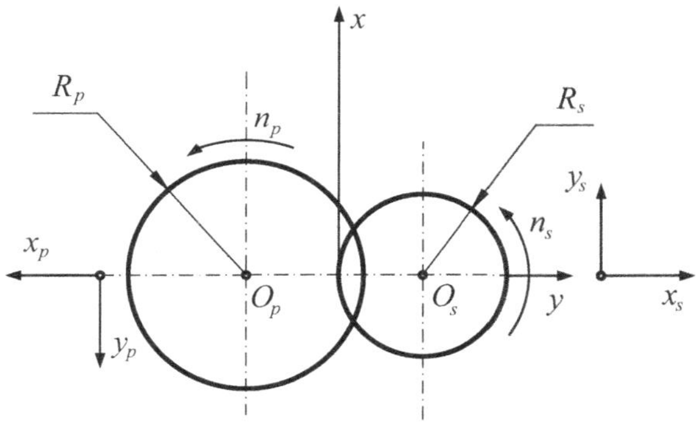

Fig. 1. Mechanical model.

It immediately results in the relations $O_pO = R_p - w_s$ and $O_sO = R_s$. The cutting process is an orthogonal one, and the cutting edge of each tooth of the milling cutter is parallel to the workpiece's surface and normal to the cutting direction. We will assume that the depth of cut is much smaller than the width of the cutting edge of the tool, so that one gets a two-dimensional cutting process (situated in the plane Oxy). The cutting edge of the milling cutter is set at the depth of the cut w_s. The cutting process may be described by two oscillators as in Fig. 2. It results in the differential equations of motion in the form.

$$m_s\ddot{x}_s + c_{sx}\dot{x}_s + k_{sx}x_s = F_x,\ m_s\ddot{y}_s + c_{sy}\dot{y}_s + k_{sy}y_s = F_y, \tag{1}$$

for the tool, and

$$m_p\ddot{x}_p + c_{px}\dot{x}_p + k_{px}x_p = F_x,\ m_p\ddot{y}_p + c_{py}\dot{y}_p + k_{py}y_p = F_y, \tag{2}$$

for the workpiece, respectively.

The cutting force is assumed to depend on only the depth of cut w and the relative velocity v between the tool and workpiece. The force F_x is the main cutting force, while the force F_y is the friction force, between them exists a relation of proportionality in the

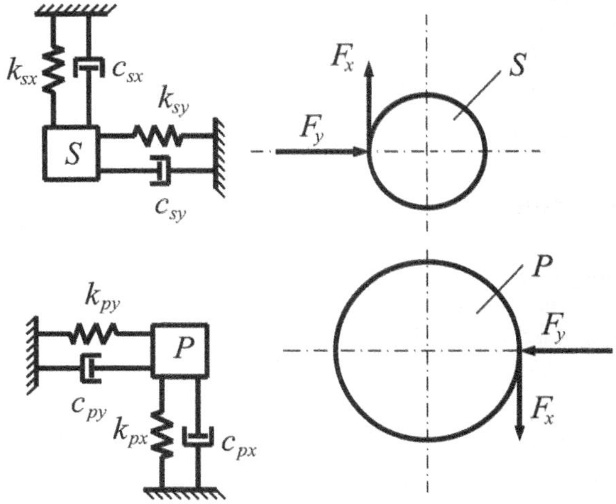

Fig. 2. The two oscillators.

form [14, 15] $F_y = KF_x$; this relation implies that the two oscillators are coupled. There exist the relations [14, 15]

$$F_x = F_{x0}\frac{w}{w_0}\left[C_1\left(\frac{v}{v_0}-1\right)^2+1\right], \tag{3}$$

$$K = K_0\left[C_2\left(\frac{v}{v_0}-1\right)^2+1\right]\left[C_3\left(\frac{w}{w_0}-1\right)^2+1\right], \tag{4}$$

where F_{x_0}, K_0, w_0, v_0, C_1, C_2, and C_3 are constants that depend on the cutting process.

The friction coefficient is described by the relative flow velocity of the workpiece's material with respect to the tool onto the x-direction. On the other hand, the friction is a consequence of the slipping of the material along the tool's surface along the y-direction; it results that, because of the plastic deformation, the slipping velocity along the y-direction is diminished and we will mark this thing by a factor R

$$v_f = \frac{v}{R}. \tag{5}$$

Dynamically, the friction coefficient is given by the relation

$$K = K_0\left[C_2\left(\frac{v_f R}{v_0}-1\right)^2+1\right]\left[C_3\left(\frac{w}{w_0}-1\right)^2+1\right]. \tag{6}$$

If we denote by φ the friction angle, then $R = \cot\varphi$;since, in general, the angle φ depends on the cutting velocity, one may write a relation in the form

$$R = R_0\left[C_4\left(\frac{v}{v_0}-1\right)^2+1\right], \tag{7}$$

where R_0 and C_4 are constants depending on the cutting conditions.

The references [14, 15] state that the previous relations are experimental ones that are valid between the average values of the dynamical variables. We will assume that these relations also hold true between the instantaneous values of the variables.

First of all, we will refer to Fig. 3, where we marked by O_p' and O_s' the positions of the centers of the workpiece and tool, respectively, at a given moment of time. From the figure, one deduces the expression of the depth of cut

$$w = R_p + R_s - O_p'O_s' = R_p + R_s - \sqrt{(R_p - w_s + R_s + y_p + y_s)^2 + (x_p + x_s)^2}. \quad (8)$$

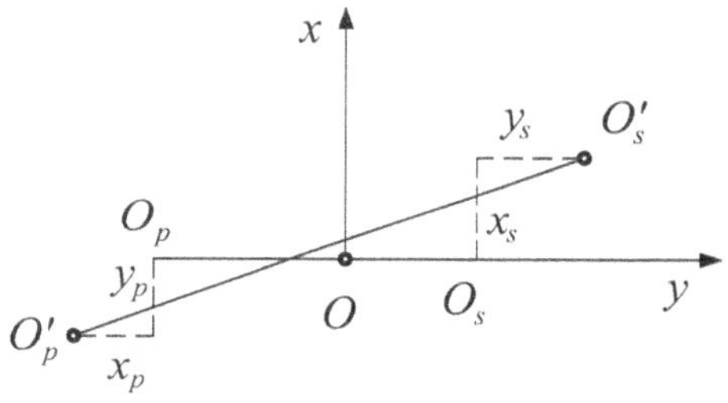

Fig. 3. The depth of cut.

In this way, one obtains the previously mentioned relations

$$w(t) = R_p + R_s - \sqrt{[R_p - w_s + R_s + y_p(t) + y_s(t)]^2 + [x_p(t) + x_s(t)]^2}, \quad (9)$$

$$v(t) = v_i - \Delta v_i, \quad (10)$$

$$v_f(t) = \frac{v(t)}{R(t)} - [\dot{y}_s(t) + \dot{y}(t)], \quad (11)$$

where $v(t)$ is the instantaneous cutting velocity, $v_i = R_s\omega_s \pm R_p\omega_p$, while $\Delta v(t)$ is the variation of the cutting velocity, $\Delta v(t) = \dot{x}_s(t) + \dot{x}_p(t)$. One takes the sign + if the tool and the workpiece have opposite rotational senses, and the sign—if they have the same rotational sense. We have denoted by ω_p and ω_s the angular velocities of the workpiece and tool, respectively; these angular velocities read $\omega_p = \frac{\pi n_p}{30}$, $\omega_s = \frac{\pi n_s}{30}$, where n_p and n_s are the revolutions of the workpiece and tool, respectively, given in rev/min.

The oscillations of the workpiece and tool may lead to situations in which $w(t) < 0$, $v(t) < 0$ or $v_f(t) < 0$. We have to add the conditions $F_x = 0$ for $w(t) < 0$ or $v(t) < 0$, and $K(-v_f) = -K(v_f)$.

The main cutting force is given by a nonlinear expression in function of the depth of cut w, that is the expression (3) must be corrected in the form

$$F_x = F_{x0}\left(\frac{w}{w_0}\right)^p \left[C_1\left(\frac{v}{v_0} - 1\right)^2 + 1\right], \quad (12)$$

where p is a polytropic exponent.

Taking into account the previous discussion, it results in the expressions

$$F_x = F_{x0}\left(\frac{w}{w_0}\right)^p\left[C_1\left(\frac{v}{v_0}-1\right)^2+1\right]\Theta(w(t))\Theta(v(t)), \tag{13}$$

$$K_F = K_{F_0}\left[C_2\left(\frac{v_f}{v_0}-1\right)^2+1\right]\left[C_3\left(\frac{w}{w_s}-1\right)^2+1\right]\Theta(F_x)\operatorname{sgn}(v_f), \tag{14}$$

where $\Theta(a)$ and $\operatorname{sgn}(a)$ are the Heaviside, and sign functions, respectively.

3 The Non-Dimensional System

We introduce the non-dimensional variables: $T = t\frac{v_0}{w_0} = t\omega_0$, $X_p = \frac{x_p}{w_0}$, $Y_p = \frac{y_p}{w_0}$, $X_s = \frac{x_s}{w_0}$, $Y_s = \frac{y_s}{w_0}$, $V_i = \frac{v_i}{v_0}$, $V = V_i - X_s' - X_p'$, $V_f = V_i - RY_p' - RY_s'$, $W_s = \frac{w_s}{w_0}$, and $\overline{R}_p = \frac{R_p}{w_0}$, $\overline{R}_s = \frac{w_s}{w_0}$, where the superior index prime signifies the derivative with respect to the non-dimensional time, $X' = \frac{\mathrm{d}X}{\mathrm{d}T}$, $X'' = \frac{\mathrm{d}^2X}{\mathrm{d}T^2}$, etc. We also define the parameter $W = W_s - \sqrt{\left(\overline{R}_p+\overline{R}_s-W_s+Y_p-Y_s\right)^2+\left(X_p+X_s\right)^2}$, and use the notations $C_{px} = \frac{c_{px}w_0}{m_pv_0} = \frac{c_{px}}{m_p\omega_0}$, $C_{py} = \frac{c_{py}w_0}{m_pv_0} = \frac{c_{py}}{m_p\omega_0}$, $C_{sx} = \frac{c_{sx}w_0}{m_sv_0} = \frac{c_{sx}}{m_s\omega_0}$, $C_{sy} = \frac{c_{sy}w_0}{m_sv_0} = \frac{c_{sy}}{m_s\omega_0}$, $K_{px} = \frac{k_{px}w_0^2}{m_pv_0^2}$, $K_{py} = \frac{k_{py}w_0^2}{m_pv_0^2}$, $K_{sx} = \frac{k_{sx}w_0^2}{m_sv_0^2}$, $K_{sy} = \frac{k_{sy}w_0^2}{m_sv_0^2}$, $F_0 = \frac{F_{x0}}{m_p\omega_0^2w_0}$, $F_1 = \frac{F_{x0}}{m_s\omega_0^2w_0} = F_0\frac{m_p}{m_s} = \lambda F_0$, $F = F_0W^p\left[C_1(V-1)^2+1\right]\Theta(W)\Theta(V)$, $F_s = \lambda F_0W^p\left[C_1(V-1)^2+1\right]\Theta(W)\Theta(V)$, $R = R_0\left[C_4(V-1)^2+1\right]$, and $K_F = K_{F_0}\left[C_2\left(V_f-1\right)^2+1\right]\left[C_3(W-1)^2+1\right]\operatorname{sgn}\left(V_f\right)\Theta(F)$, in which m_p and m_s are the masses of the workpiece and tool, respectively.

One obtains the non-dimensional system

$$X_p'' + C_{px}X_p' + K_{px}X_p = F_0W^p\left[C_1(V-1)^2+1\right]\Theta(W)\Theta(V), \tag{15}$$

$$\begin{gathered} Y_p'' + C_{py}Y_p' + K_{py}Y_p \\ = K_{F_0}F_0W^p\left[C_1(V-1)^2+1\right]\left[C_2\left(V_f-1\right)^2+1\right]\left[C_3(W-1)^2+1\right] \\ \Theta(w)\Theta(V)\Theta(F)\operatorname{sgn}\left(V_f\right), \end{gathered} \tag{16}$$

$$X_s'' + C_{sx}X_s' + K_{sx}X_s = \lambda F_0W^p\left[C_1(V-1)^2+1\right]\Theta(W)\Theta(V), \tag{17}$$

$$\begin{gathered} Y_s'' + C_{sy}Y_p' + K_{sy}Y_p \\ = \lambda K_{F_0}F_0W^p\left[C_1(V-1)^2+1\right]\left[C_2\left(V_f-1\right)^2+1\right]\left[C_3(W-1)^2+1\right] \\ \Theta(w)\Theta(V)\Theta(F)\operatorname{sgn}\left(V_f\right). \end{gathered} \tag{18}$$

The system formed by Eqs. (15), (16), (17), and (18) is a system of four second-order nonlinear differential equations. For the numerical integration, it must be transformed in a system of first-order differential equations using the notations $Z_1 = X_p$, $Z_2 = Y_p$, $Z_3 = X_s$, $Z_4 = Y_s$, $Z_5 = X_p'$, $Z_6 = Y_p'$, $Z_7 = X_s'$, $Z_8 = Y_s'$.

4 Numerical Example

As a numerical example, we consider the milling of a cylindrical workpiece in three cases. The common parameters are [14, 15]: $p = 0.8$, $F_0 = 0.5$, $C_1 = 0.3$, $C_2 = 0.7$, $C_3 = 1.5$, $C_4 = 1.2$, $K_{F_0} = 0.36$, $\lambda = 1$, $R_0 = 2.2$, $w_0 = 0.25$, $v_0 = 6.6\,\text{m/s}$, $\omega_0 = 2.64 \times 10^4\,\text{rad/s}$, $R_s = 40$ mm, $R_p = 25$ mm, $w_s = 0.25$ mm, $n_p = 10\,\text{rev/min}$, the initial conditions $X_p = Y_p = X_s = Y_s = X_p' = Y_p' = X_s' = Y_s' = 0$, $t_{\max} = 10$ s (time of simulation), $dt = 10^{-3}$ s (increment of time), $C_{px} = C_{py} = C_{sx} = C_{sy} = 0$. In addition, the first case is characterized by $K_{sx} = 2$, $K_{px} = 1.9$, $K_{sy} = 1.25$, $K_{py} = 1.2$, and $n_s = 3000\,\text{rev/min}$; the second one by $K_{sx} = 1$, $K_{px} = 0.9$, $K_{sy} = 0.25$, $K_{py} = 0.2$, and $n_s = 1500\,\text{rev/min}$, while the third one is defined by $K_{sx} = 1$, $K_{px} = 0.9$, $K_{sy} = 0.25$, $K_{py} = 0.2$, and $n_s = 3000\,\text{rev/min}$. The results of simulations are captured below (X_p in blue, Y_p in green, X_s in red, and Y_s in cyan). The reader may see the complex dynamics involved by this model. One may state the absence of any period in the motion (Fig. 4).

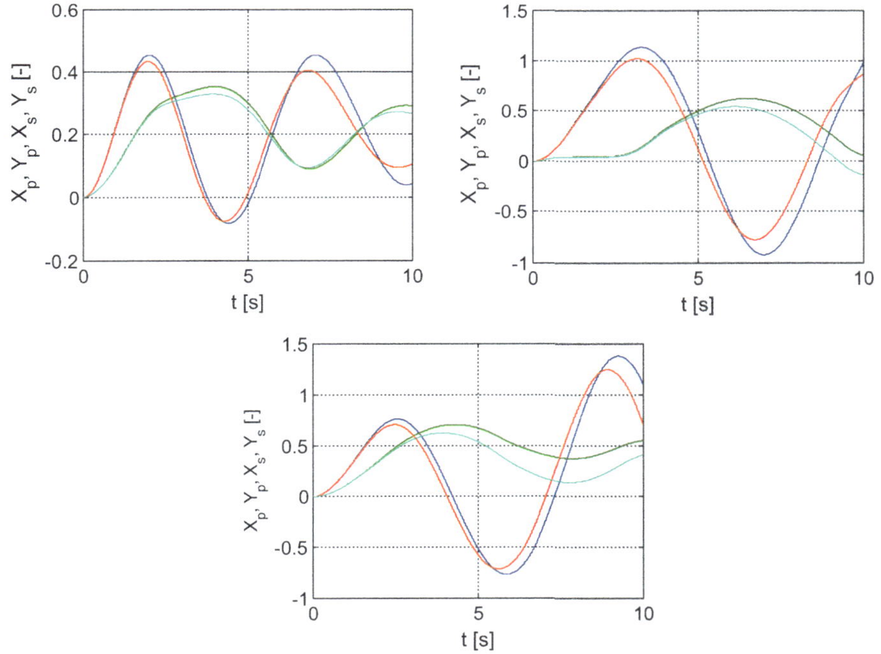

Fig. 4. Time histories in the three cases.

5 Conclusions

In this paper, we present a new mechanical model for the study of the vibrations which occur during the milling process. The system has four degrees of freedom. The equations are highly nonlinear implying a complex behavior of the technological process, validated by the shapes of the curves obtained by simulation. The next steps of analysis must include the phase portrait, Poincaré section, entropy, cutting force, etc., for a better description of the dynamics. This will be the goal of our future work.

References

1. Sellmeier, V., Denkena, B.: Stable islands in the stability chart of milling processes due to unequal tooth pitch. Int. J. Mach. Tools Manuf. **51**, 152–164 (2011)
2. Parus, A., Powałka, B., Marchelek, K., Domek, S., Hoffmann, M.: Active control vibration in milling flexible workpieces. J. Vib. Contr. **19**(7), 1103–1120 (2012)
3. Balachandran, B., Zhao, M., X.: A Mechanics based model for the study of dynamics of milling operation, Meccanica **35**, 89–109 (2000)
4. Susanto, A., Liu, C.-H., Yamada, K., Hwang, Y.-R., Tanaka, R., Sekiya, K.: Application of Hilbert-Huang transform for vibration signal analysis in end-milling. Precis. Eng. **53**, 263–277 (2018)
5. Bąk, P.A., Jemelniak, K.: Self-excited vibrations avoidance methodology in non-linear numerical simulation environment. Procedia CIRP **62**, 245–249 (2017)
6. Bachrathy, D., Munoa, J., Stepan, G.: Experimental validation of appropriate axial immersions for helical mills. Int. J. Adv. Manuf. Technol. **84**, 1295–1302 (2016)
7. Mann, B.P., Garg, N.K., Young, K.A., Helvey, A.M.: Milling bifurcations from structural asymmetry and nonlinear regeneration. Nonlinear Dyn. **42**, 319–337 (2005)
8. Dombovari, Z., Altintas, Y., Stepan, G.: The effect of serration on mechanics and stability of milling cutters. Int. J. Mach. Tools Manuf. **50**, 511–520 (2010)
9. Liu, M.-K., Halfmann, E.B., Suh, C.S.: Multi-dimensional time frequency control of micro-milling instability. J. Vib. Contr. **20**(5), 643–660 (2014)
10. Banihasan, M., Bakhtiari-Nejad, F.: Chaotic vibrations in high-speed milling. Nonlinear Dyn. **66**, 557–574 (2011)
11. Liu, X., Li, R., Yang, L., Yue, C.: A prediction method of milling chatter stability for complex surface mold. Int. J. Adv. Manuf. Technol. **89**, 2637–2648 (2017)
12. Geng, Z., Ridgway, K., Turner, S.: Linear improvement of the machining stability lobes and application in milling process prediction. Proc. Int. Inst. Mech. Eng. Part B: J. Eng. Manuf. **223**, 369–378 (2009)
13. Jin, G., Zhang, Q., Hao, S., Xie, Q.: Stability prediction of milling process with variable pitch and variable helix cutters. Proc. Int. Inst. Mech. Eng., Part C J. Mech. Eng. Sci. **228**(2), 281–293 (2014)
14. Grabec, I.: Chaotic dynamics of the cutting process. Int. J. Mach. Tools Manuf. **28**(1), 19–32 (1988)
15. Stanescu, N.-D., Pandrea, N., Fressengeas, C.: Can grinding be chaotic? 13th international congress on sound and vibration. ICSV 2006 **7**, 5417–5424 (2006)

Vibrations of a Planar System of Cantilever Beams Cylindrically Jointed. Part I: Theory

Maria-Luiza Beşliu-Gherghescu and Nicolae-Doru Stănescu(✉)

University of Pitesti, 110040 Pitesti, Romania

Abstract. This paper is an extension and generalization of a previous paper. The system considered has an arbitrary number of cantilever beams. We will prove that the parameters of the system, for arbitrary forces and large elastic deformations, may be determined only if additional conditions are considered. These supplementary conditions cannot be arbitrary and they must take into account some particularities of the mechanical system. It is also possible that some obvious supplementary conditions be redundant.

Keywords: Cantilever beams · Cylindrical joint · Large deformations

1 Introduction

In a previous paper [11], we discussed the case of a simple system consisting of two cantilever beams, cylindrically jointed one to another and we proved that the solution may be obtained only considering some supplementary conditions. Now we make a step forward considering arbitrary systems of planar cantilever beams acted by forces situated in the same plane and we ask about their deformations and the forces that appear in the system.

The theory presented here considers that the deformations of the cantilever beams are high, but elastic ones, so the classical approximation used in the strength of materials is no longer valid [1–11]. The practical situation corresponds to the parts obtained from plastic materials by Fused Deposition of Materials (FDM) technology. For the case of small deformations in which one considers only the bending moments and assumes that the axial forces are negligible, the problem reduces to a linear system and the solution becomes trivial.

2 Mechanical Model

We consider the system in Fig. 1 consisting of cantilever beams of type $O_{i_k}A$ and cantilever beams of type $O_{j_l}A$. At the point A , the beams are cylindrical jointed one to another. The system is a planar one and at the point A it is acted by a system of planar forces reduced at the force **P**. We denote by OXY the general reference system (Fig. 2).

N. Herisanu and V. Marinca (eds.), *Acoustics and Vibration of Mechanical Structures – AVMS-2021*, Springer Proceedings in Physics 274,
https://doi.org/10.1007/978-3-030-96787-1_10

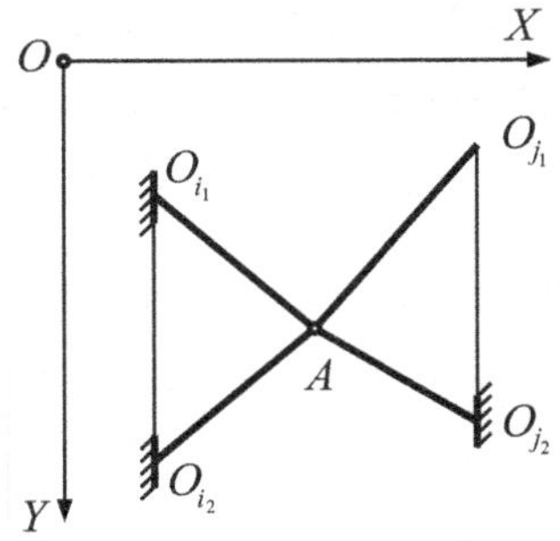

Fig. 1. Mechanical model

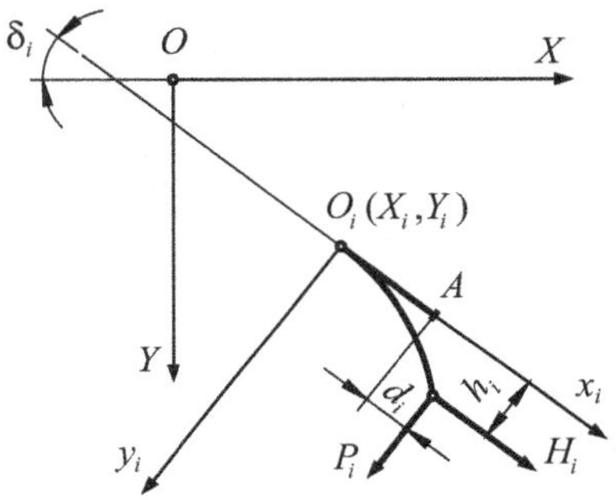

Fig. 2. The case of the O_iA beam

In the beginning, we isolate a beam of type O_iA to which we attach the local reference $O_ix_iy_i$ and we denote by δ_i the angle between the O_ix_i and OX axes. Let X_i, Y_i be the coordinates of the point O_i with respect to the system OXY. The passing from the local system $O_ix_iy_i$ to the system OXY is performed with the aid of the formula

$$[\mathbf{R}_i] = \left[\mathbf{R}_{O_i}\right] + [\mathbf{A}_i]\begin{bmatrix} x_i \\ y_i \end{bmatrix}, \tag{1}$$

where

$$[\mathbf{R}_i] = \begin{bmatrix} X \\ Y \end{bmatrix}, \ \left[\mathbf{R}_{O_i}\right] = \begin{bmatrix} X_i \\ Y_i \end{bmatrix}, \tag{2}$$

X, Y being the coordinates relative to the reference system OXY, x_i, y_i being the coordinates in the reference system $O_ix_iy_i$, while $[\mathbf{A}_i]$ is the rotation matrix of the system $O_ix_iy_i$ with respect to the system OXY; in our case

$$[\mathbf{A}_i] = \begin{bmatrix} \cos\delta_i & -\sin\delta_i \\ \sin\delta_i & \cos\delta_i \end{bmatrix} \tag{3}$$

After the application of the external forces, the beam O_iA of initial length L_{i0} deforms, the end A reaching at the position A'; that is, it has a displacement d_i along the direction O_ix_i and a displacement h_i along the direction O_iy_i.

The new length of the bar O_iA reads

$$L_i = \int_0^{L_{i0}+d_i} \sqrt{1 + (y_i')^2}\mathrm{d}x_i \tag{4}$$

or, equivalently,

$$\begin{aligned} L_i &= L_{i0} + \frac{1}{E_i(x_i)A_i(x_i)} \int_0^{L_{i0}} (H_i \cos\theta_i + P_i \sin\theta_i)\mathrm{d}x_i \\ &= L_{i0} + \frac{1}{E_i(x_i)A_i(x_i)} \int_0^{L_{i0}} \frac{H_i + P_i y_i'}{\sqrt{1 + (y_i')^2}}\mathrm{d}x_i, \end{aligned} \tag{5}$$

where we assumed that at the end A', we have the forces H_i along the direction O_ix_i, and P_i along the direction O_iy_i; in addition (Fig. 3).

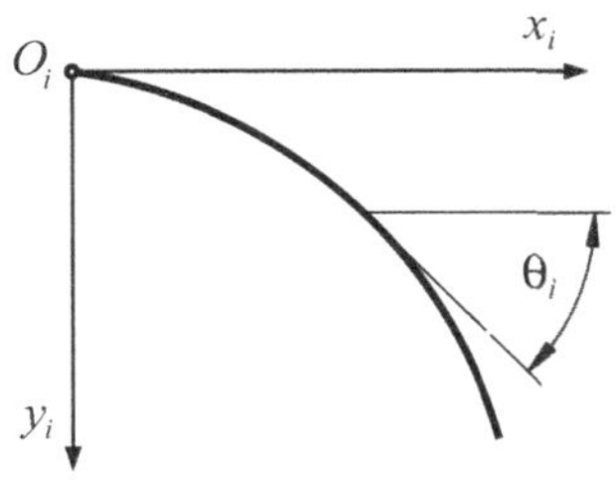

Fig. 3. Curvature of the beam

$$\tan\theta_i = \frac{\mathrm{d}y_i}{\mathrm{d}x_i} = y_i', \ \cos\theta_i = \frac{1}{\sqrt{1 + (y_i')^2}}, \ \sin\theta_i = \frac{y_i'}{\sqrt{1 + (y_i')^2}} \frac{-b \pm \sqrt{b^2 - 4ac}}{2a}. \tag{6}$$

We may also write

$$\frac{y_i''}{\left[1 + (y_i')^2\right]^{\frac{3}{2}}} = \frac{P_i(L_{i0} - x_i) - H_i(h_i - y_i)}{E_i(x_i)I_i(x_i)} \tag{7}$$

and

$$h_i = y_i(L_{i0} + d_i). \tag{8}$$

From the Eqs. (1) and (3), one gets

$$\begin{bmatrix} X_{A'} \\ Y_{A'} \end{bmatrix} = \begin{bmatrix} X_{O_i} \\ Y_{O_i} \end{bmatrix} + \begin{bmatrix} \cos\delta_i & -\sin\delta_i \\ \sin\delta_i & \cos\delta_i \end{bmatrix} \begin{bmatrix} L_{i0} + d_i \\ h_i \end{bmatrix}, \tag{9}$$

while the projections of the forces H_i and P_i onto the directions OX and OY are

$$H_X = H_i \cos\delta_i,\ H_Y = H_i \sin\delta_i,\ P_X = -P_i \sin\delta_i,\ P_Y = P_i \cos\delta_i. \tag{10}$$

We now proceed analogically for the beams of type O_jA for which (Fig. 4) the passing to the system OXY being performed with the aid of the intermediary system $O_jX'Y'$ and using the formulae

$$[\mathbf{R}'_j] = \begin{bmatrix} X'_j \\ Y'_j \end{bmatrix} = [\mathbf{A}_j] \begin{bmatrix} x_j \\ y_j \end{bmatrix}, \tag{11}$$

$$[\mathbf{R}_j] = \begin{bmatrix} D_j \\ E_j \end{bmatrix} + \begin{bmatrix} -X'_j \\ Y'_j \end{bmatrix} = \begin{bmatrix} D_j - X'_j \\ E_j + Y'_j \end{bmatrix}. \tag{12}$$

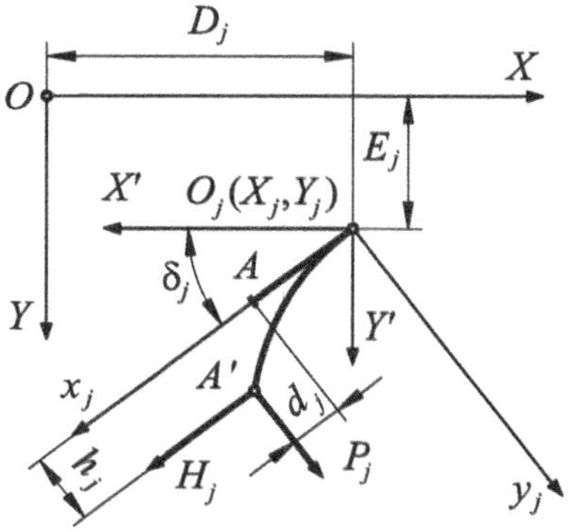

Fig. 4. The case of the O_jA beam

One may also write the relations

$$L_j = L_{j0} + \frac{1}{E_j(x_j)A_j(x_j)} \int_0^{L_{j0}} \frac{H_j + P_j y'_j}{\sqrt{1 + \left(y'_j\right)^2}} \mathrm{d}x_j, \tag{13}$$

$$\tan\theta_j = \frac{\mathrm{d}y_j}{\mathrm{d}x_j} = y'_j,\ \cos\theta_j = \frac{1}{\sqrt{1 + \left(y'_j\right)^2}},\ \sin\theta_j = \frac{y'_j}{\sqrt{1 + \left(y'_j\right)^2}}, \tag{14}$$

$$\frac{y_j''}{\left[1+\left(y_j'\right)^2\right]^{\frac{3}{2}}}=\frac{P_j\left(L_{j0}-x_j\right)-H_j\left(h_j-y_j\right)}{E_j\left(x_j\right)I_j\left(x_j\right)}, \tag{15}$$

$$h_j=y_j\left(L_{j0}+d_j\right). \tag{16}$$

3 Algorithm and Comments

The algorithm consists of the following steps:

Input data

- the components P and H of the forces that act at the point A;
- the laws of deformations $y_i = y_i(x_i)$.

We assume as known the displacement h and d of the point A.

Step 1: One determines the displacements h_i and d_i;

Step 2: One determines the new lengths of the beams;

Step 3: One calculates:

$$\alpha_i=\int_0^{L_{i0}}\frac{\mathrm{d}x_i}{\sqrt{1+\left(y_i'\right)^2}},\quad \beta_i=\int_0^{L_{i0}}\frac{y_i'}{\sqrt{1+\left(y_i'\right)^2}}\mathrm{d}x_i, \tag{17}$$

resulting in the relations

$$\alpha_i H_i+\beta_i P_i=E_i A_i(L_i-L_{i0}), \tag{18}$$

where $i=\overline{1,m}$, m being the number of beams.

Step 4: One writes

$$\sum_{i=1}^{m}\mathbf{H}_i+\sum_{i=1}^{m}\mathbf{P}_i=\mathbf{P}+\mathbf{H} \tag{19}$$

resulting in two new relations.

At this moment, we have m equations in the form (18) and 2 equations in the form (19), that is, a total of $m+2$ equations. The number of unknowns is $2m+2$ (the forces H_i and P_i, $i=\overline{1,m}$, and the displacements d and h).

It results that the problem is possible if and only if $m=1$. If $m>1$, then one needs m additional relations.

4 Conclusions

The theory presented here is valid no matter the number m of cantilever beams. Moreover, the deformations of the beams are arbitrary but elastic. The reader may ask what kind of relation is needed. The answer is more complicated and it will be discussed in detail in the second part of the paper. The statement that if one knows the shapes of the beams, then the problem can be solved is not necessarily true. The statement that we have $m+2$ equations is valid if and only if these equations are independent.

References

1. Fertis, D.G.: Nonlinear Mechanics. CRC Press, Boca Raton (1999)
2. Huang, X., Yu, T.X., Lu, G., Lippmann, H.: Large deflection of elastoplastic beams with prescribed moving and rotating ends. Proc Inst. Mech. Eng., Part C: J. Mech. Eng. Sci. **217**, 1001–1013 (2003)
3. Jiao, P., Alavi, A.H., Borchani, W., Lajnef, N.: Small and large deformation models of post-buckled beams under lateral constraints. Math. Mech. Solids **24**(2), 386–405 (2019)
4. Kang, Y.A., Li, X.F.: Large deflections of a non-linear cantilever functionally graded beam. J. Reinf. Plast. Comp. **29**(12), 1761–1774 (2010)
5. Nada, A.A., Hussein, B.A., Megahed, S.M., Shabana, A.A.: Use of the floating frame of reference formulation in large deformation analysis: experimental and numerical validation. Proc. Inst. Mech. Eng., Part K: J. Multi-body Dyn. **224**, 45–58 (2009)
6. Kopmaz, O., Gündoğdu, Ö.: On the curvature of an Euler-Bernoulli beam. Int. J. Mech. Eng. Educ. **31**(2), 132–142 (2003)
7. Stănescu, N.D., Beşliu–Gherghescu, M.L., Rizea, A., Anghel, D.: Determination of the shape of a beam obtained by fused deposition material with general loads, innovative manufacturing engineering and energy (IMANEE 2019)—50 years of higher technical education at the University Of Pitesti, Book Series: IOP Conference Series-Materials Science and Engineering 564, 012057 (2019)
8. Teodorescu, P., Stănescu, N.D., Pandrea, N.: Numerical Analysis with Applications in Mechanics and Engineering. John Wiley & Sons, Hoboken (2013)
9. Yu, H., Zhao, C., Zheng, B., Wang, H.: A new higher-order locking-free beam element based on the absolute nodal coordinate formulation. Proc. Inst. Mech. Eng., Part C: J. Mech. Eng. Sci. **232**(19), 3410–3423 (2018)
10. Zhao, C., Yu, H., Zheng, B., Wang, H.: New stiffened plate elements based on the absolute nodal coordinate formulation. Proc. Inst. Mech. Eng., Part K: J. Multi-body Dyn. **231**(1), 213–229 (2017)
11. Beliu-Gherghescu, M.L., Stanescu, N.-D., Pandrea, N., Popa, D.: Study of the vibrations of a system consisting in cantilever beams, In: Herisanu N., Marinca V. (eds) Acoustics and Vibration of Mechanical Structures—AVMS 2019. Springer Proceedings in Physics **251**, 59–66, (2021)

Vibrations of a Planar System of Cantilever Beams Cylindrically Jointed. Part II: Example

Maria-Luiza Beşliu-Gherghescu and Nicolae-Doru Stănescu(✉)

University of Pitesti, 110040 Pitesti, Romania

Abstract. This paper is the second part of the paper that presents the general theory. It contains a complete solved example, which highlights all parts of the theory. It is also proved that sometimes the system of equations may not contain independent ones and some additional conditions have to be considered. The solution must be validated by experiments.

Keywords: Cantilever beams · Cylindrical joint · Large deformations

1 Introduction

This second part of the paper is dedicated to a single example. Even if it looks simple, the solution of it involves some technical difficulties. Some assumptions may look obvious. However, the solution has to be tested by practice.

As we already proved in the first part [12], we need some additional conditions. The obvious supplementary conditions are those involving the shapes of the deformed beams [1–11], which are considered here to be governed by the same expression, the values of the coefficients being different.

2 Numerical Example

Let us consider the system of four cantilever beams O_1A, O_2A, O_3A, and O_4A captured in Fig. 1 in which $O_1O_2O_3O_4$ is a square, and the beams are identical. At point A the four beams are cylindrically joined to one another. The system is acted by the vertical force P. We have $O_1O_2 = O_1O_3 = O_2O_4 = O_3O_4 = 2l$. In addition, we assume that the displacements of the point A are d along the OX axis and h along the OY axis. Using the reference systems in Fig. 5 and the notations previously described, one may write:

$$\begin{bmatrix} x_1 \\ y_1 \end{bmatrix} = \frac{\sqrt{2}}{2}\begin{bmatrix} 1 & 1 \\ -1 & 1 \end{bmatrix}\begin{bmatrix} l+d \\ l+h \end{bmatrix}, \begin{bmatrix} x_2 \\ y_2 \end{bmatrix} = \frac{\sqrt{2}}{2}\begin{bmatrix} 1 & -1 \\ 1 & 1 \end{bmatrix}\begin{bmatrix} l+d \\ -l+h \end{bmatrix},$$
$$\begin{bmatrix} x_3 \\ y_3 \end{bmatrix} = \frac{\sqrt{2}}{2}\begin{bmatrix} 1 & 1 \\ -1 & 1 \end{bmatrix}\begin{bmatrix} l-d \\ l+h \end{bmatrix}, \begin{bmatrix} x_4 \\ y_4 \end{bmatrix} = \frac{\sqrt{2}}{2}\begin{bmatrix} 1 & -1 \\ 1 & 1 \end{bmatrix}\begin{bmatrix} l-d \\ -l+h \end{bmatrix}, \quad (1)$$

N. Herisanu and V. Marinca (eds.), *Acoustics and Vibration of Mechanical Structures – AVMS-2021*, Springer Proceedings in Physics 274,
https://doi.org/10.1007/978-3-030-96787-1_11

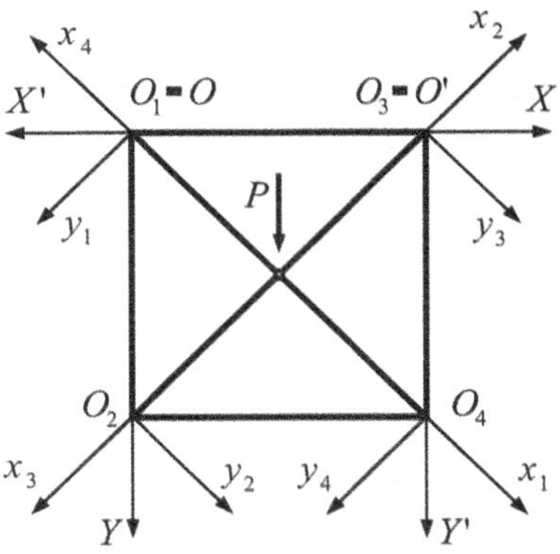

Fig. 1. Mechanical system.

where from $d_1 = (d+h)\frac{\sqrt{2}}{2}$, $h_1 = (h-d)\frac{\sqrt{2}}{2}$, $d_2 = (d-h)\frac{\sqrt{2}}{2}$, $h_2 = (d+h)\frac{\sqrt{2}}{2}$, $d_3 = (h-d)\frac{\sqrt{2}}{2}$, $h_3 = (d+h)\frac{\sqrt{2}}{2}$, $d_4 = -(d+h)\frac{\sqrt{2}}{2}$, $h_4 = (h-d)\frac{\sqrt{2}}{2}$.

We approximate the shapes of the deformed beams by parabolas $y_i = a_i x_i^2$ getting

$$L_i = \frac{1}{4a_i}\ln\left(2a_i(L_0+d_i) + \sqrt{1+4a_i^2(L_0+d_i)^2}\right) + \frac{1}{2}(L_0+d_i)\sqrt{1+4a_i^2(L_0+d_i)^2}, \tag{2}$$

$$\alpha_i = \int_0^{L_{i0}} \frac{\mathrm{d}x_i}{\sqrt{1+(y_i')^2}} = \frac{1}{4a_i}\ln\left(2a_i L_{i0} + \sqrt{1+4a_i^2 L_{i0}^2}\right) + \frac{1}{2}L_{i0}\sqrt{1+4a_i^2 L_{i0}^2}, \tag{3}$$

$$\beta_i = \int_0^{L_{i0}} \frac{y_i'}{\sqrt{1+(y_i')^2}}\mathrm{d}x_i = \frac{1}{2a_i}\left(\sqrt{1+4a_i^2 L_{i0}^2} - 1\right), \tag{4}$$

where $L_{i0} = l\sqrt{2}$, $i = \overline{1,4}$.

It results in the equation

$$\alpha_i H_i + \beta_i P_i = EA\Delta L_i,\ \ i = \overline{1,4}. \tag{5}$$

From the symmetry of the mechanical system, one may state the following relation

$$P_1 = P_3,\ P_2 = P_4,\ H_1 = H_3,\ H_2 = H_4. \tag{6}$$

The equations of projections onto OX and OY axes read (Fig. 2):

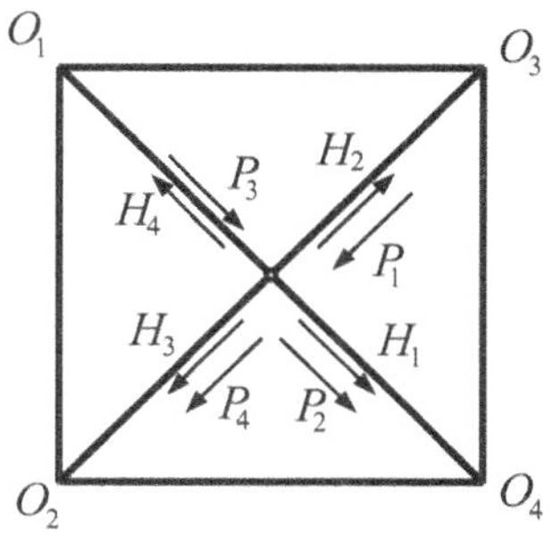

Fig. 2. The forces.

$$\begin{aligned}&(P_1 + P_3 + P_2 + P_4 - H_4 - H_2 + H_1 + H_3)\frac{\sqrt{2}}{2} = P,\\ &(-H_4 + H_2 + P_2 + P_3 - P_4 - H_3 + H_1 - P_1)\frac{\sqrt{2}}{2} = 0.\end{aligned} \tag{7}$$

It is easy to observe that the second Eq. (7) reduces to $0 = 0$, that is, the Eqs. (5), (6), and (7) form a system of 9 equations with 10 unknown (H_i, P_i, $i = \overline{1, 4}$, d, and h).

We need the tenth equation. One may state that taking into account the existence of a single vertical force, then $d = 0$ (the displacement of the point A along the OX-axis). In this situation, it results in the system.

$$\begin{aligned}&\alpha_1 H_1 + \beta_1 P_1 = EA\Delta L_1,\ \alpha_2 H_2 + \beta_2 P_2 = EA\Delta L_2,\\ &P_1\left(1 - \frac{\beta_1}{\alpha_1}\right) + P_2\left(1 + \frac{\beta_2}{\alpha_2}\right) = \frac{P}{\sqrt{2}} + EA\left(\frac{\Delta L_2}{\alpha_2} - \frac{\Delta L_1}{\alpha_1}\right),\end{aligned} \tag{8}$$

which can be solved if we add a new relation between P_1 and P_2. We state that this new relation is

$$P_1 = P_2 \tag{9}$$

resulting

$$P_1 = P_2 = \frac{\frac{P}{\sqrt{2}} + EA\left(\frac{\Delta L_2}{\alpha_2} - \frac{\Delta L_1}{\alpha_1}\right)}{2 - \frac{\beta_1}{\alpha_1} + \frac{\beta_2}{\alpha_2}},\ H_1 = \frac{EA\Delta L_1 - \beta_1 P_1}{\alpha_1},\ H_2 = \frac{EA\Delta L_2 - \beta_2 P_2}{\alpha_2}. \tag{10}$$

As we already stated, we consider that the shapes of the beams are described by the equations

$$y_i = a_i x_i^2 \tag{11}$$

obtaining

$$h_i = a_i(L_0 + d_i)^2,\ a_i = \frac{h_i}{(L_0 + d_i)^2} = \frac{h_i}{\left(l\sqrt{2} + d_i\right)^2},\ i = \overline{1, 4}. \tag{12}$$

Taking into account that

$$\frac{y_i''}{\left[1 + (y_i')^2\right]^{\frac{3}{2}}} = \frac{P_i(L_{i0} - x_i) - H_i(h_i - y_i)}{E_i(x_i)I_i(x_i)}, \tag{13}$$

(see part I), and writing this equation for $x_1 = 0$, one obtains

$$\frac{2a_1}{1} = \frac{P_1L_{10} - H_1a_1\left(L_{10} + \frac{h\sqrt{2}}{2}\right)^2}{EI}, \tag{14}$$

$$h = 2\left(\sqrt{\frac{P_1 l\sqrt{2} - 2a_1EI}{2H_1a_1}} - l\right). \tag{15}$$

The algorithm of calculation is the following one:

Input data: E, l, I, P, $\min = 10^{10}$, $b = 1$, $L_{i0} = l\sqrt{2}$, $i = \overline{1, 2}$, $h_0 = 10^{-6}$, $dh = 10^{-6}$.

Step 1: One calculates: $a_1 = \frac{h\sqrt{2}}{(2l+h)^2}$, $a_2 = a_1$, $d_1 = \frac{h\sqrt{2}}{2}$, $d_2 = -d_1$, $L_i = \frac{1}{4a_i}\ln\left[2a_i(L_0 + d_i) + \sqrt{1 + 4a_i^2(L_0 + d_i)^2}\right] + \frac{1}{2}(L_0 + d_i)\sqrt{1 + 4a_i^2(L_0 + d_i)^2}$, $\Delta L_i = L_i - L_{i0}$, $\alpha_i = \frac{1}{4a_i}\ln\left(2a_iL_{i0} + \sqrt{1 + 4a_i^2L_{i0}^2}\right) + \frac{1}{2}L_{i0}\sqrt{1 + 4a_i^2L_{i0}^2}$, $\beta_i = \frac{1}{2a_i}\left(\sqrt{1 + 4a_i^2L_{i0}^2} - 1\right)$, $i = \overline{1, 2}$, $P_1 = \frac{\frac{P}{\sqrt{2}} + EA\left(\frac{\Delta L_2}{\alpha_2} - \frac{\Delta L_1}{\alpha_1}\right)}{2 - \frac{\beta_1}{\alpha_1} - \frac{\beta_2}{\alpha_2}}$, $H_1 = \frac{EA\Delta L_1 - \beta_1P_1}{\alpha_1}$, $h^* = 2\left(\sqrt{\frac{P_1l\sqrt{2} - 2a_1EI}{2H_1a_1}} - l\right)$, $\Delta h = h^* - h$.

Step 2: If $|\Delta h| < \min$ then $\min = |\Delta h|$, $b = 1$, $h = h + dh$; else $b = 0$.

Step 3: If $b = 1$ then return to Step 1.

As numerical values we consider $l = 0.25$ m, $d = 5 \times 10^{-3}$ m (diameter of the beams), $A = \frac{\pi d^2}{4}$, $I = \frac{\pi d^4}{64}$, $t_{\max} = 3$ s (maximum time of simulation), $dt = 10^{-3}$ s (increment of time), and $P = P_0 + P_1\sin(\omega t) + P_2\sin(2\omega t)$. In the first case, we have $E = 2.1 \times 10^8\,\mathrm{N/m^2}$, $P_0 = 100$ N, $P_1 = 10$ N, $P_2 = 20$ N, and $\omega = 20\,\mathrm{rad/s}$, while in the second case, the values are $E = 3 \times 10^7\,\mathrm{N/m^2}$, $\omega = 20\,\mathrm{rad/s}$, $P_0 = 15$ N, $P_1 = 5$ N, and $P_2 = 7$ N. The results of simulations are given in the next figures (Figs. 3, 4, 5 and 6).

One may easily observe the periodicity of the variation for each parameter. Their amplitudes depend on both the amplitude of the force P and the rest of the mechanical

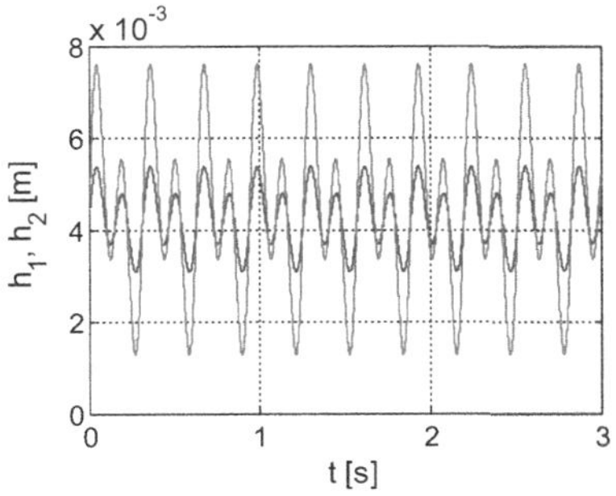

Fig. 3. Time history $h = h(t)$ for the first case (in blue), and the second case (in green).

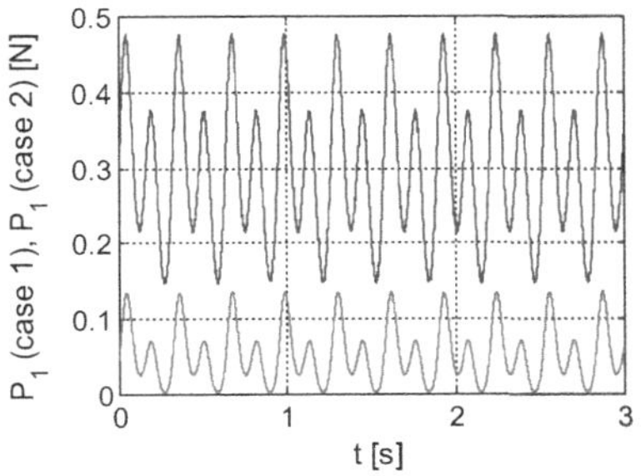

Fig. 4. Time history $P_1 = P_1(t)$ for the first case (in blue), and the second case (in green).

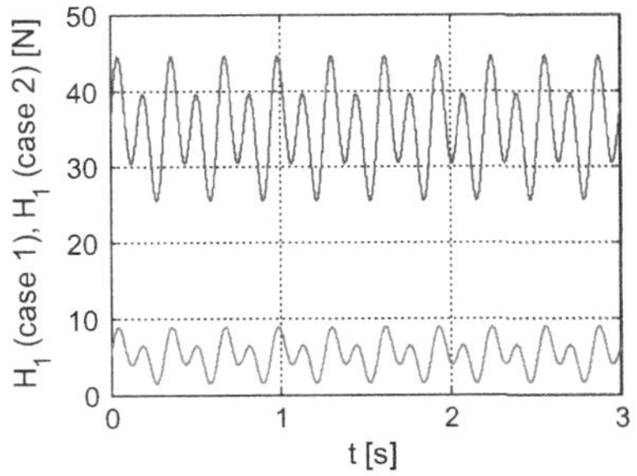

Fig. 5. Time history $H_1 = H_1(t)$ for the first case (in blue), and the second case (in green).

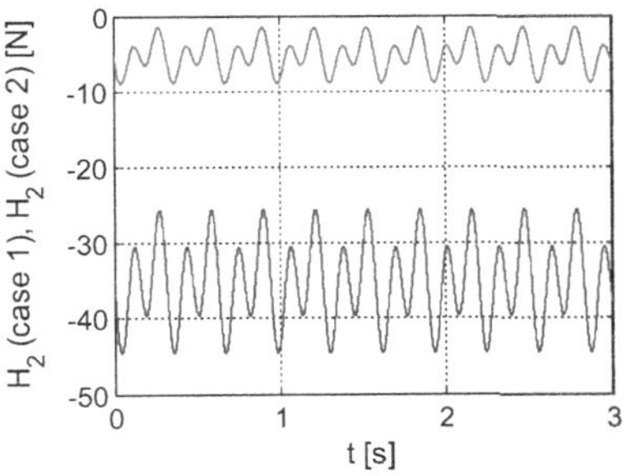

Fig. 6. Time history $H_2 = H_2(t)$ for the first case (in blue), and the second case (in green).

parameters (in this situation, the Young modulus E). These dependencies of the parameters can be stated to be direct or inversely proportional to the force P and the Young modulus E.

For instance, when the force P and the Young modulus E decrease, the amplitude h of the oscillations of the point O increases, the amplitude of the force P_1 decreases (the force becomes negative, but the same statement holds true also for the absolute value of the force P_1), the amplitude of the force P_2 decreases (with the same observation as in the case of the force P_1), while the amplitude of the force H_1 increases (the force becomes positive, but the same statement holds true also for the absolute value of the force H_1).

The periods of all parameters (h, P_1, P_2, and H_1) are the same and equal to the period of the force P.

3 Conclusions

The example presented here shows the application of the theory developed in part I of the paper. The independence of the equations from which the solution of the problem is obtained remains an open problem. Some additional conditions can be considered obvious (e.g., $d = 0$ in this example), but other ones do not have mathematical proof (e.g., $y_i = a_i x_i^2$). The practice may validate some conditions and may invalidate other ones. It results that for a better solution, one must have knowledge about this supplementary conditions.

The algorithm for obtaining the solution implies the use of numerical methods. It is known that the convergence of these methods is assured only if some hypotheses are fulfilled, that is, one has to know some information about the solution.

References

1. Fertis, D.G.: Nonlinear Mechanics. CRC Press, Boca Raton (1999)
2. Huang, X., Yu, T., X., Lu, G., Lippmann, H.: Large deflection of elastoplastic beams with prescribed moving and rotating ends. Proc. Inst. Mech. Eng., Part C: J. Mech. Eng. Sci. **217**, 1001–1013 (2003)
3. Jiao, P., Alavi, A.H., Borchani, W., Lajnef, N.: Small and large deformation models of post-buckled beams under lateral constraints. Math. Mech. Solids **24**(2), 386–405 (2019)
4. Kang, Y., A., Li, X., F.: Large deflections of a non-linear cantilever functionally graded beam. J. Reinf. Plast. Comp. **29**(12), 1761–1774 (2010)
5. Nada, A.A., Hussein, B.A., Megahed, S.M., Shabana, A.A.: Use of the floating frame of reference formulation in large deformation analysis: experimental and numerical validation. Proc. Inst. Mech. Eng., Part K: J. Multi-body Dyn. **224**, 45–58 (2009)
6. Kopmaz, O., Gündoğdu, Ö.: On the curvature of an Euler-Bernoulli beam. Int. J. Mech. Eng. Educ. **31**(2), 132–142 (2003)
7. Stănescu, N.D., Beşliu–Gherghescu, M.L., Rizea, A., Anghel, D.: Determination of the shape of a beam obtained by fused deposition material with general loads, innovative manufacturing engineering and energy (IMANEE 2019) - 50 years of higher technical education at the University of Pitesti, Book Series: IOP Conference Series-Materials Science and Engineering 564, 012057, (2019)

8. Teodorescu, P., Stănescu, N., D., Pandrea, N.: Numerical Analysis with Applications in Mechanics and Engineering. John Wiley & Sons, Hoboken (2013)
9. Yu, H., Zhao, C., Zheng, B., Wang, H.: A new higher-order locking-free beam element based on the absolute nodal coordinate formulation. Proc. Inst. Mech. Eng., Part C: J. Mech. Eng. Sci. **232**(19), 3410–3423 (2018)
10. Zhao, C., Yu, H., Zheng, B., Wang, H.: New stiffened plate elements based on the absolute nodal coordinate formulation. Proc. Inst. Mech. Eng., Part K: J. Multi-body Dyn. **231**(1), 213–229 (2017)
11. Besliu-Gherghescu, M.L., Stanescu, N.-D., Pandrea, N., Popa, D.: Study of the vibrations of a system consisting in cantilever beams, In: Herisanu N., Marinca V. (eds) Acoustics and Vibration of Mechanical Structures—AVMS 2019. Springer Proceedings in Physics **251**, 59–66 (2021)
12. Besliu-Gherghescu, M.L., Stanescu, N.-D.: Vibrations of a planar system of cantilever beams cylindrically jointed. Part I: Theory, AVMS 2021 (in press)

Oscillations of a Vehicle Equipped with Hydraulic Shock Absorbers

Bogdan Marinca(✉) and Eugen Tudor

Politehnica University Timişoara, Bd. M. Viteazul Nr. 1, 300222 Timişoara, România
bogdan.marinca@upt.ro

Abstract. In the present paper, the nonlinear oscillations of the suspension of a vehicle equipped with hydraulic shock absorbers is studied using Optimal Auxiliary Functions Method (OAFM). By neglecting the elasticity of the tires and the coupling between the solutions of the front and rear axles, the vibrations of the vehicle that are symmetrical with respect to the longitudinal axis may be studied on a single degree of freedom. Analytical solutions and natural frequencies of the system are calculated. Our results obtained through this procedure are in very good agreement with numerical results, which provides the accuracy of the method.

1 Introduction

Suspensions with nonlinear damping characteristics are frequently utilized, because the nonlinearity limits displacements and velocities, reduces the extreme values of the acceleration, and leads to a more uniform dynamic loading of the suspension. The damping nonlinearity is usually achieved by using hydraulic shock absorbers.

The oscillator with symmetric quadratic damping characteristics has been minutely studied. Magnus [1] obtained some analytical solutions for the differential equations of the oscillator with strong quadratic damping and Andronov et al. [2] provided a qualitative analysis of the oscillator with quadratic damping. The trajectories are plotted in the phase plane. Dincă and Teodosiu [3] considered the successive approximations to study the dissipative systems with dry friction and with quadratic damping. Cvetičanin [4] analyzed exact solutions by using the Jacobi elliptic function and approximate analytical by using the elliptic-harmonic-balance method. For other details, see [5–13].

In this work, we propose a relative novel approach, the Optimal Auxiliary Functions Method [14–21] to investigate the nonlinear oscillations of a vehicle equipped with hydraulic shock absorbers. Unlike other solutions, the method applied to find approximate analytical solutions to nonlinear dynamical systems, the proposed procedure is based upon the original construction of a solution using a moderate number of convergence-control parameters on every subdomain. The conditions of the continuity and derivability of the approximate solutions of two adjacent subintervals are fulfilled.

These parameters are basic components of the original auxiliary functions introduced in our paper, which lead to a high precision, comparing our approximate solutions with numerical ones.

The accuracy of obtained results is proved by numerical investigations, with validated analytical results.

N. Herisanu and V. Marinca (eds.), *Acoustics and Vibration of Mechanical Structures – AVMS-2021*, Springer Proceedings in Physics 274,
https://doi.org/10.1007/978-3-030-96787-1_12

2 The Optimal Auxiliary Functions Method (OAFM)

In order to develop an application of the OAFM, let us consider the nonlinear differential equation

$$L[x(t)] + N[x(t)] = 0, \quad t \in D, \tag{1}$$

where L is the linear operator, N is the nonlinear operator, t being an independent variable, $x(t)$ an unknown function at this stage, and D is the domain of interest. The initial boundary conditions are known:

$$B\left(x(t), \frac{dx(t)}{dt}\right) = 0. \tag{2}$$

An exact solution of a strongly nonlinear equation of types (1) and (2) is often very hard to find [14]. To find an approximate solution $\overline{x}(t)$, we suppose that this can be expressed in the form:

$$\begin{gathered}\overline{x}(t, C_i) = x_0(t) + x_1(t, C_i), \quad i = 1, 2, ..., p, \quad t \in D_j; j = 1, 2, ..., s \\ D_1 \cup D_2 \cup ... \cup D_s = D; \quad D_j \cap D_k = \emptyset.\end{gathered} \tag{3}$$

If we substitute Eq. (3) into Eq. (1), we obtain:

$$L[x_0(t)] + L[x_1(t, C_i)] + N[x_0(t) + x_1(t, C_i)] = 0, \tag{4}$$

where C_i, $i = 1, 2, ..., p$ are the convergence-control parameters, which will be rigorously determined. The initial approximation may be determined from the linear differential equation:

$$L[x_0(t)] = 0, \quad B\left[x_0(t), \frac{dx_0(t)}{dt}\right] = 0, \tag{5}$$

while the first approximation x_1 is obtained from Eqs. (4) and (5):

$$L[x_1(t)] + N[x_0(t) + x_1(t, C_i)] = 0, \quad B\left[x_1(t, C_i), \frac{dx_1(t, C_i)}{dt}\right] = 0. \tag{6}$$

The nonlinear term Eq. (6) is expanded as:

$$N[x_0(t) + x_1(t, C_i)] = N[x_0(t)] + \sum_{k=1}^{\infty} \frac{x_1^k(t, C_i)}{k!} N^{(k)}[x_0(t)]. \tag{7}$$

In order to avoid the difficulties appearing in solving Eq. (7) and also to accelerate the convergence of the solution $\overline{x}(t, C_i)$, instead of the last term, one can suggest another expression, so that Eq. (7) can be rewritten as:

$$L[x_1(t, C_i)] + \sum_{i=1}^{p} C_i f_i^*(t) = 0, \quad B\left[x_1(t, C_i), \frac{dx_1(t, C_i)}{dt}\right] = 0, \tag{8}$$

where C_i are p arbitrary unknown parameters, $f_i^*(t)$ are functions depending on the initial approximation $x_0(t)$, on the functions which appear in $N[x_0(t)]$ and $N[x(t)]$ or are combinations of such expressions. These auxiliary functions $f_i^*(t)$ are very important and evidently are not unique, and it should be emphasized that we have much freedom to choose such of auxiliary functions. From Eq. (8) we can deduce the first approximation $x_1(t, C_i)$ in the form:

$$x_1(t, C_i) = \sum_{i=1}^{p} C_i f_i(t), \quad B\left[x_1(t, C_i), \frac{dx_1(t, C_i)}{dt}\right] = 0. \tag{9}$$

In conclusion, the auxiliary functions are of the same form like $x_0(t)$, $N[x_0(t)]$, or $N[x(t)]$. In other words, $x_0(t)$, $N[x_0(t)]$, and $N[x(t)]$ are the "source" for the auxiliary functions.

Now, the unknown parameter C_i is optimally identified via rigorous methods, and we have much freedom to choose between the last square method, Ritz method, collocation method, Galerkin method, or Kantorovich method or by minimizing the square residual error.

3 Application of the OAFM for Oscillations of a Vehicle Equipped with Hydraulic Shock Absorbers

The equation of motion of this oscillator is [3, 5, 9–13]:

$$\ddot{x}(t) + \alpha\dot{x}(t)|\dot{x}(t)| + \varphi[x(t)] = 0, \tag{10}$$

where α is a positive constant (damping coefficient) and $\varphi[x(t)]$ is the elastic restoring force.

In the present paper, we consider $\alpha = 1$ and $\varphi(x) = x$. Initial conditions for (10) are:

$$x(0) = \mathbf{A}, \quad \dot{x}(0) = 0. \tag{11}$$

The linear operator for (10) can be chosen in the form:

$$L[x(t)] = \ddot{x} + 2\lambda\dot{x} + \left(\lambda^2 + \omega^2\right)x, \tag{12}$$

where the point denotes differentiation with respect to time t and λ and ω are unknown parameters at this moment with $\lambda > 0$. The corresponding nonlinear operator can be obtained from (10) and (12):

$$N[x(t)] = -2\lambda\dot{x}(t) + \alpha\dot{x}(t)|\dot{x}(t)| + x(t) - \left(\lambda^2 + \omega^2\right)x(t). \tag{13}$$

The approximate solution of (10) and (11) is of the following form:

$$\overline{x}(t) = x_0(t) + x_1(t, C_i). \tag{14}$$

From (5) and (12), we obtain:

$$x_0(t) = Ae^{-\lambda t}\left(\cos\omega t + \frac{\lambda}{\omega}\sin\omega t\right), \tag{15}$$

such that from (13) and (15), it is clear that

$$\begin{aligned} N[x_0(t)] = \frac{2\lambda A}{\omega}\left(\lambda^2+\omega^2\right)e^{-\lambda t}\sin\omega t - \frac{\alpha A^2}{\omega^2}\left(\lambda^2+\omega^2\right)^2 e^{-\lambda t}\sin\omega t + \\ +\left(1-\lambda^2-\omega^2\right)Ae^{-\lambda t}\left(\cos\omega t + \frac{\lambda}{\omega}\sin\omega t\right). \end{aligned} \tag{16}$$

Taking into consideration that:

$$\begin{aligned} |\sin\omega t| = \sin\omega t sign(\sin\omega t) = \sin\omega t \cdot \frac{4}{\pi} \\ \left(\sin\omega t + \frac{1}{3}\sin 3\omega t + \frac{1}{5}\sin 5\omega t + \frac{1}{7}\sin 7\omega t + \ldots\right). \end{aligned} \tag{17}$$

Equation (16) becomes

$$\begin{aligned} N[x_0(t)] = -\frac{2\alpha A^2}{\pi\omega^2}\left(\lambda^2+\omega^2\right)e^{-\lambda t} \\ \left(1-\frac{2}{3}\cos 2\omega t - \frac{2}{15}\cos 4\omega t - \frac{3}{35}\cos 6\omega t - \frac{2}{63}\cos 8\omega t\ldots\right) + \\ +\frac{2\lambda A}{\omega}\left(\lambda^2+\omega^2\right)e^{-\lambda t}\sin\omega t + \left(1-\lambda^2-\omega^2\right)Ae^{-\lambda t}\left(\cos\omega t + \frac{\lambda}{\omega}\sin\omega t\right). \end{aligned} \tag{18}$$

The "source" for the auxiliary function are (15) and (18), such that, (9) can be written as:

$$x_1(t, C_i) = e^{-\lambda t}[C_1\cos\omega t + C_2\sin\omega t + C_3\cos 2\omega t + C_4\sin 2\omega t]. \tag{19}$$

The initial conditions for x_1 are $x_1(0) = \dot{x}_1(0) = 0$. Considering that the first approximation is given by Eq. (19), we obtain:

$$x_1(t, C_1, C_2) = e^{-\lambda t}\left[C_1(\cos\omega t - \cos 2\omega t) + C_2\left(\sin\omega t - \frac{1}{2}\sin 2\omega t\right)\right]. \tag{20}$$

Finally, the approximate solution of (10) is obtained from Eqs. (14), (15), and (20):

$$\begin{aligned} \overline{x}(t, C_1, C_2) = Ae^{-\lambda t}\left[A\cos\omega t + \frac{\lambda A}{\omega}\sin\omega t + C_1(\cos\omega t - \cos 2\omega t) + \right] \\ \left[+C_2\left(\sin\omega t - \frac{1}{2}\sin 2\omega t\right)\right]. \end{aligned} \tag{21}$$

The convergence-control parameters C_1, C_2 and the parameters λ and ω can be obtained by collocation method.

4 Numerical Results

4.1 For subdomain [0, 3.487) for $A = 0.5$, we obtain $\lambda = 0.1449147$ and $\omega = 0.9010218$. The approximate solution (21), becomes:

$$x(t) = 0.5e^{-0.1449197t}(0.991952921595\cos\omega t + 0.025869203\sin\omega t+) \\ \left(+6.471422708 \cdot 10^{-4}\sin 2\omega t + 0.0080470815\cos 2\omega t\right). \tag{22}$$

4.2 For subdomain [3.487, 6.626) one get get $\lambda = 0.04397666$ and $\omega = 1$. The approximate solution can be written as:

$$\bar{x}(t) = -0.2968e^{-0.043970 66(t-t_1)}[\cos(t - t_1) + 0.04397666\sin(t - t_1)+ \\ +0.91004341071908 \cdot (\cos(t - t_1) - \cos 2(t - t_1))+ \\ +0.0231957721(\sin 2(t - t_1) - \sin(t - t_1))],\ t_1 = 3.487. \tag{23}$$

4.3 For subdomain [6.626, 9.893) we have:

$$\bar{x}(t) = -0.21197e^{-0.03765712(t-t_2)}[\cos\omega(t - t_2) - 0.03917111991\sin(t - t_2)+ \\ +0.845061185223(\cos\omega(t - t_2) - \cos 2\omega(t - t_2))- \\ -0.01043143193(\sin 2\omega(t - t_2) - 2\sin\omega(t - t_2))],\quad t_2 = 6.626, \\ \omega = 0.961349078. \tag{24}$$

Figure 1 illustrates the approximate graph of the displacement obtained by OAFM and comparison with the numerical solution obtained by means of Runge–Kutta approach. It is observed a very good accuracy of the purposed approximate results compared to numerical integration ones.

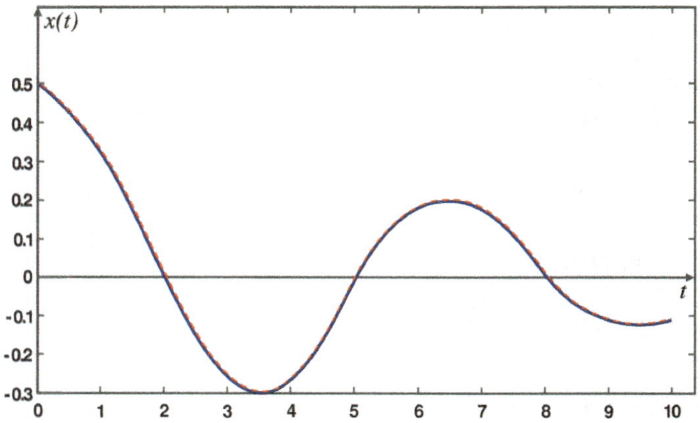

Fig. 1. Comparison between the appropriate relation and numerical results. Numerical OAFM

5 Conclusions

The present paper proposes a relative novel analytical approximate procedure to solve the strongly nonlinear differential equations to oscillations of a vehicle equipped with hydraulic shock absorbers. The nonlinear equation is analytically solved by means of optimal auxiliary functions method (OAFM) and extensive numerical simulations were performed in order to validate the proposed technique. OAFM accelerates the convergence of the approximate analytical solution and leads to very good values of the parameters. A major advantage of this procedure is that it is independent of any small or large parameters and the convergence-control parameters C_i, λ and ω are rigorously identified, using only the first iteration, which underlines the validly of the proposed procedure and on the other hand, it proves that this procedure is very efficient in practice.

References

1. Magnus, K.: Schwingungen. Teubner, Stuttgart (1997)
2. Andronov, A.A., Vitt, A.A., Haikin, S.E.: Teorija Kolebanij. Nauka, Moskow (1981)
3. Dincă, F., Teodosiu, C.: Nonlinear and random vibrations. Academic Press (1973)
4. Cvetičanin, L.: Oscillator with strong quadratic damping force. Publications de l'Institut Mathématique, 119–150 (2009)
5. Wallashek, J.: Dynamics of nonlinear automobile shock-absorbers. Int. J. Nonlinear Mech. **2**, 299–308 (1990)
6. Fay, T.H.: Quadratic damping. Int. J. Math. Educ. Sci. Technol. 4316, 789–803 (2012)
7. Zhu, J.W.: A new exact solutions of a damped quadratic nonlinear oscillators. Appl. Math. Model. **38**, 5986–5993 (2014)
8. Elliott. S.J., Tehrani, M.G., Langley, R.S.: Nonlinear damping of quasi-linear modeling. Philos. Trans., R. Soc. A. **373**. https://doi.org/10.1098/rsta2014.0402 (2014)
9. Zhang, C.H., Zhao, J.H., Wang, Y.: Effect of quadratic damping in a shock isolation. J. Skip Mech. **18**(7), 834–840 (2014)
10. Pandey, A.,Ghose Choudhury, A., Guha, P.: Quadratically damped oscillation with nonlinear restoring force. arxiv:1610.07821v1[nlin.C.D,J.] (2016)
11. Guo, F.: Global analysis of a Liénard system with quadratic damping. Discrete Dynamics in Nature and Society. Article ID1249620 (2018)
12. Fangnou, R., Ainamon, A.V., Monwanon, C.H., Chabi Orou, J.B.: Nonlinear dynamic of the quadratic damping Helmholtz oscillator. Complexity. Article 10:8822534, 17 pages (2020)
13. Demina, M.V., Kuznetsov, N.S.: Liouville Integrability of the Poincare Problem for nonlinear oscillator with quadratic damping and polynomial forces. J. Dyn. Contr. Syst. (2020). https://doi.org/10.1007/s10883-020-09513-2
14. Marinca, V., Herişanu, N., Marinca, B.: Optimal Auxiliary Functions Method for Nonlinear Dynamical Systems, Springer (2021)
15. Marinca, B., Marinca, V.: Some exact solution of MHD flow and heat transfer to modified second grade fluid with variable thermal conductivity in the presence of thermal radiation and heat generation/absorption. Comput. Math. Appl. **76**, 1515–1524 (2018)
16. Marinca, V., Marinca, B.: Optimal auxiliary functions method for nonlinear thin film flow of a third grade fluid on a moving belt. Proc. Roman. Acad. Series A **19**, 575–580 (2018)
17. Marinca, V., Herişanu, N., Bota, C., Marinca, B.: An optimal homotopy asymptotic method applied to the steady flow of a fourth grade fluid past a porous plate. Appl. Math. Lett. **22**, 245–251 (2009)

18. Marinca, V., Herişanu, N.: Nonlinear Dynamical Systems in Engineering. Some Approximate Approaches. Spinger (2011)
19. Marinca, B., Marinca, V., Bogdan, C.: Dynamics of SEIR epidemic model by optimal auxiliary functions method. Chaos, Solut. Fractals **14**, 110949 (2021)
20. Herişanu, N., Marinca, V.: An efficient analytical approach to investigate the dynamics of a misaligned multirotor system. Mathematics **8**. https://doi.org/10.3390/math8071983 (2020)
21. Marinca, V., Herişanu, N.: Construction of analytic solution to axisymmetric flow and heat transfer on a moving cylinder. Symmetry **12**. https://doi.org/10.3390/sym12081335 (2020)

Environmental and Occupational Noise

Determination of Annual Environmental Noise Indicators Using a Series of Short-Term Measurements

Darko Mihajlov[1(✉)], Momir Praščević[1], Marko Ličanin[1], Miomir Raos[1], and Branko Radičević[2]

[1] Faculty of Occupational Safety in Niš, University of Niš, Čarnojevića 10A, 18000 Niš, Serbia
{darko.mihajlov,momir.prascevic,marko.licanin, miomir.raos}@znrfak.ni.ac.rs

[2] Faculty of of Mechanical and Civil Engineering in Kraljevo, University of Kragujevac, Dositejeva 19, 36000 Kraljevo, Serbia
radicevic.b@mfkv.kg.ac.rs

Abstract. Completely reliable annual values of environmental noise indicators can be obtained only through continuous annual noise measurements. This paper examines the justification of using a series of short-term measurements to determine the annual values of environmental noise indicators. The examination involves the application of a proposed measurement program to the results of continuous annual noise monitoring at two locations in the city of Niš, Serbia, in which road traffic is the dominant noise source. Monitoring results include the period when the state of emergency was in effect because of the Covid-19 pandemic. The obtained results indicate that using the proposed program is fully justified if the goal is to define annual environmental noise indicator values for specific cases only.

Keywords: Road traffic noise · Noise monitoring · Short-term measurements

1 Introduction

The existing road infrastructure in large cities is becoming increasingly burdened by everyday vehicle flow. The main reasons include population growth, increased population density, increased number of cars per capita, increased activity of public utility companies, and so forth. Another reason is that, owing to the current lot coverage, terrain conditions, and other limitations, it is impossible to make any significant changes to the transport network in most of the already developed urban areas, for instance, to expand the existing roads or to build new ones. Under current conditions, measures such as public transportation electrification, subsidies for electric car purchases, driving bans for specific vehicle categories in specific parts of a city, construction of roundabouts, introduction of one-way traffic, etc. do not contribute much to noise level reduction at numerous urban locations that are particularly vulnerable to road traffic noise [1]. Such

N. Herisanu and V. Marinca (eds.), *Acoustics and Vibration of Mechanical Structures – AVMS-2021*, Springer Proceedings in Physics 274,
https://doi.org/10.1007/978-3-030-96787-1_13

a state of affairs has environmental pollution as its direct consequence, accompanied by the effects on the performance of everyday tasks and the health of people living in such an environment [2, 3].

In order to control environmental noise and to take proper reduction measures, it is necessary to obtain the information on the current status of noise. The data can be obtained in two ways and they refer to the annual values of environmental noise indicators [4]. The first way is to assess noise indicator values using specialised software tools. The results of the assessment are then presented by means of strategic noise maps and are produced every five years, according to the current regulations [5]. In addition to noise indicator values, strategic noise maps also provide a plethora of other useful information [6] that can be used to develop noise control action plans. The second way to obtain the data on annual noise indicators is to use various noise monitoring programs [7], while bearing in mind that the use of different programs may affect the reliability of the results. They allow the monitoring of noise changes in a time–amplitude domain and a subsequent statistical analysis of the measurement results [8]. If necessary, this method allows the monitoring of noise level changes due to specific activities at a given location, in addition to regular or altered traffic conditions, such as construction works, different public events, etc. Finally, this method can be used to exactly determine the impact of decreased traffic flow on the status of environmental noise, as exemplified by the measures of restricted movement imposed in Serbia from 15 March to 6 May 2020 because of the Covid-19 pandemic.

Determination of annual environmental noise indicators through monitoring requires the use of specific measuring equipment and procedures [9]. However, it is necessary to understand that a monitoring program (selection of measurement duration, frequency, and time intervals) has to encompass all operating regimes of noise sources and all the various weather conditions in the course of a calendar year. In terms of noise, the main properties of road traffic include the following: composition or structure of traffic flow, traffic flow size, infrastructure, and road capacity and topography. In addition to the said factors, the status of noise level is also affected by the road base type [10] as well as the weather conditions at the time of measurement. With the current traffic flow structure and the road capacity, topography, and base type, the noise level status is predominantly influenced by traffic flow size, primarily the vehicle flow, flow density and speed, as well as the temporal variability and complexity of the traffic flow, as its main characteristics. The ideal approach to the analysis of the said influencing factors involves continuous monitoring over the course of one calendar year using the equipment specifically intended for that purpose. All other monitoring programs are attempts to rationally solve the problem more or less successfully in terms of reliability of obtained results [11]. The rationale and justification for using such noise monitoring programs can be analyzed through comparison of their results with the results of continuous annual monitoring.

2 Examples of Results of Continuous Annual Noise Monitoring

In the city of Niš, Serbia, continuous annual environmental noise monitoring has been performed since 2014 [12] using an automated sound monitoring system [13, 14]. The basic report format for the status of noise at a measurement point is the value of equivalent noise level for a 15-min measurement time interval, which is used to create hourly, daily, monthly, and annual reports on the value of equivalent noise level, as the standard report format. The system can be adjusted to create reports on the values of main and additional environmental noise indicators in accordance with Directive 2002/49/EC [5].

Table 1 shows the results of continuous annual noise monitoring at two measurement points (MP1 and MP2) in the city of Niš from 1 November 2019 to 31 October 2020. The annual values of certain quantities are the result of energy averaged monthly values.

Table 1. Results of continuous annual noise monitoring at measurement points MP1 and MP2—monthly and annual noise indicator values in dB.

	L_{day}		$L_{evening}$		L_{night}		L_{den}		$L_{eq,\text{total}}$	
	MP1	MP2	MP1	MP2	MP1	MP2	MP1	MP2	MP1	MP2
Nov. 19	67.3	64.7	66.7	64.2	62.0	60.8	70.2	68.4	66.0	63.6
Dec. 19	68.3	64.9	67.6	64.6	62.9	61.3	71.1	68.8	67.0	63.9
Jan. 20	68.8	64.1	67.2	63.9	62.4	61.9	70.9	68.9	67.2	62.5
Feb. 20	67.9	64.2	67.3	64.1	62.5	60.7	70.7	68.2	66.6	63.3
Mar. 20	67.2	64.2	64.8	62.1	60.2	58.5	68.8	66.6	65.4	62.6
Apr. 20	65.4	63.0	56.6	55.5	54.3	52.7	64.6	62.6	62.8	60.5
May. 20	67.2	64.0	65.4	62.7	60.5	58.6	69.1	66.7	65.5	62.5
Jun. 20	67.6	64.1	67.2	64.2	62.6	62.1	70.6	69.0	66.4	63.5
Jul. 20	66.5	64.0	66.8	64.2	61.9	60.2	70.0	67.9	65.5	63.1
Aug. 20	66.8	64.0	66.8	64.3	62.1	60.7	70.1	68.2	65.7	63.2
Sep. 20	67.2	63.9	66.6	63.8	61.8	60.4	70.0	67.9	65.9	63.0
Oct. 20	68.1	64.4	67.3	64.3	62.4	60.5	70.8	68.1	66.8	63.4
Annual value	**67.4**	**64.1**	**65.9**	**63.2**	**61.3**	**59.9**	**69.7**	**67.6**	**65.9**	**62.9**
Std. dev	0.86	0.45	2.90	2.41	2.25	2.40	1.69	1.68	1.11	0.84
L_{max}	68.8	64.9	67.6	64.6	62.9	62.1	71.1	69.0	67.2	63.9
L_{min}	65.4	63.0	56.6	55.5	54.3	52.7	64.6	62.6	62.8	60.5

Traffic flows at the two locations are independent. Both traffic flows are characterised by a relative regularity and a complex structure on a daily basis. The traffic flow at measurement point MP1 has a volume of over 1600 vehicles per hour during the day/evening and over 600 vehicles per hour during the night. The road has three lanes in each direction [15]. The traffic flow at measurement point MP2 has the volume of over

1200 vehicles per hour during the day/evening and over 400 vehicles per hour during the night. The road has one lane in each direction [15].

The current traffic flows generate a time-variable noise over a broad frequency range. The time profile of the noise is also affected by the occasional short-burst noise events, which are usually due to emergency vehicle sirens but its occurrence and duration are impossible to predict.

Table 2 shows the deviations of monthly values from the annual values of specific quantities expressed in decibels for measurement points MP1 and MP2. The bottom row shows the absolute values of maximum deviations of monthly from annual values of specific quantities $|\Delta L_{max}|$ in decibels.

Table 2. Deviations of monthly from annual noise indicator values in dB for measurement points MP1 and MP2.

	L_{day}		$L_{evening}$		L_{night}		L_{den}		$L_{eq,\text{total}}$	
	MP1	MP2	MP1	MP2	MP1	MP2	MP1	MP2	MP1	MP2
Nov. 19	−0.1	0.6	0.8	1.0	0.7	0.9	0.5	0.8	0.1	0.7
Dec. 19	0.9	0.8	1.7	1.4	1.6	1.4	1.4	1.2	1.1	1.0
Jan. 20	1.4	0.0	1.3	0.7	1.1	2.0	1.2	1.3	1.3	−0.4
Feb. 20	0.5	0.1	1.4	0.9	1.2	0.8	1.0	0.6	0.7	0.4
Mar. 20	−0.2	0.1	−1.1	−1.1	−1.1	−1.4	−0.9	−1.0	−0.5	−0.3
Apr. 20	**−2.0**	**−1.1**	**−9.3**	**−7.7**	**−7.0**	**−7.2**	**−5.1**	**−5.0**	**−3.1**	**−2.4**
May. 20	−0.2	−0.1	−0.5	−0.5	−0.8	−1.3	−0.6	−0.9	−0.4	−0.4
Jun. 20	0.2	0.0	1.3	1.0	1.3	2.2	0.9	1.4	0.5	0.6
Jul. 20	−0.9	−0.1	0.9	1.0	0.6	0.3	0.3	0.3	−0.4	0.2
Aug. 20	−0.6	−0.1	0.9	1.1	0.8	0.8	0.4	0.6	−0.2	0.3
Sep. 20	−0.2	−0.2	0.7	0.6	0.5	0.5	0.3	0.3	0.0	0.1
Oct. 20	0.7	0.3	1.4	1.1	1.1	0.6	1.1	0.5	0.9	0.5
$\mathbf{|\Delta L_{max}|}$	**2.0**	**1.1**	**9.3**	**7.7**	**7.0**	**7.2**	**5.1**	**5.0**	**3.1**	**2.4**

The values of $|\Delta L_{max}|$ in Table 2 for every quantity are actually the values for April 2020. The reason for this occurrence is the state of emergency declared in Serbia from 15 March to 6 May 2020 to prevent contact and the spread of the coronavirus among the population during the pandemic. One of the measures under those circumstances was the overall curfew (excluding official purposes) during specific periods of the day, especially during the night. The measure was repeatedly imposed for periods of two to five days even after the state of emergency had been lifted. The change in the regime of road traffic and other outdoor activities was the direct result of restricted movement of the urban population, which, among other things, directly affected the status of environmental noise.

3 Methodology

Based on potential unplanned changes of traffic flow quantities and characteristics in urban environments during a year, this paper analyses the justification of using the proposed noise monitoring program, which can be used to obtain both monthly reports on the status of noise and reliable data on annual noise indictor values. The proposed program has to consider scenarios that can unfold once or repeatedly over the course of a year with a specific duration, involving the following occurrences:

- Vehicle flow change;
- Traffic flow density change;
- Traffic flow composition (structure) change;
- Traffic flow time irregularity;
- Unplanned traffic-induced transient noise events.

Considering that it is impossible to predict the above occurrences when adopting the annual noise monitoring program because it is impossible to know their onset time and duration, special attention must be given to the following elements when defining the noise monitoring program involving a series of short-term measurements:

- Choice of the observation time interval during which the series of measurements is performed;
- Choice of the measurement time interval during which individual measurements are performed;
- Choice of the frequency of short-term noise measurements;
- Choice of measured quantities.

The model noise monitoring program analysed in this paper is a program modified according to the Serbian national standard SRPS U.J6.205:2007 [16]. The program is adjusted to meet the needs of annual environmental noise monitoring for a specific case. The following are taken as the basic characteristics of the program:

- Observation time interval: one calendar year (12 months);
- Measurement time interval: $t = 15$ min;
- Measurement frequency: 24 measurements per month during weekdays (Monday to Friday), one 15-min measurement per hour (00:00–01:00, 01:00–02:00, ... 23:00–24:00) according to a freely chosen schedule;
- Measured quantity: equivalent noise level during the measurement time interval, $L_{\mathrm{Aeq},t}$, in dB.

The results of such monitoring consist of monthly and annual noise indicator levels. The monthly indicator values L_{day}, $L_{evening}$, and L_{night} are obtained via energy averaging of the measured 15-min values of equivalent noise levels in specific reference periods (day, evening, and night) during one month. The monthly value of the total noise indicator L_{den} is obtained via energy averaging of L_{day}, $L_{evening}$, and L_{night} values, while taking into account the duration of specific reference periods (day: 12 h, evening: 4 h, and night: 8 h) and dB corrections due to more prominent people's reaction to noise during the evening (+5 dB) and night (+10 dB) [5].

The annual indicator values L_{day}, $L_{evening}$, and L_{night} are obtained via energy averaging of the twelve-monthly values of specific noise indicators. The annual value of the total noise indicator L_{den} is obtained via energy averaging of annual L_{day}, $L_{evening}$, and L_{night} values, analogously to the determination of monthly L_{den} [5].

The reliability of thus obtained results depends on the degree of agreement with the results that would have been obtained from continuous annual noise monitoring. Accordingly, this study demonstrates the application of the proposed program to the results of continuous annual noise monitoring at measurement points MP1 and MP2. Methodologically, the testing of the proposed program's reliability is performed in five phases.

The first phase is the organization of a database containing the results of continuous annual noise monitoring for each measurement point. Based on the presets, the database comprises monthly reports on the measured $L_{\mathrm{Aeq},15}$ min values on weekdays, chronologically ordered from 00:00 to 24:00.

The second phase is the calculation of K monthly and annual L_{day}, $L_{evening}$, L_{night}, and L_{den} values by applying the proposed program to the results of continuous annual noise monitoring using randomly selected samples from the phase-one database. The random selection of samples and the calculation of required data K times are performed using Python 3.7.

The third phase involves the calculation of the deviations of the calculated from the reference (measured) monthly and annual ΔL noise indicator values.

The fourth phase involves the statistical analysis of the results from the previous phase. This includes the calculation of the mean value of absolute deviations of the calculated from the reference monthly and annual $|\Delta L_{\mathrm{avr}}|$ indicator values as well as the standard deviation of the deviations of the calculated from the reference values of σ_L noise indicator.

The fifth phase is the evaluation of the proposed program's reliability according to the values of quantities calculated in the previous phase. The mean value of absolute deviations of the calculated noise indicators from their reference values is the indicator of the program's accuracy. On the other hand, the standard deviation of the deviations of the calculated noise indicators from their reference values indicates the program's precision, whereby lower values of these quantities indicate a higher precision of the program.

4 Results and Discussion

Reliability testing of the proposed program using the described methodology was performed based on $K = 50$ sets of samples for every month of the year. A single set comprises 24 $L_{\mathrm{Aeq},15\ \mathrm{min}}$ values for each hour during a day, randomly selected from the database containing the monthly results of continuous noise monitoring at measurement points MP1 and MP2.

For every set k ($k = 1 \div 50$), monthly values of $L_{day,k}$, $L_{evening,k}$, $L_{night,k}$, and $L_{den,k}$ and their deviations from the reference (measured) noise indicator values (Table 1),

ΔL_k, were calculated. The ΔL_k values were then used to calculate the mean values of absolute deviations of the calculated from the reference values of noise indicators as well as specific standard deviations.

The results of the application of the proposed program on the results of continuous annual noise monitoring at measurement points MP1 and MP2 are shown in Figs. 1, 2, 3 and 4.

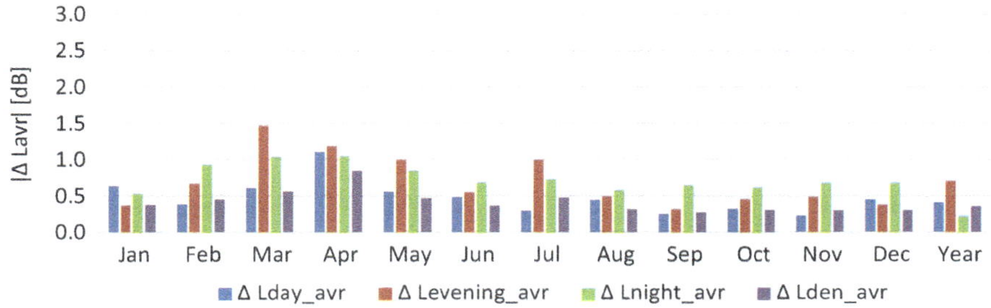

Fig. 1. MP1—Mean value of absolute deviations of the calculated noise indicators from their reference values based on 50 randomly selected sets of samples.

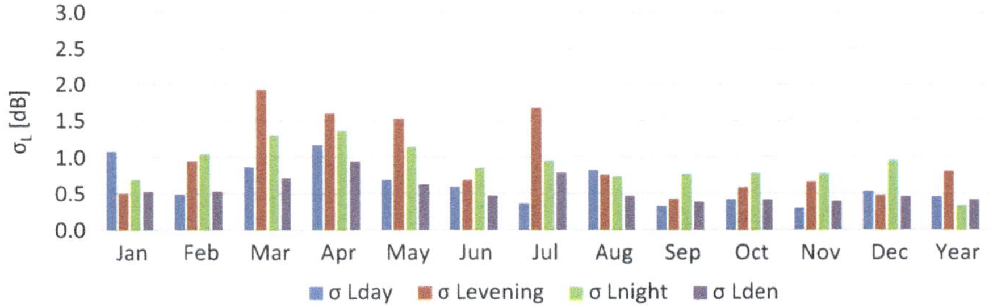

Fig. 2. MP1—Standard deviation of the deviations of the calculated noise indicators from their reference values based on 50 randomly selected sets of samples.

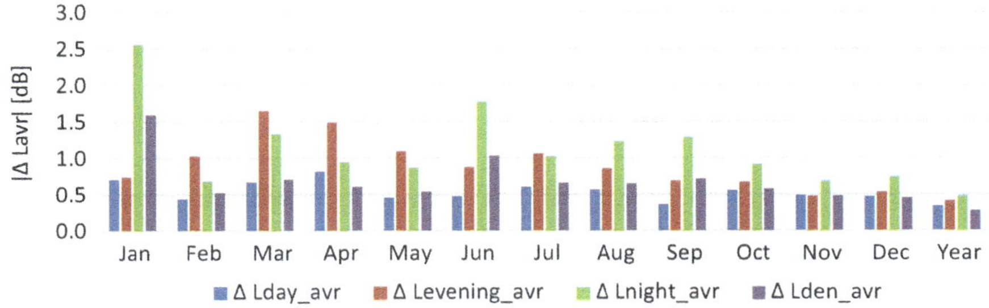

Fig. 3. MP2—Mean value of absolute deviations of the calculated noise indicators from their reference values based on 50 randomly selected sets of samples.

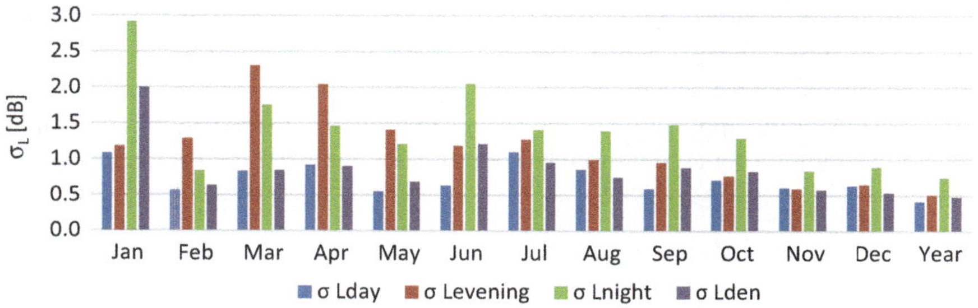

Fig. 4. MP2—Standard deviation of the deviations of the calculated noise indicators from their reference values based on 50 randomly selected sets of samples.

The results of the application of the proposed program on the results of continuous annual noise monitoring at measurement points MP1 and MP2 can be analysed according to the value range of quantities that indicate the program's reliability, as shown in Table 3.

Table 3. Range of $|\Delta L_{avr}|$ and σ_L values in dB.

	$\|\Delta L_{day\_avr}\|$		σ_{Lday}		$\|\Delta L_{evening\_avr}\|$		$\sigma_{Levening}$		$\|\Delta L_{night\_avr}\|$		σ_{Lnight}	
	MP1	MP2	MP1	MP2	MP1	MP2	MP1	MP2	MP1	MP2	MP1	MP2
Min	0.24	0.38	0.31	0.56	0.33	0.49	0.44	0.59	0.55	0.70	0.72	0.85
Max	1.12	0.83	1.19	1.11	1.48	1.66	1.94	2.31	1.07	2.58	1.38	2.92

The values shown in Table 3 indicate that the reliability of the proposed program is higher when vehicle flow is higher and when traffic flow has a more regular time, as is the case with MP1.

The results of the application of the proposed monitoring program on the results of continuous annual noise monitoring indicate that the values of $|\Delta L_{\_avr}|$ and σ_L were significantly higher for the evening and night periods during March and April 2020, when traffic was maximally reduced because of the imposed state of emergency on account of the pandemic. Over the subsequent months, the movement was restricted on specific days during the evening and night, so the randomly selected samples outside of those periods contributed to the prominent deviations of the calculated from the reference values and to the increase in standard deviation. More prominent values of $|\Delta L_{avr}|$ and σ_L in the rest of the cases, especially in the night periods, were due to samples with reduced vehicle flow and the consequent time irregularity of the traffic flow.

5 Conclusions

The reliability of noise monitoring results based on short-term measurements is reflected in the deviations from the results obtained via continuous noise monitoring over the same observed period. In conditions of variable vehicle flow, variable flow speed and density, variable composition, and traffic flow time irregularity, the reliability of short-term measurement results is crucially influenced by the program elements such as the measurement time interval and the noise parameter measurement frequency. The increase of the measurement time interval and the number of measurements asymptotically tends to continuous monitoring and provides more reliable results.

Upon the application of the proposed monitoring program on the results of continuous annual monitoring of environmental noise originating predominantly from road traffic, it can be concluded that the proposed program can be used reliably, especially in stable traffic flow characteristics.

If it is not feasible or even necessary to resort to continuous monitoring, the proposed noise monitoring program is a satisfactory alternative solution for providing insight into the noise level status in urban environments.

Acknowledgements. This paper was written as a result of the research financially supported by the Ministry of Education, Science and Technological Development of the Republic of Serbia in accordance with the agreement no. 451-03-9/2021-14/200148.

References

1. Papadimitriou, E.A., Papageorgiou, G.P., Alamanis, N., Diakosavva, T.N.: Road noise levels in urban environment compared to specification limits. The case of the City of Larissa, Greece. Periodica Polytechnica Civil Eng. **64**(4), 964–974. https://doi.org/10.3311/PPci.14868 (2020)
2. World Health Organization: Environmental noise guidelines for the European region. https://euro.who.int/_data/assets/pdf_file/0008/383921/noise-guidelines-eng.pdf?ua=1 (2018)
3. Felcyn, J.: The influence of a signal's time structure on the perceived noise annoyance of road traffic noise. J. Environ. Health Sci. Eng. (2021). https://doi.org/10.1007/s40201-021-00655-4
4. Zambon, G., Roman, H.E., Smiraglia, M., Benocci, R.: Monitoring and prediction of traffic noise in large urban areas. Appl. Sci. **8**(2), 251 (2018). https://doi.org/10.3390/app8020251
5. Directive 2002/49/EC: Assessment and Management of Environmental Noise. European Parliament, Council of the European Union. https://eur-lex.europa.eu/legal-content/EN/TXT/PDF/?uri=CELEX:32002L0049&from=EN
6. Alam, P., Ahmad, K., Afsar, S. S., Akhtar, N.: Noise monitoring, mapping, and modelling studies–A review. J. Ecol. Eng. **21**(4) (2020). https://doi.org/10.12911/22998993/119804
7. Jagniatinskis, A., Fiks, B., Zaporozhets, O., Mickaitis, M.: Annual assessment of noise generated by road traffic using measurements. Procedia Eng. **187**, 614–619 (2017). https://doi.org/10.1016/j.proeng.2017.04.421
8. Prascevic, M.R., Mihajlov, D.I., Cvetkovic, D.S.: Measurement and evaluation of the environmental noise levels in the urban areas of the city of Nis (Serbia). Environ. Monit. Assess. **186**(2), 1157–1165 (2014). https://doi.org/10.1007/s10661-013-3446-2
9. International Organization for Standardization: Acoustics - Description, measurement and assessment of environmental noise - Part 2: Determination of sound pressure levels (ISO Standard No. 1996–2:2017)
10. Stančerić, I., Dragčević, V., Ahac, S.: Toward environmental noise estimation according to the road surface characteristics and traffic volume. Technical Gazette **17**(2), 191–197 (2010). https://hrcak.srce.hr/55484

11. Aumond, P., Can, A., Mallet, V., Gauvreau, B., Guillaume, G.: Global sensitivity analysis for road traffic noise modelling. Appl. Acoust. **176**, 107899 (2021). https://doi.org/10.1016/j.apacoust.2020.107899
12. Mihajlov, D.I., Prascevic, M.R.: Permanent and semi-permanent road traffic noise monitoring in the city of Nis (Serbia). J. Low Freq. Noise, Vibr. Active Cont. **34**(3), 251–268 (2015). https://doi.org/10.1260/0263-0923.34.3.251
13. Technical documentation - Noise Monitoring terminal Types 3639-A, 3639-B and 3639-C with Hand-held Analyzer Type 2250-N or Hand-held Analyzer Type 2250-N-D00, version 4.1.1, Bruel&Kjaer, BE 1818–17 (2013)
14. Technical documentation - Environmental Noise Management System Software Type 7843, version 2.8.1, Bruel&Kjaer, BE 1767–15, (2010)
15. COWI d.o.o. Belgrade: Study of checking the capacity of the traffic network of the city of Niš for the development of the General Plan of City of Niš, Book 1: Research of the characteristics of traffic flows (2010)
16. Institute for Standardization of Serbia: Building acoustic - Acoustical zoning (SRPS U.J6.205:2007).

Repeatability of Impulse Response Measurements Using Raspberry Pi

Marko Licanin(✉), Darko Mihajlov, Momir Prascevic, and Dejan Ciric

University of Nis, 18000 Nis, Serbia
{marko.licanin,darko.mihajlov,momir.prascevic}@znrfak.ni.ac.rs, dejan.ciric@elfak.ni.ac.rs

Abstract. Advancement in computing power of the Single Board Computers (SBC) has enabled hobbyists and researchers to be creative even without expensive equipment. When it comes to audio and noise, the precise measuring equipment often does not bring enough flexibility, and it is limited by options enabled by manufacturer. For this reason, the authors of this paper were interested in the possibilities of using a Raspberry Pi SBC as a main audio recording and analysis device for the integration in more complex real-time system. Audio recording performance of the Raspberry Pi (RPi) has been evaluated through repeatability of the impulse response measurements. Two setups have been used, the first with RPI paired with MEMS microphone, while second included the RPi, a studio soundcard, and studio measurement microphone. Two states of the RPi were analyzed (GUI turned ON and GUI turned OFF). Data have shown that the configuration with MEMS microphone has a surprisingly good repeatability, both of the latency and amplitude. Impact of the GUI is highly observable in repeatability of the latency. Results obtain in the paper show good promise when it comes to using RPi in audio data acquisition applications.

Keywords: Impulse response · Latency · Raspberry pi · Audio hardware

1 Introduction

Professional equipment in audio recording, acoustical, and noise measurement is widely used in many applications. This equipment includes different soundcards, studio and measurement microphones, sound sources, time and frequency domain sound analyzers, monitoring equipment etc. It is very common that these devices are very expensive, thus available to a small number of professionals who can afford them. On the opposite side, during the last decade semiconductor components became significantly cheaper, and electronic devices available to a broad population. The computing power and a form factor of cheap electronics became formidable, while at the same time, innovation in measurement equipment progress at much slower rate. For instance, when looking at sound analyzer device of a very known manufacturer during the period of 5 years,

N. Herisanu and V. Marinca (eds.), *Acoustics and Vibration of Mechanical Structures – AVMS-2021*, Springer Proceedings in Physics 274,
https://doi.org/10.1007/978-3-030-96787-1_14

other than software, there is very little change in hardware design. Latest flagship smartphone device, which now possesses the processing power that slowly approaches PC performance, have changed significantly in the last couple of years.

Along the smartphones and PC industry, another type of affordable devices has attracted hobbyists and professionals in electronics. These devices are known as Single Board Computers (SBCs) and are widely used in open-source applications. Modern SBCs combine the processing power of typical smartphone, flexibility of a PC (peripherals and operating systems), and hardware expandability of a microcontroller. One of the most popular SBC is the Raspberry Pi (RPi), recently available as 4th generation model. This device has a decent number of peripherals, wireless and ethernet capabilities, and serious computing power. This makes Raspberry Pi interesting for audio data acquisition, processing, and communication of systems in real-time [1].

This paper addresses the possibility of using RPi in audio applications. This has been done through the analysis of the recorded room Impulse Responses (IRs) [2], more precisely repeatability of the latency [2, 3] and amplitude of the IRs. Insight into these parameters can give a better clue whether is it possible to use the RPi in developing complex audio systems (for instance, low-cost noise monitoring stations) that rely on the real-time audio data transfer. To analyze the performance of RPi in the context of IRs repeatability, another audio system has been introduced for comparison. This system is composed of studio soundcard and studio measurement microphone. Recording of the IRs has been performed using MEMS (Micro Electro Mechanical System) microphone, then repeated by the second audio system. Measurements were performed in two RPi states, first when GUI of the RPi was running, while at the second state, commands were issued from Linux terminal with GUI of the RPi turned off. Processing of the data has been done in Python, and results are presented graphically.

2 Methodology

To evaluate the audio performance of the RPi, a series of measurements has been taken in the Laboratory for Noise and Vibration (Room 2) at the Faculty of Occupational Safety, University of Nis. As an acoustical excitation signal, a 12-s sine signal which frequency changes in time (swept sine or sweep) [4, 5] was generated in Python, together with the inverse function of this waveform. Signal was reproduced over the reference loudspeaker and recorded in the memory card of the RPi. Two different setups were used:

- Setup 1—using MEMS microphone;
- Setup 2—using studio microphone.

The reason for the second setup was to compare the recording quality of MEMS microphone to the studio microphone. It is important to mention that this is the most affordable studio microphone pricing around 80€. The cost of the single PCB with mounted MEMS microphone is 2€, while typical class I measurement microphone costs more than 1000€.

Measurements were performed in two different RPI states:

- State 1—RPi GUI was turned on;
- State 2—RPi GUI was turned off.

During the state, 2 commands were issued only from the terminal in Linux. This was done to observe the impact of the operating system on the stability of audio recording. Process of the entire measurement was done in the following steps (combinations):

- C1 (Setup1-State1): Connected Setup 1, GUI turned ON, 10 recordings performed;
- C2 (Setup1-State2): GUI turned OFF, 10 recordings performed;
- C3 (Setup2-State1): Connected Setup 2, GUI turned ON, 10 recordings performed;
- C4 (Setup2-State2): GUI turned OFF, 10 recordings performed.

2.1 Laboratory Setup

During the measurement, the following equipment has been used:

- MEMS microphone InvenSense I2S INMP441;
- Measurement condenser microphone Beyerdynamic MM1;
- Sound card Tascam US2 $\times$ 1;
- Brüel & Kjær Power amplifier type 2734;
- Brüel & Kjær Omnidirectional reference loudspeaker type 4292-L;
- Raspberry Pi 4, 8 GB RAM, USB 3.0, Broadcom BCM2711.

MEMS Microphone and loudspeaker are placed 1 m apart at the height of 1.5 m. Room has been cleared so there was no object at distances less than 2 m horizontally and 1.5 m vertically from the elements of the experimental setup.

Raspberry Pi is a central component of the measurement system. It does not have analog audio input and can receive audio through HDMI port and USB devices. Another way to input the audio from the microphone is to use the I2S protocol that has to be implemented in software. It is fast to download and compile kernel for the I2S audio module, but it takes some time to adjust the recording volume level of the module. In essence, this software module acts as a virtual sound card and runs the I2S driver. If I2S device is connected to the dedicated General Purpose Input Output (GPIO) pins, driver accepts the 24-bit audio from external device to memory buffer. As a I2S device, MEMS microphone InvenSense INMP441 has been used. This device has been connected to the RPi and placed in the 3D printed casing, resembling the shape of the regular microphone enclosure. Figure 1a describes the Setup 1 of the recording system.

Acoustical measurement signal is stored at the memory card on the RPi. On the device itself, recording software has been written in Python. Software utilizes parallel processing in python, where separate cores of the processor handles recording and reproduction, enabling these operations to run simultaneously. This way, the load of the third-party recording/reproduction software (which are sparse for this SBC) on the processor was eliminated. When executed by ARM processor, process 1 plays the signal. The digital signal converted to analog in Digital-to-Analog Converter (DAC), amplified in power amplifier Brüel & Kjær type 2734, then fed to the loudspeaker Brüel & Kjær type 4292-L. The audio signal is picked up by MEMS microphone and through GPIO directly led to the ARM processor which buffers data save it to wav file.

Setup 2 is composed of the USB audio interface (soundcard) Tascam US2 $\times$ 1, studio measurement microphone Beyerdynamic MM1, power amplifier Brüel & Kjær type

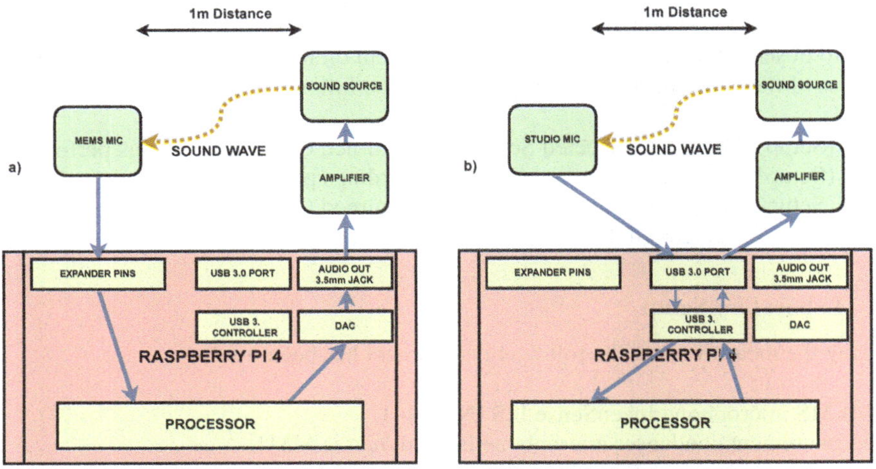

Fig. 1. Simplified schematic of the Setup 1 (**a**) and Setup 2 (**b**) of the recording system. Blue arrow describes the travel direction of electrical signal from input to processor, as well as from the processor to output.

2734 and omnidirectional reference loudspeaker Brüel & Kjær type 4292-L. Simplified schematic of the Setup 2 can be seen in Fig. 1b. Difference from the previous setup is that signals travel in both ways through USB 3.0 controller. This controller could increase the latency of the Impulse Response since it is another digital circuit that handles data. Knowing this information could prove very useful in the future when making a decision whether to use USB interface for audio input or consider other options.

3 Results

After all of the measurement has been completed, wav signals have been transferred from the RPi memory card to PC and post-processing was performed. Data have been imported to python, where recorded sweep signal has been convolved with the inverse filter generated at the same time as the original sweep that was reproduced during measurements. As a result of this operation, impulse responses of all measurements have been extracted. For each of the different combinations (C1–C4) index of the least delayed IR has been found. This index has been set to 0, trimming the unnecessary data before the index. The rest of signals within each combination was presented as the relative latency compared to the least delayed IR (which is set to 0 latency). This can be seen in Fig. 2. This representation is separated into 4 figures, where each figure represents one of the mentioned combinations. The least delayed IR for each C is visible, green color for C1 and C3 (Fig. 1a and c), gray color for C1 (Fig. 1b), and blue color for C4 (Fig. 1b).

Impact of the choice of Raspberry Pi state is particularly visible in Fig. 2. Latencies in samples of the IRs are much closer grouped in C2 and C4, than in C1 and C3. This implies that Linux operating system GUI is the most influential factor in increasing the latency of the IR. When looking at the latency for the State 2 and combination C2 and

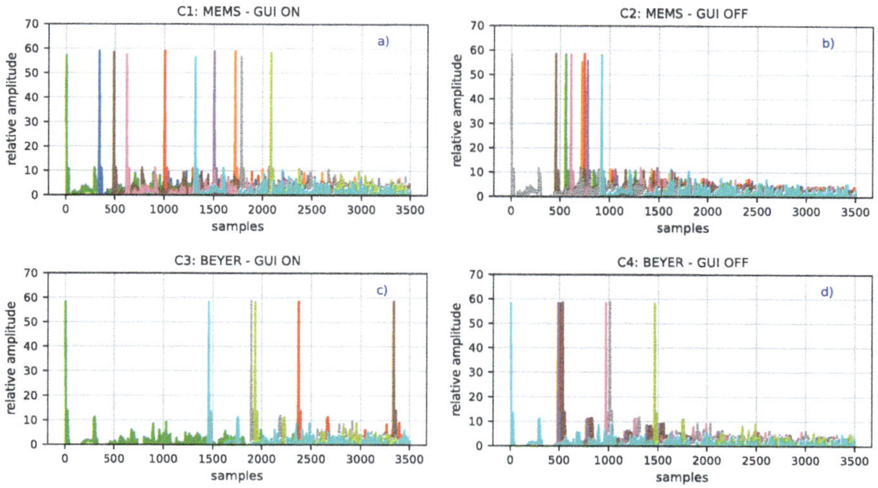

Fig. 2. Distribution of the IRs presented in relation to the least delayed IR (latency set to 0). Results are given for measured combination C1 (**a**), C2 (**b**), C3 (**c**), and C4 (**d**).

C4 (Fig. 2b and d), Setup 1 with the MEMS microphone shows better results. This is opposite to the expectation, having in mind the price of the MEMS microphone, and given the fact the studio equipment is often marketed as zero-latency devices. The reason that MEMS is better in the terms of latency may lie in the previous suspicion that in Setup 2, additional hardware in USB 3.0 controller and Tascam soundcard increase the latency. On the other hand, in Setup 1, MEMS microphone is practically directly digitally connected to the RPi ARM processor which significantly reduces the number of circuits on data path. This discussion can be better observed in Fig. 3. On y-ordinates, latency is presented in milliseconds. Distribution of results has been presented in the same way as in Fig. 4. For all of the combinations. Mean values of the latencies have been calculated for each of the C (1–4) and presented by red line, as well as by red labels. In each of the C (1–4), there is one IR with a zero latency, which corresponds to the least delayed IR. It can be seen that mean value of the latency for the State 1 is almost twice higher than for State 2, in case of Setup 1 (Fig. 3a versus Fig. 3b). In the same way, mean latency for the State 1 is more than three times higher than for State 2, in case of Setup 2 (Fig. 3c versus Fig. 3d).

The last analysis is related to the repeatability of the IR amplitude. This has been calculated as the maximum value of each IR in any of the combinations. For each of the C (1–4), maximum amplitudes composed an array containing the same number of values as the number of measurements. From these arrays, mean values of the amplitude have been calculated. To determine the variance, values in the arrays for all of the C (1–4) have been then subtracted from the corresponding calculated mean, and presented in Fig. 4. Distribution of the results for com combinations has been done in the same way as in Figs. 2 and 3. Standard deviation has been determined and given on Fig. 4 in the form of red line and labels for $\pm\sigma$. It is important to note that axis for the Setup 2

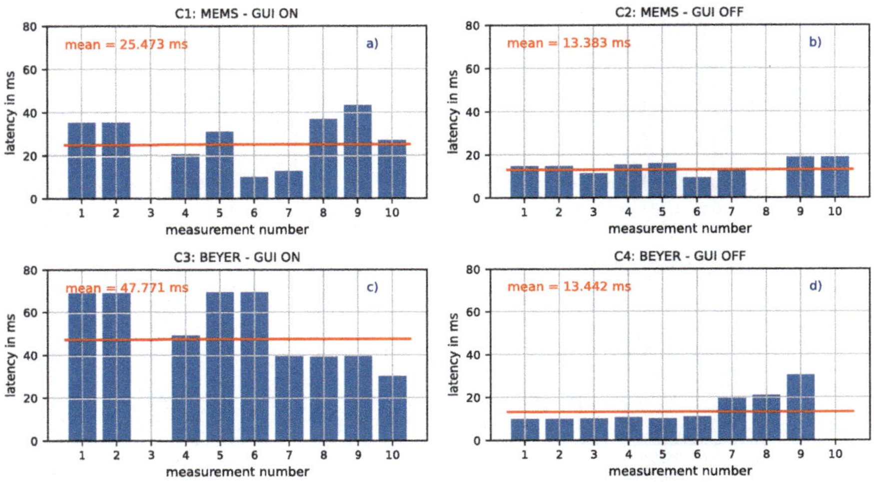

Fig. 3. Latency expressed in milliseconds, calculated as a relative time delay from the least delayed IR. Results are given for measured combination C1 (**a**), C2 (**b**), C3 (**c**), and C4 (**d**).

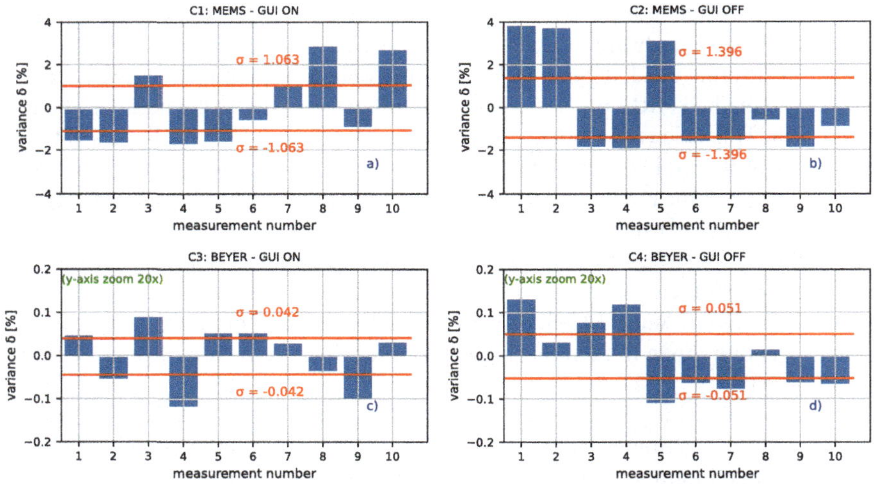

Fig. 4. Repeatability of the impulse response amplitude. Results are given for measured combination C1 (**a**), C2 (**b**), C3 (**c**), and C4 (**d**).

(Fig. 4c and d) has been enlarged 20 times in comparison to Setup 1 (Fig. 4a and b), so data can be visible.

When analyzing the results in Fig. 4, it is easily observable that the clear winner in the category of amplitude repeatability is Setup 2 for all of the RPi States (Fig. 4c and d). Given the fact that results are presented as the variance for particular C (1–4), and considering the cost of the MEMS microphone, Setup 1 (Fig. 4a and b) does give a respectable result regarding repeatability of IR since the maximum variance is less than 4% in any of the Setup 1 measurements (see Fig. 1b). When comparing State 1 to the results of State 2 (Fig. 4a and c versus Fig. 4b and d) it seems that turning GUI ON or OFF does not significantly affect the repeatability of the IR measurement.

4 Conclusion

Availability and affordability of the SBC systems enable engineer hobbyists and scientists to put them to good use in many types of projects. Realtime applications require that hardware is capable of sampling, processing, and outputting the data minimizing the loss and signal degradation. This is especially true for the audio real-time applications where SBCs can play an important role. This paper addresses the possibility of using Raspberry Pi for audio recording.

Analysis was done through investigation of the IR measurements, that is, through repeatability of amplitude latency of the IR. Different setups were composed, using MEMS microphone for recording at first (Setup 1), then in Setup 2 using studio microphone and sound card. Raspberry Pi was observed in State 1 when Linux OS GUI was turned ON, then repeating measurements when Linux OS GUI was turned OFF. Results were processed in Python and graphically presented. Results showed that repeatability of the latency is better in the case of MEMS microphone in comparison to the studio equipment. It was also noted that by OS GUI running, latency is increased more than twice which suggests that running GUI in real-time audio applications is not recommended.

Repeatability of amplitude showed better results when using studio equipment, which is to be expected. Turning the OS GUI ON does not influence the repeatability of the amplitude. Even though studio equipment is better, using the cheap MEMS microphone does seems to be valid, since maximum of the variance in the amplitude was 4%. In the future, it would be interesting to perform much larger set of measurements to see if MEMS amplitudes will hold the repeatability.

References

1. Ličanin, M., Praščević, M., Mihajlov, D.: Realization of the low-cost noise measurement monitoring station using MEMS microphone technology and micro-PC. In: Proceedings of 26th International Conference: Noise and Vibration, pp. 133–136. ISBN: 978-86-6093-088-2, COBISS.SR-ID 270804748, Niš (2018)
2. Müller, S., Massarani, P.: Transfer-Function Measurement with Sweeps. J. Audio Eng. Soc. **49**(6), 443–471 (2001)
3. Stan, G., Embrechts, J., Archambeau, D.: Comparison of different impulse response measurement techniques. J. Audio Eng. Soc. **50**(4), 249–262 (2002)

4. Farina, A.: Simultaneous Measurement of Impulse Response and Distrortion with a Swept-sine Technique. 108th Convention Audio Eng. Soc., abstract in J. Audio Eng. Soc., **48**(4), 350 (2000)
5. Jelenković, M., Ćirić, D., Ličanin, M., Đorđević, A.: Ponovljivost impulsnog odziva prostorije. In: Zbornik radova konferencije ETRAN, pp. AK.2.7-1–AK.2.7-4. Društvo za ETRAN, Banja Vrućica (2011)

Protecting Residential Areas Against Noise Generated by Car Wash Units. A Case Study

Vasile Bacria, Nicolae Herisanu(✉), and Cristina Chilibaru-Opritescu

University Politehnica Timisoara, Timişoara, Romania
nicolae.herisanu@upt.ro

Abstract. In car washes, which are often located near residential areas, technical installations and noise-generating activities can be identified. This noise can affect people who carry out their activities in car washes and in neighboring areas. The paper presents the results obtained in the investigation of the acoustic fields in the space of a car wash and of the one in the neighboring inhabited area. In this sense, the noise sources, its propagation mode, its frequency spectrum, the statistical distribution, the characteristic parameters, the harmful effects, and the methods of eliminating the undesirable effects are underlined. The passive shielding method was used to protect the neighboring inhabited areas against the noise generated by the car wash. The efficiency of the implemented method is highlighted by the measurements performed before and after the installation of the screens.

Keywords: Noise protection · Residential area · Car wash units

1 Introduction

Car washes have an important role in maintaining cleanliness and reducing pollutants that worsen air quality. Within them there are installations and activities that generate noise. It can affect the life and health of people who work on the car wash platform or living in neighboring areas.

In this sense, the water jet installation for the washing operation of the vehicles, the suction installation used after washing, the movement of the vehicles before and after washing are identified as noise sources.

The noise generated by these sources in the car wash unit can be characterized by the equivalent noise level defined by [1–3]:

$$L_{Aeq,T} = 10\lg\left[\frac{1}{t_2 - t_1}\int_{t_1}^{t_2}\frac{p_A^2(t)}{p_0^2}dt\right] \quad (1)$$

where L is the equivalent continuous level of sound pressure A-weighted, measured in dB, determined in a time interval T, which begins at time t_1 and ends at time t_2, $p_0 = 2 \times$

N. Herisanu and V. Marinca (eds.), *Acoustics and Vibration of Mechanical Structures – AVMS-2021*, Springer Proceedings in Physics 274,
https://doi.org/10.1007/978-3-030-96787-1_15

10^{-5} [N/m^2] is the reference acoustic pressure and $p_A(t)$ is the A-weighted instantaneous acoustic pressure of the acoustic signal.

Indeed, the equivalent continuous noise level can be obtained by calculation using Eq. (1) or can be measured directly and recorded using appropriate equipment [4–7]. It allows the measurement and recording of several characteristic sound parameters as well as the spectral analysis in frequency bands and the statistical distribution of the acoustic levels, as well.

In the paper, we present the results obtained in the investigation of noise in the car wash unit and near the neighboring residential area as well as the measures established to protect it against the noise generated by the car wash unit. In this sense, the passive shielding method was used. The effectiveness of the implemented method was verified by measurements.

2 Harmful Noise Generated by Car Wash

The technical facilities and activities in the car wash unit generate noise which is very harmful to people who work here as well as for those who live in the neighboring area.

Thus, the diseases of various organs and apparatus of the human body are produced by noises that exceed the level of 40 dB. Due to the action of noise, the blood pressure changes, the pulse accelerates, the intracranial vascular tension increases even three times, the visual acuity decreases, the breathing rhythm changes, nervous irritations occur, the fatigue process is accentuated, and attention and psychic reactions weaken.

Asthenia and even nerve disease can also occur. Research has shown that sound vibrations influence the contractions of the stomach wall, the noise makes them more pronounced and less regular. It also increases blood pressure in the blood vessels that run through the stomach walls. It turns out that noise affects the human nervous system producing psychophysiological changes, sleep disorders, and negatively affects the visual function and endocrine glands producing biochemical disorders. Noise also causes hearing fatigue and noise trauma. The action of noise on the auditory organ is more harmful and more pronounced if it is discontinuous, occurs suddenly, has a wide spectrum of frequencies and is accompanied by mechanical vibrations. A month of activity in a noisy environment with a level of 85 dB produces a loss of acoustic sensitivity of up to 40–50 dB at a frequency of 4000 Hz.

If the activity is continued in this noisy environment, the hearing loss is accentuated, widening the frequency band in which the sounds are no longer received. Noise causes general fatigue in the body. It affects physical work, especially that which requires a focus of attention and especially intellectual work. Working in a noisy environment increases the risk of accidents because it reduces the possibility of acoustic signaling, orientation after various characteristic noises and reduces the intelligibility of speech at work.

Thus, according to in-force rules, the maximum permissible limit of noise exposure of workers at work is 87 dB. In those places where the daily personal noise exposure of an employee exceeds 80 dB, the employer must provide employees with individual noise protection equipment, and in places where the daily personal exposure exceeds 84 dB employees are required to wear personal protective equipment against noise.

At the same time, in order to protect people living in the adjacent car wash area, the maximum permissible limit of continuous noise level equivalent to the limit of the functional area of the car wash is set by the Romanian standard [8] at 65 dB. Moreover, the location of units with noise and vibration sources and the dimensioning of sanitary protection areas will be done in such a way that in protected territories the equivalent noise level measured at 3 m from the outer wall of the house at 1.5 m height from the ground should not exceed 50 dB and the noise curve Cz45.

By investigating the acoustic fields generated by the sources on the car wash platform within it and near the residential area, it can be ascertained whether the environment is comfortable or not for humans. If the noise level exceeds the allowed values, measures can be established to reduce it.

3 Noise Investigation

There are a significant number of various noise sources in the car wash unit. These by their operation generate acoustic waves. Due to the obstacles encountered, they are reflected and overlap over the direct ones so that complex acoustic fields are formed on the car wash platform and in the neighboring area. The sound pressure at a point in the fields is obtained by summing the sound pressures corresponding to each type of wave. Given the large number and diversity of sources that contribute to the generation of noise as well as the nature of the acoustic waves generated, the acoustic fields that form in the car wash unit and in the surrounding area are extremely complex and their investigation is indicated to be carried out experimentally by measurements.

The measurements were performed in this case using Bruel&Kjaer 2250 analyzers. These allowed the measurement and recording of the characteristic noise parameters. The assessment of the environmental conditions was carried out by using a weather station that allowed the measurement of air speed, temperature, and humidity. Measurements were made during the operation of the car wash using two Bruel&Kjaer 2250 analyzers at the boundary of the functional area as well as in the adjacent residential area, the microphone being mounted at a distance of 2 m from the wall of the nearest building and at a height of 1.3 m from sol.

The measurements were performed during the normal operation of the car wash unit with a degree of occupancy of all five available wash boxes of about 97% of the measurement time. The location of the measuring points is shown in Fig. 1. Measuring points 1 and 2 are located inside the car wash property at the boundary of the enclosure, and measuring points 3 and 4 are arranged behind the car wash fence near the nearest apartment building.

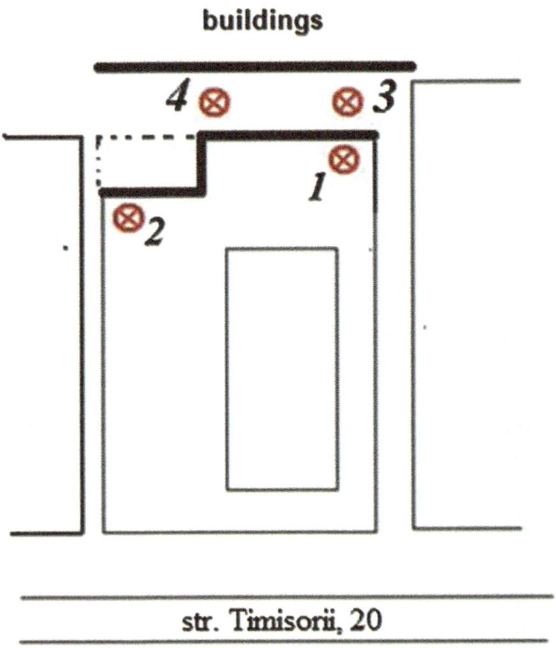

Fig. 1. The arrangement of the measuring points

Before and after performing the series of measurements the calibration status of the entire measuring system was checked using an acoustic calibrator, with a result of 93.9 for a calibrated signal of 94 dB at the frequency of 1000 Hz. In parallel with the recording of noise data, a monitoring of the environmental conditions was performed: wind speed, relative humidity, temperature, and pressure. The method for determining the characteristic parameters of noise according to [8–10] was used. During the tests, there were normal weather conditions: clear sky, absence of wind, rain or other extreme weather events, being monitored parameters that could have influenced the conditions of sound propagation.

The values of the different characteristic parameters of the recorded noise are shown in Tables 1 and 2. The following coding was used: 1-background noise location 1; 2-background noise location 2; 3-noise wash car, location 1; 4-noise wash car location 2; 5-noise vacuum cleaner location 1; 6-noise vacuum cleaner location 2; 7-noise near the apartment building location 3; 8-noise near the apartment building location 4. In Table 1 are shown the values of the continuous A-weighted equivalent noise level measured in the four points with the mentioned coding.

Table 1. Values of the equivalent noise levels

Sample number	1	2	3	4	5	6	7	8
L_{Aeq} (dB)	53.38	56.17	64.34	64.96	73.83	60.77	52.53	55.00

In addition to the equivalent sound level, the average noise-weighted and time-weighted maximum and minimum noise levels and the C-weighted noise level were determined. These are stored in the database of the investigation carried out.

The spectral distribution in frequency bands of 1/1 octave of the unweighted noise in the measurement points is presented in Table 2.

Table 2. Spectral distribution in frequency bands (case 1)

Sample number	16 Hz	31.5 Hz	63 Hz	125 Hz	250 Hz	500 Hz	1 kHz	2 kHz	4 kHz	8 kHz	16 kHz
Cz60	–	96.3	82.9	74.2	67.8	63.2	60	57.4	55.4	53.8	–
Cz45	–	86.0	71.0	61.1	53.6	48.6	45	42.2	40.0	38.3	–
1	62.88	62.25	62.78	55.33	51.3	49.26	48.99	46.06	40.39	34.44	25.08
2	2.06	10.88	4.8	0	64.56	60.78	59.06	52.75	46.82	45.59	42.66
3	61.6	65.38	63.79	58.56	59.22	59.58	59.96	57.63	53.53	50.84	42.23
4	66.13	67.46	64.7	61.44	60.95	60.87	61.03	59.23	56.05	53.39	46.21
5	62.45	63.85	64.73	58.06	64.9	71.51	64.86	69.03	61.78	48.99	33.97
6	62.71	66.94	64.03	57.7	57.25	58.13	53.79	55.26	48.34	36.94	25.35
7	59.97	67.76	62.13	51.04	50.55	49.62	46.95	45.39	40.49	35.9	27.77
8	62.6	66.14	63.46	56.29	50.44	50.52	47.86	49.27	48.16	42.76	31.37

Diagrams of the variation of the noise level over time, of the spectral distribution of noise in frequency bands and of the statistical distribution of noise levels for measuring point 4 are shown in Figs. 2, 3, and 4 and for measuring point 8 in Figs. 5, 6, and 7.

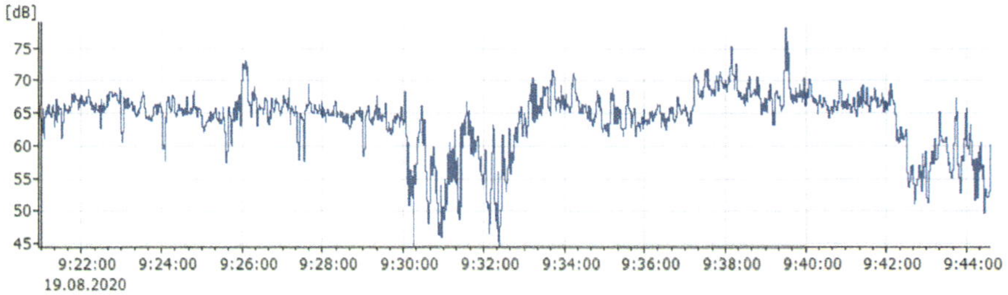

Fig. 2. Noise level time variation diagram for point 4

Analyzing the results of the measurements, it is observed that the values of the continuous equivalent noise level to the limit of the functional area of the car wash unit are very close to the allowed limit: 64.34 dB and 64.99 dB, respectively, and its values near inhabited buildings exceed 2.53 dB, respectively, by 5 dB the allowed value. Regarding the spectral distribution in bands of 1/1 octave of noise, the values recorded on the car wash platform exceed the values of the Cz60 curve only for frequencies higher than 1 kHz, and in the case of values recorded near inhabited buildings, the values of the Cz45 curve are exceeded for frequencies higher than 500 Hz. In this situation, the noise

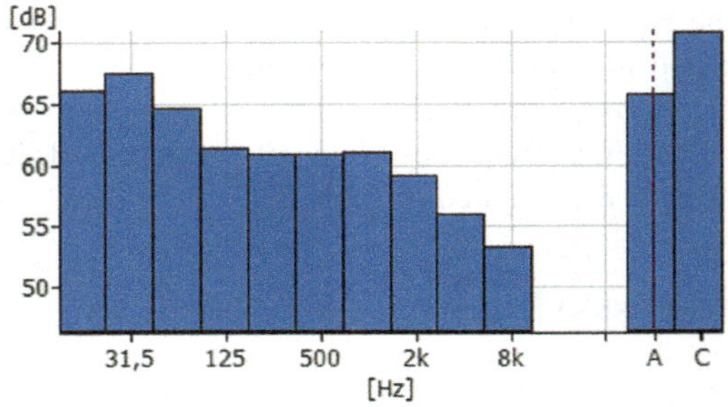

Fig. 3. Diagram of the spectral distribution of unweighted noise in 1/1 octave frequency bands for point 4

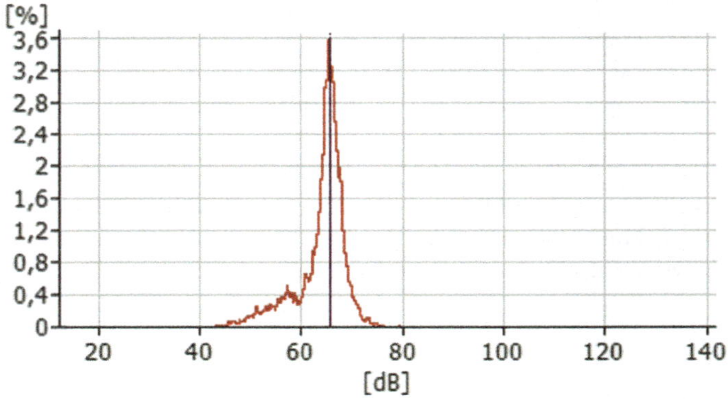

Fig. 4. Diagram of the statistical distribution of noise for point 4

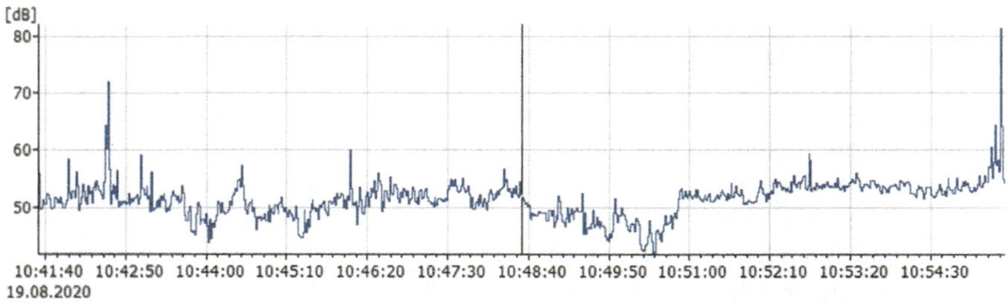

Fig. 5. Noise level time variation diagram for point 8

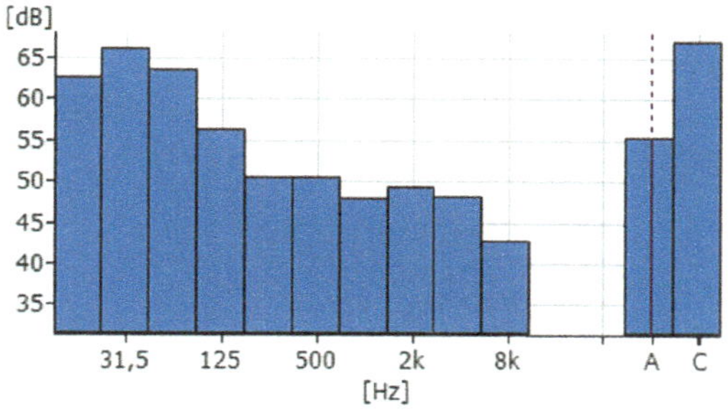

Fig. 6. Diagram of the spectral distribution of unweighted noise in 1/1 octave frequency bands for point 8

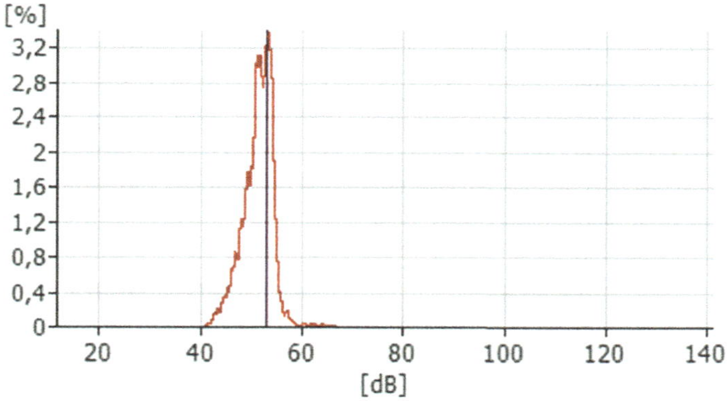

Fig. 7. Diagram of the statistical distribution of noise for point 8

on the car wash platform does not significantly affect the staff inside it, but it is harmful to people in the neighboring inhabited areas.

Due to this fact, it is necessary to establish and implement measures to protect the inhabited area against the existing noise on the car wash platform.

4 Protecting Residential Areas from Noise

To protect the inhabited areas against noise, the passive protection method was used by making a screen with dimensions $L = 20$ m, $l = 0.1$ m, $h = 4.4$ m, made of basalt mineral wool panels with sound-absorbing and sound-insulating properties. The panels used have the following dimensional characteristics: length 2500–13,500 mm, width 1000 mm, and thickness 100 mm. These are sound-absorbing insulating sandwich panels with both sides made of metal sheet, with mineral wool core, designed to achieve a sound insulation of 34 dB with a sound absorption coefficient close to the unit and are impermeable to water, air, and vapors.

On the other hand, it is known that the attenuation achieved by a protective screen depends on the height H, the wavelength λ of the disturbing sound, the distance R between the screen and the source, and D between the screen and the receiver, as well as the ground heights of source and receiver. Assuming that the source and receiver are at the same height, the ΔL attenuation achieved with a screen having a very long length in relation to the sound wavelength has the expression [1]

$$\Delta L = 10\lg 20X \tag{2}$$

where

$$X = \frac{2\left[R\sqrt{1+\left(\frac{H}{R}\right)^2}-1\right]+D\left[\sqrt{1+\left(\frac{H}{R}\right)^2}-1\right]}{\lambda\left[1+\left(\frac{H}{2}\right)^2\right]} \tag{3}$$

Generally, with such a screen raised 1–5 m above the noise source and the receiver located at a distance of 3–6 m from the source allows obtaining attenuations of 5–10 dB in case of low frequencies and 15–25 dB in the case of high frequencies. If the screen is made of sound-absorbing insulating panels, its operating qualities increase. To determine how this screen protects buildings in the surrounding area from noise generated on the car wash platform, a new set of measurements was performed after mounting the screen. The location of the measuring point is shown in Fig. 8.

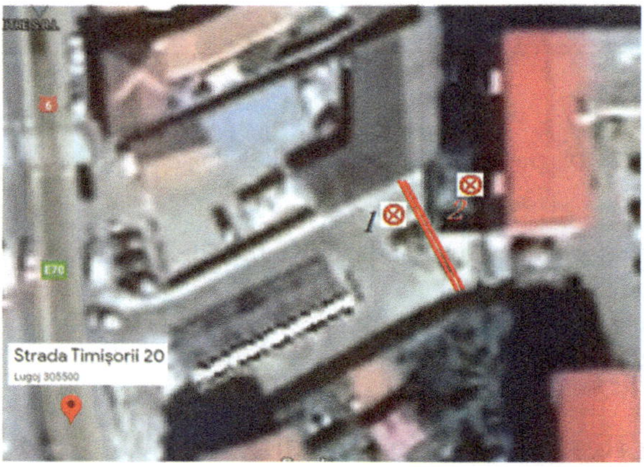

Fig. 8. The arrangement of the measurement points.

In order to evaluate the efficiency of the implemented acoustic screen and to estimate the compliance with the admissible limits provided by SR 10009:2017 for the external noise level at the facade of the residential building which is most exposed to the action of the source, a representative measurement point was chosen near the apartment block. A second measuring point was placed at the edge of the enclosure inside the car wash, the recordings being made simultaneously, the microphone being arranged at a height of 1.3 m from the ground. The measurements were performed during the normal operation of the car wash unit with a degree of occupancy of all five washrooms of approximately

90% of the measurement time, except for the period in which the background noise was measured, when the car wash activity was stopped. Measurements were made separately to determine the background noise coming largely from road traffic. The measuring points were placed in location 1 at the boundary of the car wash area and in location 2 near the building (Fig. 8). The same test equipment was used as in the initial measurements, and the same measurement procedure was employed.

The values of the different recorded noise parameters are presented in Tables 3 and 4. The following coding of the measuring points was used: 1—background noise location 1; 2—background noise location 2, series 1; 3—background noise location 2, series 2; 4—background noise location 2 Series 3; 5—noise operation laundry location 1; 6—noise operation laundry location 2, series 1; 7—laundry operation noise location 2, series 2; and 8—noise operating laundry location 2, series 3. In Table 3 are presented the values of the A-weighted equivalent noise level, measured in the two measuring points with the mentioned codifications.

Table 3. Values of the equivalent noise levels

Sample number	1	2	3	4	5	6	7	8
L_{Aeq} (dB)	53.28	46.85	46.57	46.36	68.21	48.7	48.34	48.77

Table 4. Spectral distribution in frequency bands (case 2)

Sample number	16 Hz	31.5 Hz	63 Hz	125 Hz	250 Hz	500 Hz	1 kHz	2 kHz	4 kHz	8 kHz	16 kHz
Cz60	–	96.3	82.9	74.2	67.8	63.2	60	57.4	55.4	53.8	–
Cz45	–	86.0	71.0	61.1	53.6	48.6	45	42.2	40.0	38.3	–
1	65.53	67.99	64.15	58.7	53.16	48.97	48.77	45.41	38.51	29.29	19.91
2	61.66	64.38	58.94	45.98	42.95	43.38	42.12	40.33	31.19	22.88	14.94
3	66.57	67.79	60.93	48.85	45.75	42.23	41.47	39.48	30.76	22.29	14.91
4	61.09	65.2	59.72	49.04	47.63	42.41	41.13	38.68	29.89	22.18	15.33
5	69.02	72.57	67.56	62.19	61.69	63.33	62.32	62.47	59.14	53.99	45.33
6	61.03	66.12	61.05	50.42	48.17	47.27	45.53	42.5	38.41	32.09	21.9
7	60.61	69.1	61.11	50.96	46.49	45.19	42.86	41.26	35.85	28.89	18.51
8	61.27	66.45	60.21	51.52	46.22	45.29	44.08	41.11	36.35	30.18	20.79

The spectral distribution in frequency bands of 1/1 octave of the unweighted noise in the considered measurement points is presented in Table 4.

Diagrams of variation of noise level over time, spectral distribution of noise in frequency bands, and statistical distribution of noise levels for location 1 (sample 5) are shown in Figs. 9, 10, and 11, and for location 2 (sample 7) are shown in Figs. 12, 13, and 14.

Following the results of the measurement performed after the realization and implementation of the protection screen made of basalt mineral wool panels, with sound-absorbing and sound-insulating properties positioned between the car wash platform

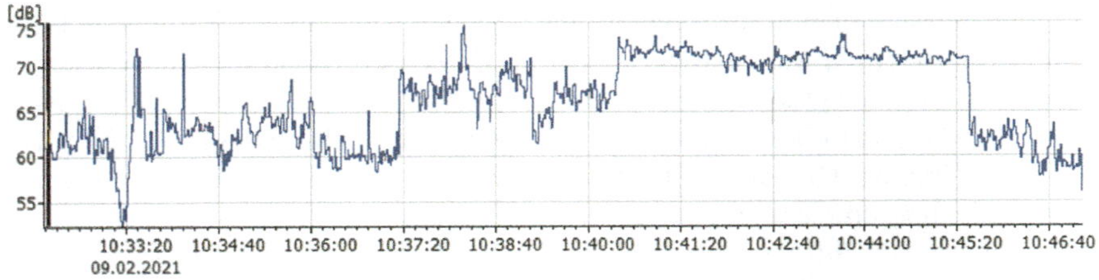

Fig. 9. Noise level time variation diagram for location 1 (sample 5)

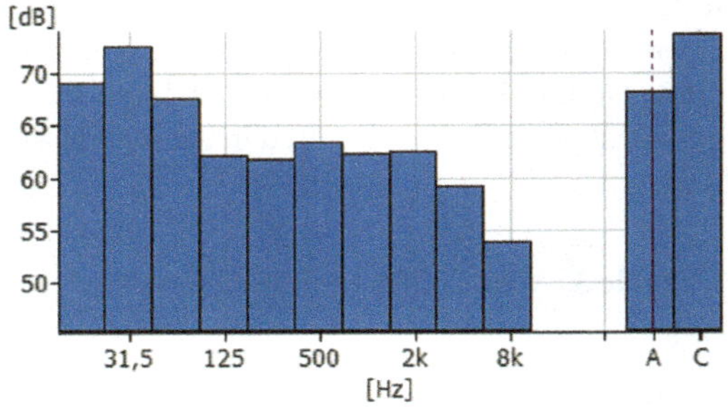

Fig. 10. Diagram of the spectral distribution of unweighted noise in 1/1 octave frequency bands for location 1 (sample 5)

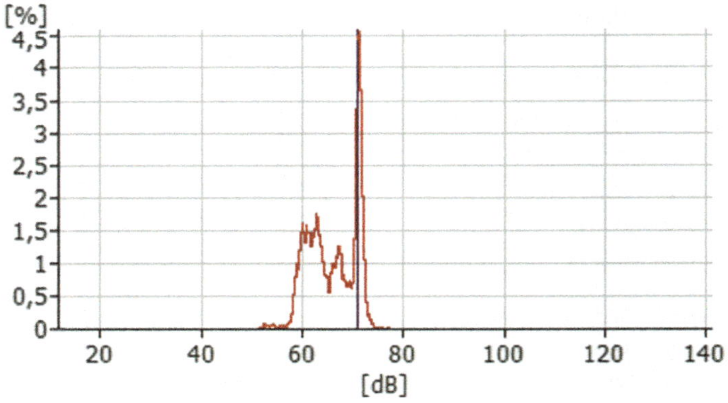

Fig. 11. Diagram of the statistical distribution of noise for location 1 (sample 5)

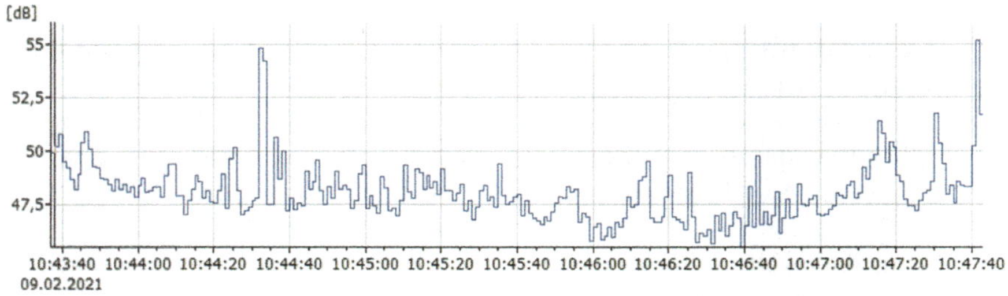

Fig. 12. Noise level time variation diagram for location 2 (sample 7)

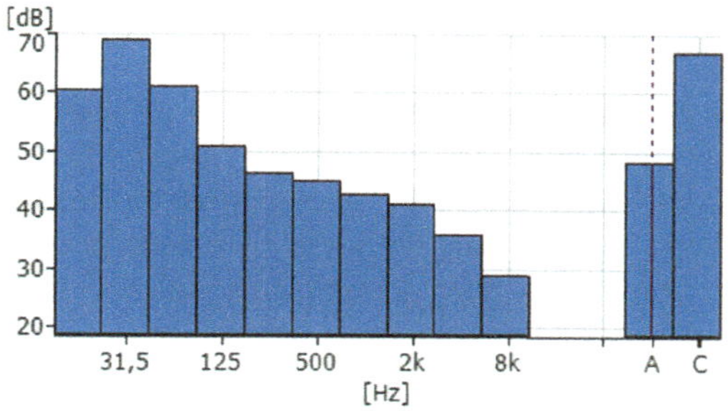

Fig. 13. Diagram of the spectral distribution of unweighted noise in 1/1 octave frequency bands for location 2 (sample 7)

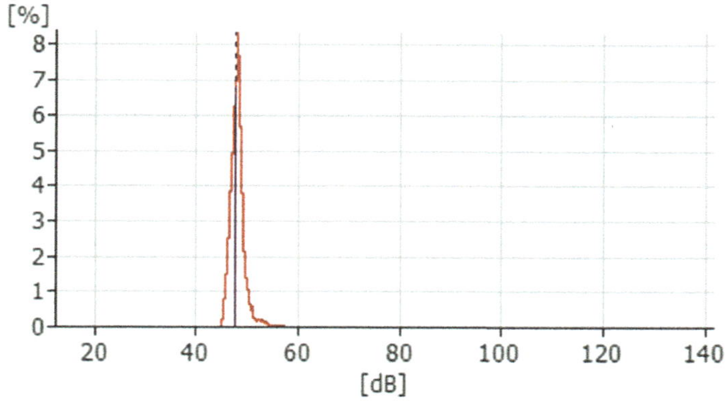

Fig. 14. Diagram of the statistical distribution of noise for location 2 (sample 7)

and the neighboring apartment block shown in Tables 3 and 4; it is observed that for a noise level of 68.21 dB registered at the limit of the functional area of the car wash unit, next to the block of flats, values of the equivalent continuous noise level of 48.7, 48.34, and 48.77 dB were registered, which fall within the allowed limits.

So, by using the designed protection screen, noise level reductions of 19.51 dB, 19.87 dB, and 19.44 dB, corresponding to the three samples, are obtained. At the same time, analyzing the spectral distribution in frequency bands of 1/1 octave of the unweighted noise presented in Table 4, it is observed that the values corresponding to the three samples near the block of flats fall almost entirely within the allowed limits, being below the corresponding values of Cz45 curve.

In this way, the protection of the residential area against the noise generated by the car wash unit by the passive method of shielding proves its validity and efficiency.

5 Conclusions

In order to find the solution to protect the inhabited areas against the noise generated by a car wash unit, it is necessary to investigate the acoustic fields on its platform and near the residential areas. The results of the investigations carried out are compared with the allowed values. From this comparison, it can be determined whether the inhabited areas are affected or not by noise. After the implementation of the protection measure, its effectiveness can be verified through new investigations.

References

1. Grumazescu, M., Stan, A., Wegener, N., Marinescu, V.: Combaterea zgomotului si vibratiilor. Ed. Tehnica, Bucuresti (1964)
2. Iudin, E.Ia.: Izolarea impotriva zgomotului, Ed. Tehnica, Bucuresti (1968)
3. Enescu, N., Magheti, I., Sarbu, M.A.: Acustica tehnica, Ed. ICPA, Bucuresti (1998)
4. Mihajlov, D., Prascevic, M., Herisanu, N.: Selection of measurement strategy for the assessment of long-term environmental noise indicators using multi-criteria optimization. Springer Proc. Phys. **198**, 77–82 (2018)
5. Bacria, V., Ghita, E., Herisanu, N.: On acoustic comfort in urban transport on rails. Springer Proc. Phys. **198**, 83–90 (2018)
6. Bacria, V., Herisanu, N.: Noise control in an industrial hall. Appl. Mech. Mater. **430**, 251–256 (2013)
7. Herisanu, N., Bacria, V.: Some effects of rubberized asphalt on decreasing the phonic pollution. Appl. Mech. Mater. **430**, 257–261 (2013)
8. SR 10009:2017 Acustică. Limite admisibile ale nivelului de zgomot din mediul ambiant
9. ISO 1996-2:2017 Acoustics - Description, measurement and assessment of environmental noise—Part 2: Determination of sound pressure levels
10. SR 6161-3:2020 Acustica în construcţii. Partea 3: Determinarea nivelului de zgomot în localităţile urbane. Metodă de determinare

Workers' Occupational Exposure to Noise in Steel Industry. A Case Study

Simion Sorin(✉), Angelica Calamar, Simion Alexandru, Toth Lorand, and Kovacs Marius

National Institute for Research and Development in Mine Safety and Protection to Explosion, INSEMEX Petroşani, 32-34 G-ral Vasile Milea Street, 332047 Petroşani, Romania
sorin.simion@insemex.ro

Abstract. A large number of workers suffer from hearing problems caused by exposure to high levels of noise, with noise being a pervasive occupational noxae in most industrial workplaces. Thus, exposure to this noxious substance is subject to legal regulations in the field of occupational health and safety, which limit exposure level, in order to reduce the risk of occupational disease generated by exposure to noxious noise. Low efficiency caused by exposure to noise, mistakes made in various operations, masking safety audible signals, distraction, worker's fatigue, and overwork, requires the implementation of technical and organizational measures that influence both the worker, by using additional protective equipment (hearing protective equipment) as well as the economic agent, by applying technical measures for modernization/redesign of technological flows (allotting material / financial resources). Assessment of occupational risks for each workplace involves determining and quantifying exposure to toxins, being the main objective of accidents and occupational diseases prevention, in order to apply the most effective measures to reduce exposure to these harmful factors. The current paper analyses the level of noise generated by technological processes in steel industry, over a monitoring period of 5 years, for the main activities. Prevention of exposure to noise must be based on both the determination of noise level and its evolution over time, to measure the effectiveness of mitigation measures. In this sense, the paper aims to analyze noise exposure evolution over time for the main workplaces in a steel plant, in order to assess the dynamics of exposure, during monitoring.

Keywords: Occupational disease · Occupational exposure · Steel industry · Occupational hazards · Noise

1 Introduction

Occupational diseases are "diseases which result from practicing a trade or profession, caused by harmful physical, chemical or biological factors characterising the workplace and overload of various body organs or systems in the process of work" [1, 2].

N. Herisanu and V. Marinca (eds.), *Acoustics and Vibration of Mechanical Structures – AVMS-2021*, Springer Proceedings in Physics 274,
https://doi.org/10.1007/978-3-030-96787-1_16

In order to be considered an "occupational disease" or "occupation related disease" the condition must be the result of practicing a trade or profession and the triggering factors (physical, chemical, or biological) must be characteristics of the work environment.

The most important feature of the causal relationship between exposure to toxins and occupational disease is the complex nature of factors that generate them, most often diseases not being caused by accidental exposure to physical and chemical toxins but by long-time exposure or decades of exposure to harmful factors. The reaction of each worker to occupational noxious substances is attenuated or aggravated by physiological individual factors, non-exhaustively listing previously acquired diseases such as cardiovascular, neuro-vegetative, osteo-articular diseases, etc.

Technological development and changes in working technologies to obtain an increased efficiency by increasing the power and working speed of technical equipment contributes to diversification of noise sources and thus to increase the number of people exposed.

Noise is an important industrial noxae that generates hazards for workers' health, occurring as a result of specific industrial activities, respectively, as a result of damage and/or improper operation of work equipment [3].

The activity carried out in the steel manufacturing industry generates noise levels that can affect health and safety of workers. In order to improve working conditions, special directives have been developed in specific areas, in order to ensure an optimal level of protection for workers.

Directive 2003/10/EC of the European Parliament and of the Council of 6th February 2003 on the minimum health and safety requirements for exposure of workers to risks arising from physical agents (noise), transposed into national legislation by Decision No. 493 of April 12, 2006, on minimum health and safety requirements regarding exposure of workers to risks generated by noise, establishes the physical parameters used as risk predictors, as follows:

- peak sound pressure (P_{peak}): the maximum value of instantaneous sound pressure, measured with the frequency weight C
- the level of daily noise exposure (Lex, 8 h) [dB (A) re. 20μPa]: the time-weighted average of noise exposure levels for a nominal 8 h working day, defined by the international standard ISO 1999:1990, point 3.6.;
- weekly noise exposure level (Lex, 8 h): the time-weighted average of daily noise exposure levels in a nominal five days working week of 8 h/day, as defined by SR ISO 1999:1996, point 3.6 [4].

National legislation also provides limits for maximum noise levels for workplaces with increased and special neuropsychologic and psychosensory demands (attention, responsibility, decision, and time constraint), respectively, a reduction of the maximum limit from 87 dB (A) to (GD no. 601 of 13 June 2007):

- 75 dB (A) for testing or troubleshooting laboratories, technological process monitoring booths, custom houses.

- 60 dB (A) for RTV studios, command and control booths, research, design and measurements laboratories, etc.
- 50 dB (A) for operating and treatment rooms, creation workshops, air traffic management, and information rooms.

The maximum limit of 87 dB (A) has been set so that health and safety of workers are not affected by noise exposure, respectively, the above limits have been set according to the complexity of activities carried out so as not to affect the quality of processes carried out.

2 Working Tools

Determination of noise levels was performed with type 2250 integrating sound level meters equipped with one-octave and/or 1/3 octave filters, 4447 noise dosimeters, type 4231 calibrators, and the BZ 5503 determinations' visualization software produced by Bruel & Kjaer.

The noise analyzer type 2250 (Fig. 1) is used for performing a series of individual measurements or a series of measurements for laboratory and field analysis. The acoustic calibrator (Fig. 1) is used for checking the sound level meters to obtain results with a high degree of accuracy (checking is performed both before and after a series of determinations). To ensure the quality of determinations' results, the sound level meters and calibrators that were used were metrologically verified in accordance with the requirements of legislation in force [4, 5].

Analyser type 2250 Acoustic calibrator type 4231

Fig. 1. Analyser type 2250, acoustic calibrator type 4231

If the calibration differs by more than ±1,5 dB from the initial calibration, the sound level meter/dosimeter used shall be checked for metrological accuracy.

3 Case Study on Occupational Exposure to Physical and Chemical Toxins

Exposure level determinations were performed on a steel plant that uses electric furnaces to melt scrap metal. The technological flow of manufacturing is linear, consisting of the following elements: scrap iron storage, loading kettles with scrap iron, loading furnaces with scrap iron, steel processing in furnaces, steel treatment plants in pots, continuous casting plant, and storage warehouse for finished products.

In order to monitor noise exposure, the INCD INSEMEX Petroşani team carried out determinations for the main activities carried out in the steel industry, likely to be generating noise beyond the limits provided by legislation in force [6].

Steel products manufacturing generates exposure to many professional pollutants, including noise, vibration, dust, carbon monoxide, carbon dioxide, iron oxides, etc.

The method for determining noise exposure is the direct method with numerical display measuring instruments, respectively, integrating sound level meters and noise dosimeters.

Results of the noise determinations, in accordance with regulations in force, are presented in Table 1 (Fig. 2).

Table 1. The evolution of noise level in the steel plant between 2016 and 2020

No.	Workplace	Measured noise level $L_{EX, 8\,h}$ dB (A)				
		2016	2017	2018	2019	2020
	EBT furnace—platform	106.5	105.5	107.6	103.4	106.4
	EBT cabin	80.5	80.0	80.3	80.6	80.6
	LF cabin	79.5	79.0	78.3	77.3	77.3
	Casting pot preparation sector	91.4	90.4	92.4	89.8	89.8
	LF installation and vacuum installation	91.7	91.6	90.4	87.7	88.9
	Masonry formation	94.6	95.6	89.9	97.1	96.5
	Crane cabin in the electrical steel hall	86.5	85.5	85.0	83.5	82.6
	Overhead crane cabin in the electrical steel hall	83.2	84.2	84.3	83.3	81.2
	Flue gas capture and treatment plant	87.6	87.5	87.9	87.7	87.5
	Flame cutting machine	90.1	90.2	90.4	90.4	89.4
	Level + 10,400	87.5	87.6	87.8	87.3	87.4
	Reception tables in MTC adjustment	94.8	92.8	96.3	96.3	95.3
	Crystallizing hall	92.8	90.8	90.3	91.5	91.1
	Pump station	96.5	94.5	96.5	95.7	95.5
	Crane cabin running bridge warehouse no. 1	86.3	86.3	85.3	83.3	82.3
	Crane cabin running bridge iron warehouse no.2	75.6	77.7	79.2	79.2	78.9
	Electric steel loading/unloading sector	87.5	87.6	87.9	87.9	88.9
	Diesel locomotive cabin—stationary	84.6	85.3	84.5	86.5	85.4
	Diesel locomotive cabin—in gear	90.2	91.2	90.5	91.5	90.4

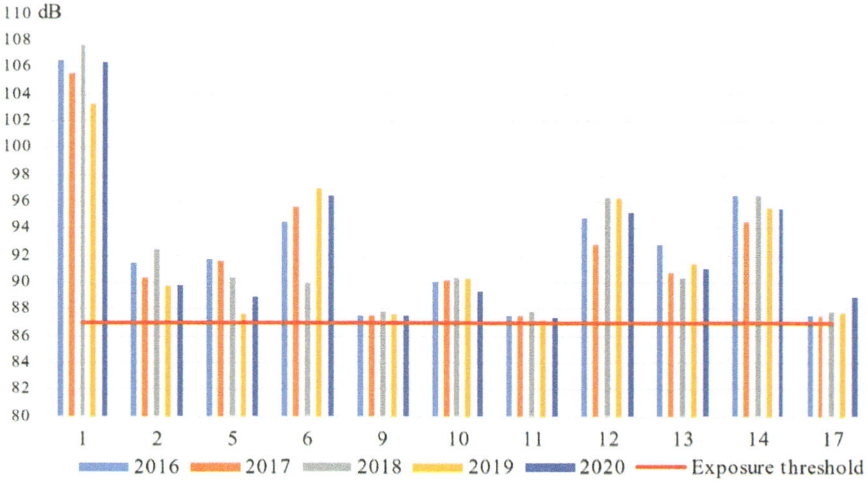

Fig. 2. Evolution of noise level in workplaces to which the exposure limit of 87 dB (A) applies

Exceedance of the exposure limit by more than 20 dB (A) at the EBT Platform Furnace is observed. Noise is generated by the action of electrode columns on scrap metal in conjunction with oxygen insufflation. This noise affects all the activities in the steel hall and it is impossible to reduce it or isolate this activity from the rest of the technological flow (Fig. 3).

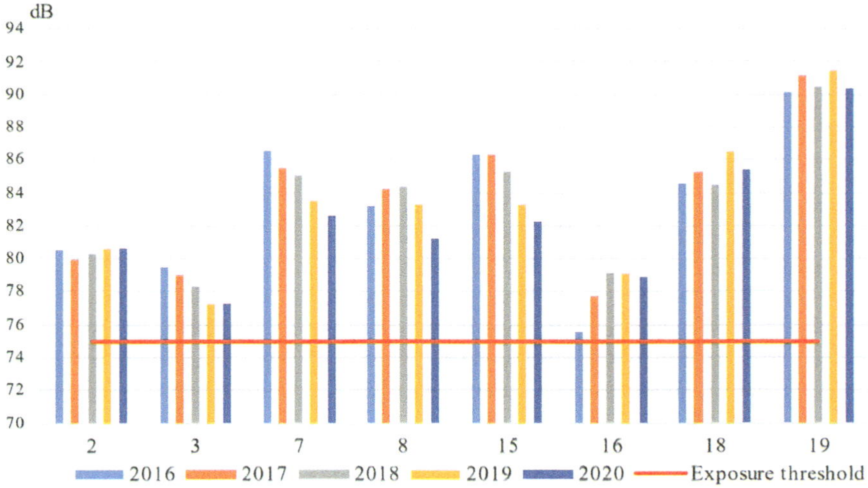

Fig. 3. Evolution of noise level in workplaces to which the exposure limit of 75 dB (A) applies

There is a decreasing trend in the level of noise exposure, in Cabin LF, Crane cabin in the electrical steel hall, overhead crane cabin in the electrical steel hall, crane cabin running bridge warehouse no. 1, crane cabin running bridge iron warehouse no. 2, these being subjected to some refurbishment and modernization processes during the monitoring period. It is also observed that the exposure limit of 75 dB (A) is exceeded by more than 16 dB (A) for diesel locomotives, regulations in the field of railway transport not allowing constructive changes to them, which leads to restriction regarding possibilities of modernization and refurbishment.

4 Conclusions

Noise determinations at the workplaces performed in order to evaluate their impact on health of personnel working in steel plants showed the following:

- Periodic determination of noise level provides the necessary data to confirm or disprove its presence in the work environment, being necessary for establishing workplaces that require special attention from health and safety at work workers. Noise exposure monitoring shall be carried out mainly to verify the level of exposure, respectively, to check compliance with the limit values set by in force legislation and to establish the plan of technical and economic measures to reduce exposure to noxae.
- Permanent medical supervision performed by occupational medicine offices within the annual medical control, identifies the cases of occupational diseases caused by exposure of workers to noise by performing audiometry examinations, respectively, by monitoring their evolution in time.
- Exceeding the maximum noise level allowed by the legal provisions in force by up to 20 dB (A), limited possibilities for modernization and re-technologization, working conditions specific to steel plants (presence of also other noxious substances), cases of occupational disease, keep the workplaces in steel plants ranked as special conditions.

In conclusion, the steel plant presented above continues to modernize equipment and optimize the production process by implementing new technologies to ensure greater protection for workers, but the limitations generated by the specifics of steel activities (high temperature, the need to use fireproof materials, etc.) significantly reduce the solutions applicable to reduce noise exposure. The production process cannot take place automatically, without direct participation of the worker, steel plants remaining a workplace in which the activity takes place in the presence of physical and chemical noxious substances. This requires medical monitoring to detect occupational diseases in short, medium, and long term.

References

1. Angelica, D., Marius, K., Lorand, T.: Occupational health and safety regarding the exposure to noxious of workers from the steel industry. Metal. Int. **18**, 2 pp, 147–153 (2013)
2. Toma, N.: Manual of professional pathology. Medical Publishing House, București (1985)

3. Simion, S., Vreme, C., Kovacs, M.: Exposure of workers to noise in mining industry. In: Herisanu, N., Marinca, V. (Eds.) Conference: 12th International Symposium Acoustics and Vibration of Mechanical Structures (AVMS 2013), Timisoara (2013)
4. Law 319/2006 Law on Safety and Health at Work
5. GD no. 493/2006 on the minimum safety and health requirements regarding exposure of workers to risks of noise pollution, provides for maximum exposure limits for workers in relation to their work
6. Dragos, V.G., Angelica, D.N., Constantin, B.: Methods for analysis and evaluation of occupational accidents and diseases risks. Environ. Eng. Manage. J. **7**(4), 443–446 (2008)

Measurement and Reduction of Underwater Noise from Anthropic Activities

Claudia Borda[1(✉)], Marinela Marinescu[1], Gabriel Gârleanu[1], Larisa Butu[1], Delia Gârleanu[1], Delicia Arsene[1], Cristina Modrogan[1], and Mihai Arsene[2]

[1] Polytehnic University of Bucharest, Bucharest, Romania
{claudia.borda,larisa.butu}@upb.ro, marinela.marinescu@upb.rocom
[2] Afico, Bucharest, Romania

Abstract. Marine water pollution does not only refer to the overflow of chemical, physical or biological harmful substances into the sea but also to the increase of environmental noise. Ambient noise results from both natural and anthropogenic acoustic sources. Noises can travel long distances underwater and can prevent marine animals from hearing prey or predators, finding their way or connecting with other animals, group members or their young. In order to assess the impact of noise pollution produced by drilling rigs on the marine environment, measurements of underwater noise produced by an exploration drilling rig were performed. The marine platform on which the study was undertaken was accompanied by two support vessels. In order to reduce the level of noise introduced into the environment by a human activity, it was proposed to create an experimental context of insertion of noise in anti-phase at the source of noise pollution. Several virtual devices have been designed to argue the idea of introducing noise in contrast to that produced by the source; two counter-phase wavefronts were emitted, which combine into a sub-VI of summation, the resulting wave having zero amplitude. Recordings taken from hydrophones can be uploaded to a virtual instrument, which provides a phase-shifted signal, amplified and played back through speakers in the same environment and at the same location as the noise source.

Keywords: Underwater · Marine pollution · Underwater noise pollution · Hydrophones · Counter-phase wavefronts · Virtual instrument

1 Introduction

The Environmental Impact Assessment (EIA) Directive on Environmental Impact Assessment states that in European waters, for most offshore construction activities, an environmental impact assessment must be carried out.

The introduction of noise into the marine environment through construction, drilling, transport and tourism activities is an important issue for marine mammals and must be properly analyzed and evaluated, taking into account that their installation is already taking place.

N. Herisanu and V. Marinca (eds.), *Acoustics and Vibration of Mechanical Structures – AVMS-2021*, Springer Proceedings in Physics 274,
https://doi.org/10.1007/978-3-030-96787-1_17

Research shows that exposure to underwater noise can cause many types of adverse effects on marine animals, from behavioral changes to their death.

Internationally, more studies are needed to establish a reliable basis for the management of underwater radiated noise from merchant ships, construction activities, activities of exploration of marine objectives; their effective implementation must be considered, including in terms of economic considerations.

Given that it is expected that most human activities that cause continuous and impulsive underwater noise will intensify in the future, it is likely that the pressure resulting from underwater noise will increase. In order to minimize their impact, consideration should be given to limiting or offsetting underwater noise emissions at an early stage when it is intended to implement the relevant technology or industrial activity.

In order to determine the impact on marine mammals, on the behavioral and physiological effects, the level of sound pressure predicted for construction and drilling activities was studied and their inclusion in the criteria of disturbance, disruption or damage to marine mammals in the Black Sea.

2 Measurement of Underwater Noise Emitted by a Drilling Rig

In order to assess the impact of noise pollution produced by drilling rigs on the marine environment, measurements were made of underwater noise produced by a drilling rig for exploration, located in the Black Sea, in Romanian waters.

In this sense, several objectives were pursued:

- the use of calibrated systems for quantifying the levels of underwater noise sources and evaluating their compliance with noise norms;
- measurement of ambient sea noise without drilling noise;
- description of sound propagation at the platform location;
- modeling the prediction of noise levels with increasing distance from the drilling site, which used the noise levels of the drilling operations, the description of the sound propagation and the influence of the ambient noise of the sea;
- assessment of the impact produced by drilling activities on marine biodiversity and fishing operations;
- measurement of underwater noise at the site, assessment of its character and whether it is influenced by drilling operations and to what extent.

The study region was located 21 km offshore 44°21′LatN, 28°72′LongE, 110 m water depth, relatively constant. Marine platforms are accompanied by ships whose role is to keep in touch with the shore, to provide supplies, transport personnel and, in general, to ensure the logistical support of the platform.

The marine platform on which the study was undertaken was accompanied by two support vessels. The two accompanying ships that have maintained their position near the platform are 11-knot ships. A research vessel was also present during the measurements. Local ambient noise measurements were also performed without any ships or noise generated by the drilling rigs.

During the study period the platform was involved in the work of maintaining the well and active drilling, with the drilling head at 2300–2400 m. The platform was anchored on eight anchors and had no active positioning systems. The major elements of the cars onboard the platform were located on the decks above the waterline.

The conditions were close to total calm during the study period so there was no masking of the ship's signatures by the sound of wind or sea. An ambient noise level of 90 dB re 1 μPa was recorded, which is considered to be close to the lowest possible level in the region.

The noise produced by the drilling rig comes from three sources. The quietest period was the work of the platform but not of drilling with the anchor support vessel.

During this period, the primary sources of noise came from the mechanical plant, the discharged fluids, the pumping systems and the various transports of the equipment on deck. As the deck of the main machines of the platform was above the waterline, the general noise level was reduced, the near field corresponded approximately to the dimensions of the platform and the highest broadband level encountered was 117 dB re 1 μPa at 125 m source. Various tones produced by machinery are found in the spectra of this noise. During this operating state and in the conditions of great calm encountered, the noise of the platform was perceptible at a distance of 1–2 km.

The second source of noise involved the drilling of the platform and the anchor support vessels. The drill string produced dominant tones, especially between 31 and 62 Hz in the 1/3 octave band. The drilling rig was linear vertical, 3.8 km long consisting of a steel tube (drilling column) rotating in a steel housing (in water) or concrete (under the seabed). Thus, two sources were active, the drilling rig itself and the drilling rig.

A sudden field transition was observed at about 400 m. During drilling and at depths less than 400 m from the drilling rig, the noise of the drilling rig was dominant, so that the measurements were similar to those of the periods when the gasket did not drill. Beyond 400 m in the received noise levels the tones of the drill string have become significant. For the drilling rig, the highest noise levels encountered were of the order of 115–117 dB re 1 μPa at 405 and 125 m, respectively, audible at distances greater than 11 km.

The third source of the noise, which far exceeds the previous two, involved stationing the support vessel for loading purposes. The ships stood on the starboard side of the bow anchor platform, activating the propeller and engines as needed. During this period strong currents were present, which caused the permanent operation of the main shaft. The use of engines or the main propeller under load produced very high levels of noise of the turbo-pump. This noise was in broadband, with the highest level measured at 137 dB re 1 μPa at 405 m from the stern of the platform and the levels of 120 dB re 1 μPa were recorded at 3–4 km and audible up to 20 km distance from the platform. The noise level produced by the accompanying ships had values of 120 dB re 1 μPa at distances between 0.5 and 1 km for ship 1 and 250 m for ship 2. The noise of each ship was composed of the noise of the installations and that of the cavitation.

Vertically placed hydrophones during the study period showed that there were strong currents throughout the water column.

3 Method and General Framework for Performing Measurements

The marine exploration platform was anchored 21 km offshore 44°21′LatN, 28°72′LongE, 110 m water depth, relatively constant. The water depth was relatively uniform near the platform. Measurement systems were also implemented 8 km north of the platform (P8) and 9 km south (P9).

The study was conducted in the spring. The research vessel was used to perform the measurements.

Recordings were made in the vicinity of the platform, ranging from 51 s to 4 h. The recordings were made 4 m from the accompanying boat, with hydrophone networks placed vertically in depth.

The hydrophone networks were located at distances between 125 m and 22.5 km from the platform and then left adrift.

The locations in the general area, the recording locations and the CTD positions are shown in Fig. 1 in descending scales.

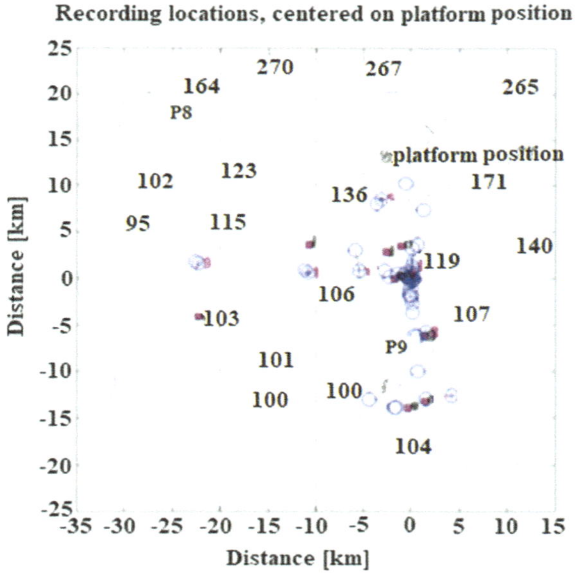

Fig. 1. Location of the study region of the underwater noise radiated from the drilling rig. The plot is centered on the drilling rig: Black Sea, 21 km offshore, 44°21′LatN, 28°72′LongE. Depth measurements are in m. The locations to be recorded are shown in open circles. For the drift set, the starting location is represented by the "+" symbol, the deviation indicated by a solid line, and each location numbered in bold. The location of system 1 is in the southern part of the platform, at P9 (9 km away from it). The throwing locations of the CTD modules are represented by solid squares, in bold letters that designate their sequence.

4 Equipment Used to Perform Measurements

All noise measurements were made with a very low wind and very calm conditions but in the presence of strong and persistent vertical sea currents. The physical noise of the sea is very much determined by the power of the wind.

The offshore platform under study is an exploratory drilling rig.

The two accompanying vessels may be considered representative of the vessels widely used for support. The movements and activities of the ships were well highlighted in the recordings made during the experiments.

Three sets of acoustic recording equipment were used for the measurement. Each was composed of a calibrated hydrophone, pre-amplifier and computer. Specific details of the equipment used are presented below.

Portable equipment—Hydrophone Brüel & Kjær 8104 deployed and monitored from 4 m from the boat. The hydrophone was equipped with a 30-m cable, spiral and with a solid casing to reduce the turbulence induced by the flow. A braking system has been implemented to help align the tilt of the boat in the direction of the current, also to reduce turbulence of the hydrophone and cable.

System 1 consisted of 3 hydrophones placed vertically in a network. The system was anchored at a depth of 70 m at the southern edge of P9 and was protected from noise by a sandbank interposed between the system and the platform and the drilling rigs.

The system operated on a recording time cycle of 3 min every 45 min from the first day at 11:30 until the third day at 19:30. Unfortunately, this system was connected with a cable of length 250 m and, from due to strong currents, there was a persistent noise of turbulence in the cable and hydrophones. Artificial pressure variations in hydrophones have led to recordings below the 150 Hz limit.

System 2 consisted of 3 hydrophones placed vertically in a network. The system operated in time cycles of 51 s every 10 min. The signal from the hydrophones was recorded on different channels. Two different recordings were made: at 420 m and at 405 m at the stern (west) of the platform, in water with a depth of 113 m. The network is anchored with a cable of 250 m length and with a diameter of 16 mm, with which are also caught buoy and floats. No flow noise was recorded in any of the implementations.

The 1A system consisted of 3 hydrophones placed vertically in a network. The system was drifted and recorded in a 60 s cycle every 15 min. Hydrophone 1 was anchored 5 m below the water surface, hydrophone 2 at 25 m and hydrophone 3 at 75 m. Depth was measured with a bathymeter.

5 Experimental Results

The latitude and longitude coordinates taken by the GPS were transferred to an x-y coordinate system.

The zero point was chosen:

- at the coordinates of the drilling rig, for the platform noise measurements,
- at the location of hydrophones for the movement of ships (which have been interpolated for drifting hydrophones).

5.1 Ambient Noise

Natural background noise levels at the location were recorded by the hydrophones located at a great distance from the platform and by the sets located in position P9.

Several ambient noise curves without any noise of the drilling rig or accompanying vessels are shown in Fig. 1. The upper set of curves A … E shows the time-mediated 1/3 octave curves (narrowband units), while the lower set is a narrowband spectral analysis (curves F and G taken with 16 averages, resolution 2.5 Hz, 2.5 kHz interval, analysis with Hanning window function), no ambient wind. Table 1 presents details of the locations of the plotted records in the curves A–G.

Table 1. Details of the locations of the plotted records in the curves A … G.

Nr. crt	Location	Hydrophone depth (m)	Equipment	Platform/support vessels movements
A	13.7 km S of platform	25	Portable	without drilling/anchoring
B	22,4 km W of platform	25	System 1a	in drilling/stationary
C	22,5 km W of platform	25	System 1a 1a	in drilling/stationary
D	13,9 km SSW of platform	70	System 1a 1	without drilling/anchoring
E	13,9 km SSW of platform	70	System 1a 1	in drilling/moving in the area
F	13,9 km SSW of platform	70	System 1a 1	in drilling/stationary
G	13,9 km SSW of platform	70	System 1a 1	in drilling/stationary

Curves B and C, elevated 22 km west of the platform, in open water reached the electronic system noise at frequencies >2 kHz. Curves D and E raised behind P9 show similar trends as in curves B and C between 100 Hz and 2 kHz but above 2 kHz show an increase in noise caused by marine life.

Curves F and G (Fig. 1) were taken with system 1 on the back of P9 and show the lowest levels of low-frequency noise (<50 Hz) encountered during the study. Most curves A–E (in Fig. 1) have a significant component of the flow noise at frequencies below 50 Hz, where curves A and E have the highest levels.

A composite curve H, which combines the portion of the low-frequency curves of the bottom hydrophone, F and G, (at frequencies < 200 Hz), all middle-frequency curves (200 Hz–2 kHz) and extended curves with the same slope over 2 kHz are shown in Fig. 2. This is the lowest level of ambient noise, which was present during the measurement period, correlated with the windless environmental conditions in the open water regions without the intervention of marine life.

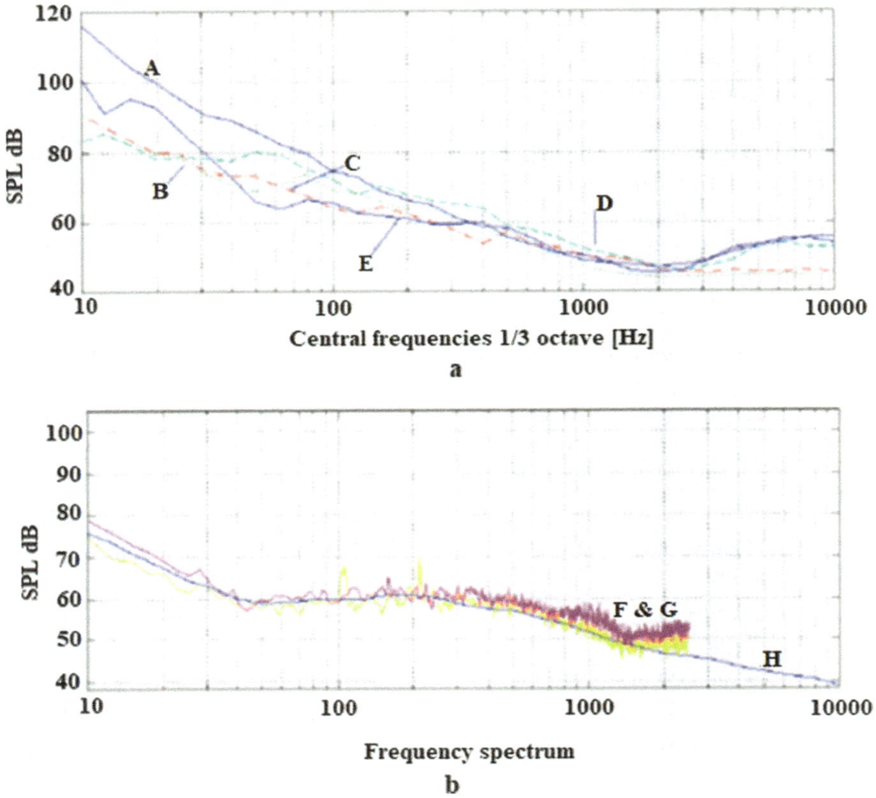

Fig. 2. Ambient marine noise curves at the platform location, without accompanying ship noise: **a** 1/3 octave marine ambient noise measurements (20 s averages): **a** drifting hydrophone at 25 m depth 13 km S of platform—displays a frequency level artificially high below ~60 Hz due to flow noise and the intervention of marine life at frequencies higher than 2 kHz; **b**, **c** set of drifting hydrophones at 25 m depth 22.3 km V distance from the platform—high levels of flow noise at frequencies lower than 50 Hz, reaches the electronic noise of the system at 1 kHz; **d**, **e** shows flow noise at frequencies lower than 50 Hz, recorded with the depth hydrophone, in water of 70 m, at S of P9; D has a slight noise produced by marine life at frequencies higher than 2 kHz; **b** frequency spectra: **f**, **g** narrowband spectra 0–2.5 kHz (2.5 Hz bandwidth, Hanning window function 40 averages) of ambient noise at the S edge of P9 with low flow noise (recorded with the depth hydrophone). A composite curve of marine ambient noise for the study period is also indicated: no wind input, no biological intervention and no ship noise, high ambient noise curve for the study period (thickened H curve).

For the predictions of the evolution of noise at long distances, the H curve presented in Fig. 2 was used. When using ship or platform noise recordings and calculating received noise levels, an appropriate ambient curve shall be used.

This is the ambient curve used in the following estimates of platform and accompanying vessel noise and which takes into account any low-frequency flow noise, as heard in all drift hydrophone recordings. For the analysis of support vessel noise, with ambient noise level 10 dB at a given frequency, the contribution of ambient noise to the received signal must be taken into account.

5.2 Noise from the Marine Drilling Rig

Marine platform noise. Variations in noise, depending on the time, produced by the exploration platform and support vessels are shown in Fig. 3 a. It shows the total noise level (broadband), over the entire frequency range of the bands 1/3 octaves with center frequencies between 10 Hz and 10 kHz. Two consecutive systems were implemented: the first at 450 m from the stern of the platform, the second at 405 m from the stern of the platform.

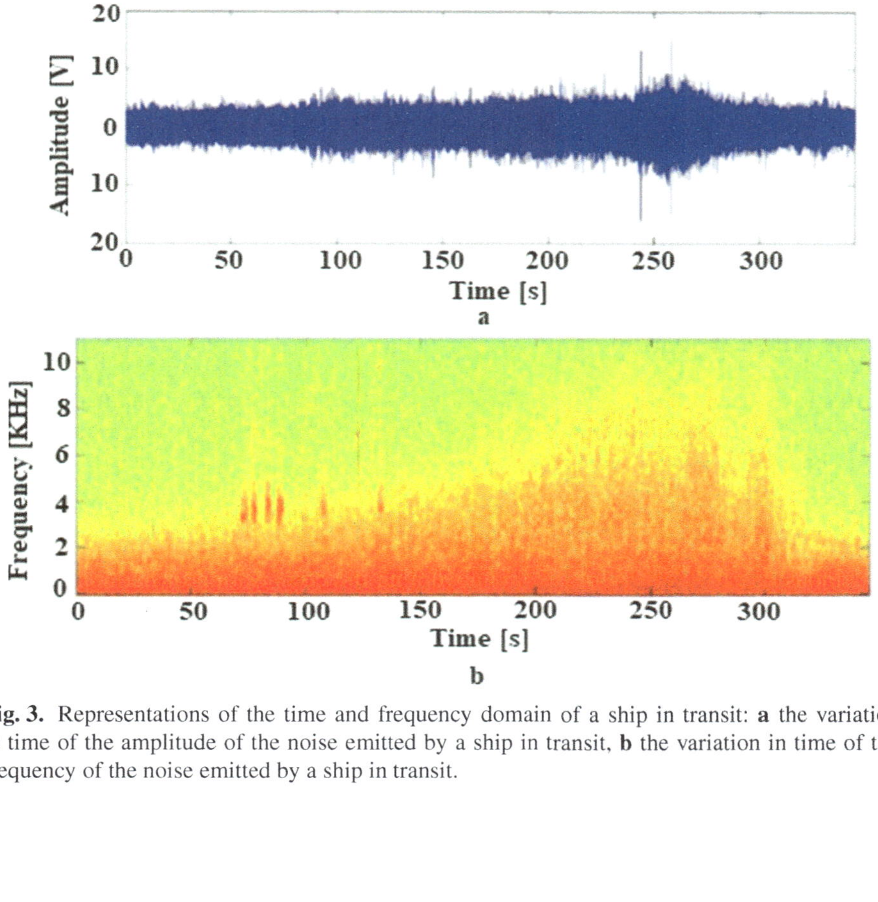

Fig. 3. Representations of the time and frequency domain of a ship in transit: **a** the variation in time of the amplitude of the noise emitted by a ship in transit, **b** the variation in time of the frequency of the noise emitted by a ship in transit.

In order to interpret the noise produced by the platform, it is necessary to detail the drilling activities on its board, as well as the movements of the support vessels.

The main noises recorded were:

- drilling noise, anchor support vessel
- drilling noise, accompanying ship at anchor.
- stationary supply ships.
- drilling noise.

Figure 4 shows the broadband measurements, performed at various distances from the platform, during the periods:

- without drilling activities and without accompanying ships;
- during drilling operations and without accompanying vessels;
- stationary accompanying ships.

In Fig. 4, the measurements performed with the depth hydrophone from the stern of the platform are represented with enlarged symbols and are determined as the average of the intensities of the noise measured at the same time with the drift measurements; these are expressed in dB. The measurements made with the hydrophone at a depth of 25 m from the drifting set 1 are analyzed and presented in a similar way.

The noise produced by the support vessel and received by the depth hydrophone, steadily increased as it approached the platform, to a distance of 405 m. The measured noise level then decreases sharply, for measurements made at distances less than 400 m. compared to the platform, at values similar to those of the period without drilling activity and with noise produced only by the platform (represented by +).

6 Solutions to Reduce the Noise Level. Virtual Instruments for Noise Injection

In order to reduce the level of noise introduced into the environment by a human activity, we proposed the realization of an experimental context of insertion of noise in anti-phase at the source of noise pollution.

6.1 Virtual Tool "Sine Signal Generator"

To demonstrate the validity of the law governing the composition of harmonic waves (with their mutual cancelation), two virtual instruments have been built that use two types of signal generators: a harmonic (sinusoidal) signal and a white noise.

The first virtual device made is a sinusoidal signal generator that generates a wave with a predetermined shape, sinusoidal and at the same time a counter-phase wave (out of phase with 180 degrees as the starting point of the wave). These waves combine in a summation VI and the signal intensity thus obtained is zero.

This virtual instrument was built to demonstrate the validity of the law governing the composition of harmonic waves (with their mutual cancelation).

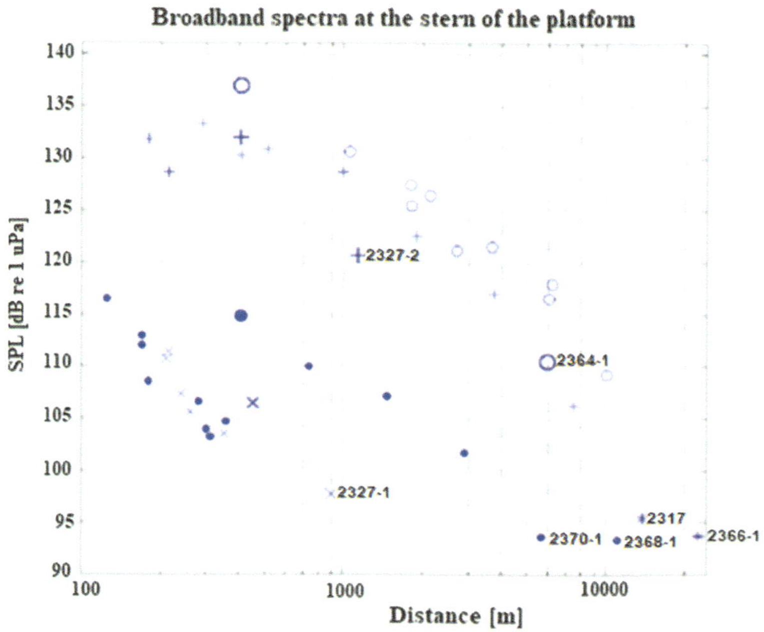

Fig. 4. Broadband ambient noise levels around the exploration platform (bands 1/3 octave central frequencies of the frequency range 40 Hz to 10 kHz): x small—periods without drilling activity and without accompanying vessels made from the boat on the first day 10:50–11:32; (X) mare-aldin—records made with system 2 at the stern of the platform on the first day at 10:50–11:32, X mare-records with the set 1a drifting on the first day at 15:30–15:48; ●—the periods with drilling activity and without accompanying vessels recordings made of the boat, on the second day 14:18–15:48; ○ small—stationary accompanying ships, recordings made of the boat on the second day 13:09–14:22 with set 2, at the stern of the platform; ○ large without labels—recordings of set 2 from the stern of the platform on the second day 13:09–14:22; ○ large with labels—recordings made with the drifting system on the second day 07:45–08:30; + small—accompanying ships in stationary, recordings made from the boat on the second day, 14:30–15:36, + large unchecked—recordings made at the stern of the platform on the second day at 14:30–15:36; + large label—records of the set drifting from the stern of the platform; *—ambient noise, recordings made from the boat on the first day at 08:30 and records of the drifting system 1a on the third day at 10:42–10:48.

6.2 Real-Time Data Acquisition with Automatic Configuration

The front panel of the virtual instrument in Fig. 5 contains only the data presentation elements corresponding to the noise signals taken from the hydrophone and the one resulting from composing with another of the same amplitude but in anti-phase.

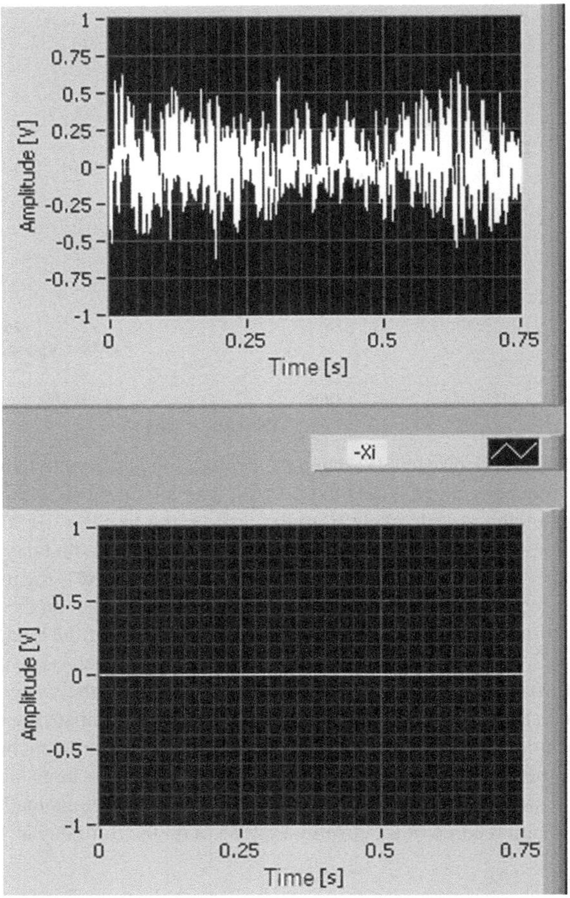

Fig. 5. Front panel of the VI "Real time acquisition".

The block diagram of this virtual tool contains:

- an "Acquire Sound Express VI" sub-VI that acquires a signal from a hydrophone; this sub-VI automatically configures the input parameters, acquires the data and resets the acquisition process;
- two "Formula" sub-VIs that process the signal (phase shift and summation);
- a sub-VI "Play Waveform Express VI" through which the virtual instrument gives signal to the power amplifier;
- two "Waveform Graphs" data presentation elements.

6.3 Virtual Tool for Uploading the File

The noise recordings presented above were made with the Bruel & Kjaer instrumentation. These recordings from hydrophones located at different depths can be set up in databases.

A recording taken from a hydrophone can be loaded into a virtual instrument and processed by it (phase reversal).

The virtual instrument provides a phase-out signal that is amplified and played back through the speakers in the same environment and at the same location as the noise source.

7 Conclusions

Noise pollution of the marine environment has increased by 20 decibels in the last 50 years, with devastating consequences for some of the inhabitants of the oceans.

For marine life, hearing is just as important as the sense of sight to the people. If there is noise pollution in the ocean, they can no longer communicate.

"Acoustic fog" affects the communication of marine mammals, which can usually send sound signals hundreds of kilometers away in order to orient themselves in their travels for reproduction and food.

Through the performed implementations were collected and processed data related to wind speed, sea situation, wave height, instrument noise, biological noise, maritime traffic, wind and wave noise, tidal noise and transmission loss.

The experimental context is original and allows the measurement of hydroacoustic parameters at different depths, on several levels:

- the highest noise level, corresponding to the criterion 120 dB re 1μPa (proposed and accepted by the Marine Directive), extended to a distance of 3–5 km from the platform;
- if the accompanying vessels were at anchor, the noise produced by the platform never exceeded the level of 120 dB re 1μPa.

The experimental results obtained consisted of a database that can be used to predict the evolution of noise over long distances.

Research on this paper has shown that the accidental or constant introduction of noise into the marine environment causes habitat and disrupts underwater life.

In order to reduce the effects of noise pollution at the source, studies were undertaken that aimed at attenuating the noise power emitted by the source by introducing counter-phase noise at the source location.

Several virtual devices have been designed to present and argue the idea of introducing noise in contrast to that produced by the source.

A "Sinusoidal Signal Generator" was developed to simply test the theory of counter-phase waveforms; two counter-phase wavefronts were emitted that combine into a sub-VI of summation; the resulting wave has zero amplitude and is displayed in the corresponding front panel window.

Therefore, the activity of the platform introduces noise in its vicinity over large areas and therefore measures must be taken to mitigate it.

The framework of this article offers openings of some study directions in order to improve the conditions of the marine habitat disturbed by the anthropic activities of construction, marine exploitation and economic.

Some study directions that can be formulated:

- defining a standard measurement procedure for underwater radiated noise to allow the creation of databases and comparable criteria;
- the application of noise reduction measures for 10% of the global fleet would lead to the achievement of the proposed objectives at international level;
- focusing on areas populated with marine life: redirecting traffic and reducing the noise of ships operating in these areas;
- identification of new measures to reduce noise by competent bodies but also by researchers in the field.

Research priorities for monitoring and evaluating underwater noise are:

- initiation of long-term monitoring programs for underwater noise;
- collecting, organizing and analyzing data on the historical evolution of underwater noise;
- development of global models for underwater noise;
- reporting the characteristics of the signals of anthropogenic noise sources;
- determining the relationship between the level of anthropogenic activity and the level of noise.

Noise levels are constantly rising, so marine noise must be managed both nationally and internationally, as a precaution, before irreversible damage to biodiversity and the marine ecosystem.

References

1. Arsene, M.: Theoretical and experimental studies on the impact of acoustic waves on the marine environment. Doctoral Thesis, UPB (2013)
2. Popovici, V., Arsene, D., Borda, C., Marinescu, M., Butu, L., Arsene, M.: Virtual laboratory for underwater noise attenuation. In: TEHNOMUS XVII, 17th International Conference on New Manufacturing Technology, Suceava, Romania, May, 17–18, 2013, P-ISSN-1224-029, E-ISSN-224-6016, pp. 375–378, indexat BDI: Copernicus, Ulrich WEB, EBSCO, SCIPIO
3. Arsene, M.: Study for reducing noise pollution of the marine environment by introducing noise at the acoustic source. U.P.B. Sci. Bull. Ser. C (4), 33–48 (2014). indexat: Ulrichs International Periodicals Directory, Inspec, Scopus
4. Laborator mobil de măsura a nivelului de zgomot la bordul navelor conform STANAG4293. Ministerul Apărării; 12/05/1008, Inovator: Arsene Mihai
5. Arsene, M., Popovici, V., Arsene, D., Borda, C., Marinescu, M., Butu, L.: Transmission loss measurement in the black sea. In: The Annual Symposium of the Institute of Solid Mechanics, SISOM, and Sess. of the Com. of Acoustics, Buc, May, 2012, ISSN 1843-5459

A Study on the Exposure to Vibration of the Whole Body and the Hand-Arm Segment of Motorcyclists

Mihaela Picu[1(✉)] and Laurentiu Picu[2]

[1] European Centre of Excellence for the Environment, "Dunărea de Jos" University of Galati, Galati, Romania
Mihaela.Picu@ugal.ro
[2] Dunarea de Jos, Galati University, Galati, Romania
Constantin.Picu@ugal.ro

Abstract. Nowadays, when traffic has become extremely crowded, the motorcycle is a much easier means of transport. This paper is an analysis of the vibrations transmitted to the whole body and to the arm-hand segment of motorcyclists and it presents the measurements of the vibration parameters transmitted to humans as well as the assessment of these parameters taking into account ISO 2631 (1997) and Directive 2002/44/EC. In order to achieve this goal, studies were performed on a number of 6 motorcyclists, men, aged between 21 and 28 years, with Body Mass Index between 19 and 24, without health problems. The subjects ran on asphalt, with speeds between 40 and 60 km/h, using the same type of motorcycle, same brand but of different manufacturing years. Accelerations were measured at foot level, coccyx, cervix, and hand with the Maestro 01 dB vibrometer and processed with "Whole-body vibration exposure computer-HSE" and "Hand-arm vibration exposure computer-HSE". For each of them were calculated: weighted r.m.s. acceleration, Vibration Dose (VDV), Time to reach EAV (T_{EAV}), Time to reach ELV (T_{ELV}), Partial Exposure, and Daily Exposure. The results were compared with the answers given on the Likert scale.

Keywords: Motorcycles · Whole-body vibration · Hand-arm vibration · Likert scale

1 Introduction

There are several occupations (policemen, paramedics, and couriers) that travel on motorcycles, to get to their destination faster, as traffic has become increasingly crowded. These people sometimes travel more than 8 h a day and the vibrations transmitted by motorcycles to the human body can become extremely dangerous to health.

Many researchers have analyzed the vibrations transmitted to the whole body (WBV) by a series of equipment (cars, trucks, tractors, ships, etc.) [1–8]. Studies of vibrations transmitted to the entire body by motorcycles are fewer and have been conducted especially in Asia, where this means of transport is well-known [9–14].

N. Herisanu and V. Marinca (eds.), *Acoustics and Vibration of Mechanical Structures – AVMS-2021*, Springer Proceedings in Physics 274,
https://doi.org/10.1007/978-3-030-96787-1_18

"The results show that newer motorcycles allow 36.3% more travel time than older motorcycles before reaching the exposure time value (EAV). Larger engine motorcycles (125 cc) allow 22.5% more travel time than smaller engine motorcycles. It is possible to travel 44% longer using fast roads than slow roads" [15].

Internationally, there is ISO 2631-1 [16] which sets out the procedure for assessing human exposure to WBV. The main purpose of IS0 2631 is to define methods for quantifying whole-body vibrations in relation to human health and comfort, the probability of vibration perception, the incidence of motion sickness.

The approximate indications of the probable reactions at different values of the total vibration are presented in Table 1 (ISO 2631-1, chapter C.2.3).

Table 1. Approximate indications of probable reactions at different values of total vibration [16]

Total vibrations (m/s^2)	Answer
<0.315	Not uncomfortable
0.315–0.63	A little uncomfortable
0.5–1.0	Quite uncomfortable
0.8–1.6	Uncomfortable
1.25–2.5	Very uncomfortable
>2.0	Extremely uncomfortable

Another extremely important document is the Directive 2002/44/EC [17] which sets out the standard values for 8 h of work in a vibrating environment:

(1) for whole-body vibration: (a) the daily exposure limit value standardized to an eight-hour reference period shall be 1.15 m/s^2 or, at the choice of the Member State concerned, a vibration dose value of 21 $m/s^{1.75}$; (b) the daily exposure action value standardized to an eight-hour reference period shall be 0.5 m/s^2 or, at the choice of the Member State concerned, a vibration dose value of 9.1 $m/s^{1.75}$.

(2) for hand-arm vibration: (a) the daily exposure limit value standardized to an eight-hour reference period shall be 5 m/s^2; (b) the daily exposure action value standardized to an eight-hour reference period shall be 2.5 m/s^2.

2 Materials and Methods

The above-mentioned standard (ISO 2631-1) states that the acceleration signal should be measured with three-axis accelerometers in the frequency range from 0.5 to 80 Hz. The signal is then subjected to weighting filters to consider the greater or lesser influence of different frequencies in the human body. The weighted global acceleration is then calculated with Eq. (1):

$$aw = \sqrt{\sum_i (W_i a_i)^2} \quad (1)$$

where a_w is the frequency-weighted acceleration, W_i is the weighting factor for the third-octave band, and a_i is the r.m.s. for the third-octave band.

Because transverse vibrations (in the case of a fixed body coordinate system for seated persons) generate more discomfort than vibrations along the z-axis fixed to the body (the axis running from the hip to the head), the transverse components of the RMS are increased by 40%, while the Oz component is left unamplified:

$$aw = \sqrt{(1.4a_x)^2 + (1.4a_y)^2 + a_z^2} \quad (2)$$

In order to study the effect of motorcycle vibrations on people, experiments were made in the real environment, running on asphalt, with speeds of 40–60 km/h, on 12 motorcycles, with an approximately equal number of kilometers on board, the same brand: 6 newer motorcycles (manufactured from 2016 to 2018) and 6 older motorcycles (manufactured from 2012 to 2014). The participants were 6 male subjects, aged between 21 and 28 years, with Body Mass Index between 19 and 24, without health problems. The duration of each determination was 15 min, on the same route. The experiment was repeated 5 times on 5 different days. The weather was good, with average temperatures (18–23 °C), no wind. The values presented in the paper are obtained through averaging. Note: The asphalt was not perfect, with "patches" and small bumps.

Accelerations were measured with the Maestro 01db vibrometer (see Figs. 1, 2, 3 and 4). At the level of the foot, the seat and the cervical were calculated for WBV: weighted r.m.s. acceleration (a_w), vibration dose value (VDV) (Eq. 3), time to reach action value (T_{EAV}), and time to reach the exposure limit value (T_{ELV}) (Eq. 4). At the level of the fingers, for HAV, the daily exposure was also calculated.

Fig. 1. Accelerometer mounted on the ankle

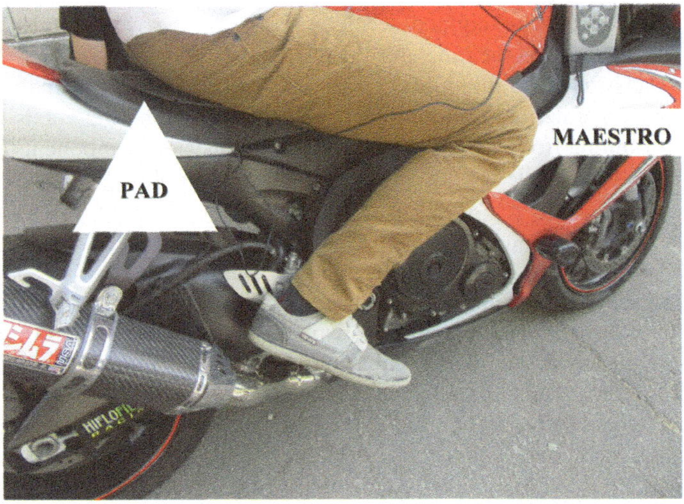

Fig. 2. Accelerometer mounted under the coccyx

Fig. 3. Accelerometer mounted on the neck

Fig. 4. Accelerometer mounted on the finger

$$\mathrm{VDV} = \left[\frac{1}{T}\int_0^T [a_w(t)]^4 dt\right]^{1/4} \left(\mathrm{m/s^{1.75}}\right) \tag{3}$$

$$T_{EAV_{A(8)}} = 8\left(\frac{0.5}{A}\right)^2 \text{ and } T_{ELV_{A(8)}} = 8\left(\frac{1.15}{A}\right)^2 (h) \tag{4}$$

Note: All subjects were informed about the experiments and gave their written consent.

3 Results and Discussions

3.1 The Impact of Vibrations Produced by Running Motorcycles on the Whole Body

The points through which the vibrations enter the body of motorcyclists are through the footpeg, through the seat (transmission through the spine up to the neck), and through the handlebars (Fig. 5).

Figure 5 shows that the total vibration magnitude decreases along the human body, according to:

- for new motorcycle (□):

$$\mathrm{a_w} = 7.546 \cdot \mathrm{x}^{-0.7945}\left(\mathrm{R}^2 = 0.9938\right) \tag{5}$$

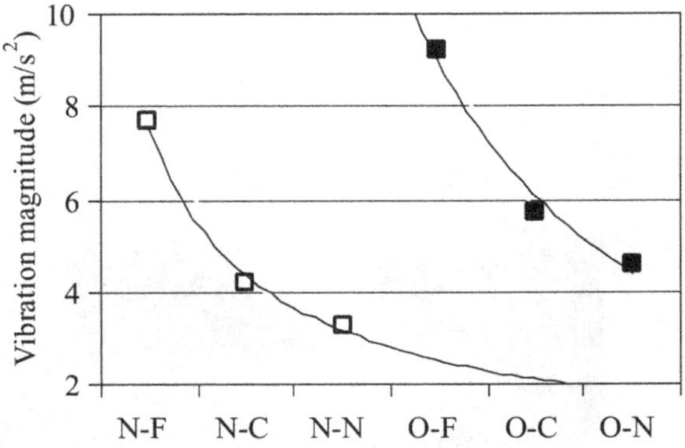

Fig. 5. Average value of vibrations at foot (F), coccyx (C), and cervical (N) level for new (N) and old (O) motorcycles

- for old motorcycle (■):

$$a_w = 101.03 \cdot x^{-1.7482} \left(R^2 = 0.9789\right) \tag{6}$$

where a_w is the weighted acceleration and x represents the position where the accelerometer is placed. This is vibration attenuation through the human body using Vibration magnitude.

Attenuation is a "power" function as:

$$a_w = A \cdot x^{-B} \tag{7}$$

with A and B two constants that depend on the characteristics of the experiment.

Next, the specific sizes were calculated with the "Whole-body vibration HSE calculator" (https://www.hse.gov.uk/vibration/wbv/calculator.htm) (Tables 2 and 3).

Total VDV decreases along the human body, according to:

- for new motorcycle (□):

$$VDV_{tot} = 124.23 \cdot x^{-1.1299} \left(R^2 = 0.9714\right) \tag{8}$$

- for old motorcycle (■):

$$VDV_{tot} = 109.5 \cdot x^{-0.6445} \left(R^2 = 0.997\right) \tag{9}$$

where x represents the position where the accelerometer is placed.

If the "Total VDV" is represented graphically, is found the same attenuation of vibrations from the legs to the cervical, given by a "power" function of the type Eq. (7) (Fig. 6).

Table 2. WBV parameters (case—new motorcycle)

Vibration magnitude	Partial VDV	Partial exposure	T_{EAV} ($A(8)$ option) 0.5 m/s^2	T_{ELV} ($A(8)$ option) 1.15 m/s^2	
m/s^2 r.m.s	m/s^2 $A(8)$	m/s$^{1.75}$	Minutes	Minutes	
7.66664	83.1	2.71	2	11	FOOT
7.65654	83.0	2.71	2	11	Total VDV 130.0 m/s$^{1.75}$
7.65124	83.0	2.71	2	11	
7.64982	83.0	2.70	2	11	
7.64951	83.0	2.70	2	11	Total exposure
7.67411	83.2	2.71	2	11	6.63 m/s^2 $A(8)$
4.17948	32.0	0.74	7	36	COCCYX
4.17895	32.0	0.74	7	36	Total VDV 50.2 m/s$^{1.75}$
4.20145	32.2	0.74	7	36	
4.20321	32.2	0.74	7	36	
4.15024	31.8	0.73	7	37	Total exposure
4.16874	32.0	0.74	7	37	1.81 m/s^2 $A(8)$
3.23519	24.8	0.57	11	61	NECK
3.25458	25.0	0.58	11	60	Total VDV 38.8 m/s$^{1.75}$
3.22667	24.7	0.57	12	61	
3.21478	24.7	0.57	12	61	
3.23755	24.8	0.57	11	61	Total exposure
3.22841	24.8	0.57	12	61	1.40 m/s^2 $A(8)$

Table 3. WBV parameters (case—old motorcycle)

Vibration magnitude	Partial VDV	Partial exposure	T_{EAV} (A(8) option) 0.5 m/s^2	T_{ELV} (A(8) option) 1.15 m/s^2	
m/s^2 r.m.s	m/s^2 A(8)	m/s$^{1.75}$	minutes	minutes	
9.19619	70.5	1.63	1	8	**FOOT**
9.18746	70.5	1.62	1	8	**Total VDV 110.4 m/s$^{1.75}$**
9.18452	70.4	1.62	1	8	
9.19547	70.5	1.63	1	8	
9.21498	70.7	1.63	1	7	**Total exposure**
9.20411	70.6	1.63	1	7	**3.98 m/s^2 A(8)**
5.71516	43.8	1.01	4	19	**COCCYX**
5.70489	43.7	1.01	4	20	**Total VDV 68.5 m/s$^{1.75}$**
5.71456	43.8	1.01	4	19	
5.68924	43.6	1.01	4	20	
5.69947	43.7	1.01	4	20	**Total exposure**
5.72045	43.9	1.01	4	19	**2.47 m/s^2 A(8)**
4.55373	34.9	0.80	6	31	**NECK**
4.55864	35.0	0.81	6	31	**Total VDV 54.7 m/s$^{1.75}$**
4.55422	34.9	0.81	6	31	
4.55632	34.9	0.81	6	31	
4.55218	34.9	0.80	6	31	**Total exposure**
4.55149	34.9	0.80	6	31	**1.97 m/s^2 A(8)**

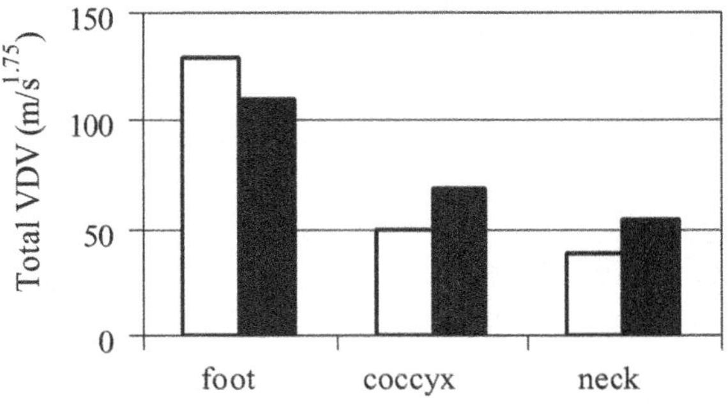

Fig. 6. Vibration attenuation through the human body using total VDV

In conclusion, Fig. 6 shows that absolutely all VDVs is much higher than the standardized daily exposure limit value for a reference period of 8 h (21 $m/s^{1.75}$), by the following percentages: from 84% (case new motorcycle, neck) to 519% (case new motorcycle, foot).

3.2 The Impact of Vibrations Generated by Running Motorcycles on the Hand-Arm Segment

In order to achieve this objective were measured the accelerations transmitted to the fingers of the motorcyclists while riding on the same motorcycles, on the same route, also 15 min; for this, the accelerometers were fixed to the subjects' fingers. The values of the r.m.s. accelerations, for each motorcycle, are shown in Fig. 7.

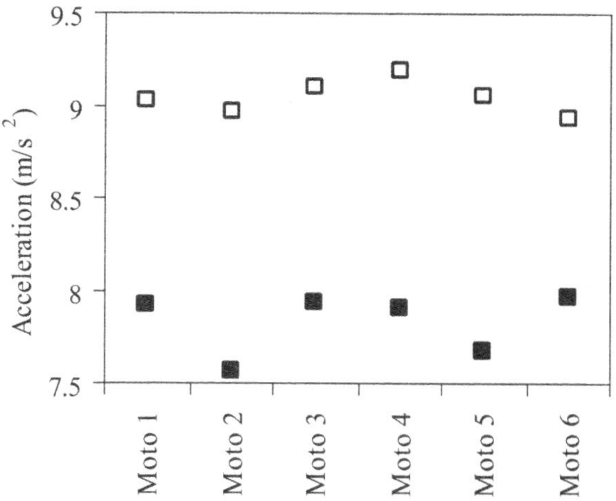

Fig. 7. Accelerations transmitted to the fingers of the 6 subjects for new motorcycles (□) and old motorcycles (■)

The calculations were made with "Hand-arm vibration exposure calculator—HSE" (https://www.hse.gov.uk/vibration/hav/vibrationcalc.htm). Given that the daily exposure limit value standardized to an eight-hour reference period should be 5 m/s^2, these values are extremely high, namely: for old motorcycles, the average acceleration exceeds by more than 80% the limit value, and for new motorcycles, with over 56%.

Time to reach EAV (T_{EAV}) and time to reach ELV (T_{ELV}) are 2 sizes that very well characterize the action of vibrations on people's hands. From Fig. 8 it can be seen that both time intervals are shorter for old motorcycles (which was to be expected) and that these intervals are very short, of the order of tens of minutes (for a 15-min run).

The most important quantities that characterize HAV are Partial Exposure and Daily Exposure (m/s^2 A(8)) which show whether the limits of 2.5 $^m/s^2$ and 5 $^m/s^2$, respectively, are exceeded (Fig. 9).

Figure 9 shows that these values fall within the limits set by Directive 2002/44/EC. But, as the experiments lasted only 15 min, these data are not useful; for this reason,

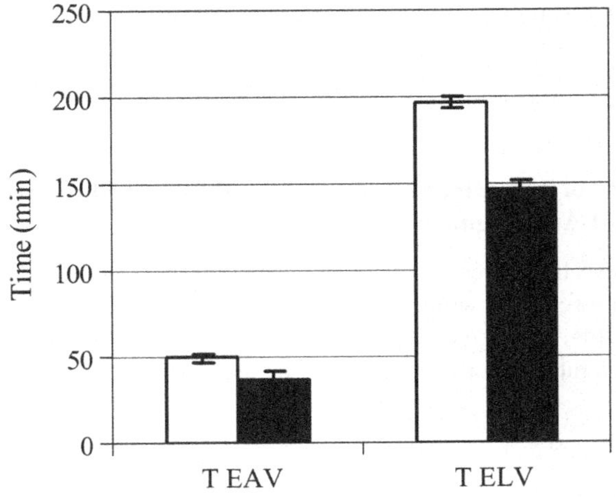

Fig. 8. T_{EAV}/T_{ELV} for new motorcycle (□) and old motorcycle (■)

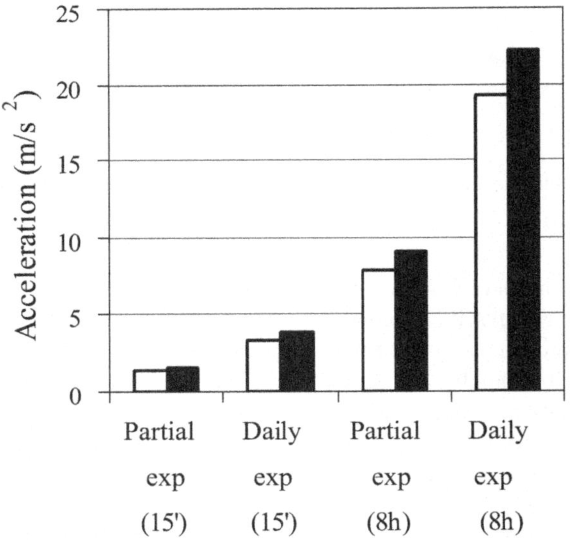

Fig. 9. Partial exposure and Daily exposure for exposure duration 15 min and for 8 h for new motorcycle (□) and old motorcycle (■)

I also compared the values corresponding to an 8-h workday. In this case, relating to the 5 m/s^2 daily exposure limit value, it is found that for new motorcycles, the daily exposure is higher by 283% and for old motorcycles by 343%.

4 Conclusions

The comfort of motorcyclists when traveling is an extremely controversial topic. Although the measurements showed values of transmitted vibrations that far exceed those indicated by Directive 2002/44/EC or ISO 2631/1997 (see Table 1), subjects do not feel this discomfort, as the pleasure of riding a motorcycle is greater than the physical pain they should feel during the transmission of vibrations to their body. To verify this, subjects were asked to quantify driving comfort according to the Likert scale which is a psychometric scale commonly involved in research that employs questionnaires (Table 4).

Table 4. The format of a typical 5-level Likert item

	Grade
Strongly disagree	1
Disagree	2
Neither agree nor disagree	3
Agree	4
Strongly agree	5

https://www.en.wikipedia.org/wiki/Likert_scale

Table 5 shows that only subject S5 wrote "Agree" for both types of motorcycles. Apart from this topic, there were 2 other subjects (S2 and S6) who wrote “Agree” only for old motorcycle. The rest of the subjects chose “Strongly agree” for all motorcycles.

Table 5. Subjects’ answers

	New motorcycle	Old motorcycle
S1	5	5
S2	5	4
S3	5	5
S4	5	5
S5	4	4
S6	5	4

As we stated, the subjects graded a maximum of 5, even if the vibrations transmitted by the motorcycles were extremely high. In this context, we are talking about people who do not earn a living using the motorcycle 8 h a day or more, but about motorcyclists who do this sport for pure pleasure.

References

1. Hazell, T.J., Jakobi, J.M., Kenno, K.A.: The effects of whole-body vibration on upper- and lower-body EMG during static and dynamic contractions. Appl. Physiol. Nutr. Metab. **32**(6), 1156–1163 (2007)
2. Kiiski, J., Heinonen, A., Järvinen, T.L., Kannus, P., Sievänen, H.: Transmission of vertical whole-body vibration to the human body. J. Bone Miner. Res. **23**(8), 1318–1325 (2008)
3. Liang, C.C., Chiang, C.F.: A study on biodynamic models of seated human subjects exposed to vertical vibration. Int. J. Ind. Ergon. **36**(10), 869–890 (2006)
4. Picu, L.: A study of the vibrations transmitted to the staff by the structures of the river ships on the romanian Danube segment. Ann. "Dunarea de Jos" Univ. Galati Fascicle XIV Mech. Eng. **24**(1), 11–16 (2017)
5. Picu, M.: Vibrations study of a naval ship in different operating modes on the Danube. Int. J. Mod. Manuf. Technol. **VII**(1), 53–60 (2015)
6. Picu, M., Picu, L.: Particular aspects regarding the effects of whole-body vibration exposure, MATEC Web of Conferences. In: Manoach, E., Stoykov, S., Wiercigroch, M. (Eds.) International Conference on Engineering Vibration (ICOEV 2017), Sofia, Bulgaria, vol. 148, Art. No. 09005. EDP Sciences (2017)
7. Picu, M.: An analysis of vibration transmission to handlers trucks. Ann. "Dunarea de Jos" Univ. Galati Fascicle XIV Mech. Eng. **20**(1), 11–16 (2013)
8. Ritzmann, R., Gollhofer, A., Kramer, A.: The influence of vibration type, frequency, body position and additional load on the neuromuscular activity during whole body vibration. Eur. J. Appl. Physiol. **113**(1), 1–11 (2013)
9. Chena, H.C., Chena, W.C.: Whole-body vibration exposure experienced by motorcycle riders – An evaluation according to ISO 2631–1 and ISO 2631–5 standards. Int. J. Ind. Ergon. **39**(5), 708–718 (2009)
10. Figlus, T., Szafraniec, P., Skrúcaný, T.: Methods of measuring and processing signals during tests of the exposure of a motorcycle driver to vibration and noise. Int. J. Environ. Res. Public Health **16**(17), 3145 (2019)
11. Chen, H.-C., Chen, W.-C., Liu, Y.-P., Chen, C.-Y., Pan, Y.-T.: Whole-body vibration exposure experienced by motorcycle riders—an evaluation according to ISO 2631–1 and ISO 2631–5 standards. Int. J. Ind. Ergon. **39**(5), 708–718 (2009)
12. Mohamad, D., Md Deros, B., Daruis, D.D.I., Khamis, N.K., Tahir, N.H.M.: Assessment of hand-arm vibration exposure among motorcyclist in Malaysia. Appl. Mech. Mater. **663**, 395–399 (2014)
13. Roseiroab, L.M., Neto, M.A., Amaro, A.M., Alcobia, C.J., Paulino, M.F.: Hand-arm and whole-body vibrations induced in cross motorcycle and bicycle drivers. Int. J. Ind. Ergon. **56**, 150–160 (2016)
14. Shivakumara, B.S., Sridhar, V.: Study of vibration and its effect on health of the motorcycle rider. Online J. Health Allied Sci. **9**(2), 1–9 (2010)
15. Moreno, R., Cardona, J., Pintado, P., Chicharro, J.: Estimadores de la vibraci6n de cuerpo entero para motociclistas (Predictors of whole-body vibration exposure in motorcycle riders). Rev. Fac. Ing. Univ. Antioquia **61**, 93–103 (2011)
16. ISO 2631-1: Mechanical vibration and shock—evaluation of human exposure to wholebody vibration—Part 1: General requirements (1997)
17. European Parliament and of the Council, Directive 2002/44/EC on the minimum health and safety requirements regarding the exposure of workers to the risks arising from physical agents (vibration) (2002)

Biomechanics and Bioacoustics

Three Segments Model for Computing Angular Momentum of Lower Limb in Gait

Andreea Stoia(✉), Dan Ioan Stoia, Cosmina Vigaru, and Nicolae Herisanu

Department of Mechanics and Strength of Materials, Politehnica University Timisoara, Timisoara, Romania
{dan.stoia,cosmina.vigaru}@upt.ro

Abstract. The study presents an analytical model for computing the angular momentum of the lower limb segments, during gait phases. The model describes the problem in the lateral plane of the human body, and it relays on Newton-Euler formulation for rotational motion. The input parameters are kinematical measurements of the human gait, recorded on a treadmill. The anthropometric, mass and inertia data were computed using the ratios founded in the literature. The total angular momentum was computed by summing the two individual components of it: spin angular momentum and orbital angular momentum. This kinetic parameter gives us a measure of net torque developed by the muscles in order to generate the movement.

Keywords: Spin angular momentum · Orbital angular momentum · Gait · Analytical model · Kinematics

1 Introduction

Nowadays, the recording and processing of the human movement are regular practice for universities, clinics and research centers. The development of motion analysis systems of all classes (video, ultrasounds, accelerometric, infrared and gyroscopic) makes kinematical parameters of the human body very accessible [1–4]. On the other hand, the kinetic parameters are a bit different to determine due to the necessity of taking into consideration the masses and moments of inertia of individual body segments. In addition, simplified or more complex analytical models are required for applying the basic principles of theoretical mechanics, which leads to kinetic data [5–9].

The angular momentum is a kinetic parameter that can be determined for any object in motion, respecting to a point. It conserves if there are no external torques are acting on the object [10]. Since it's computing depends on a reference point, when applying the basic mechanical principles to human body models four special points can be identified: the total body center of mass (CoM), the center of mass of every individual segment (Com), the joint centers (JC) and the ground contact point (GcP). Taking into consideration the position differences between these points, the kinetic parameters will be consequently different [11, 12].

N. Herisanu and V. Marinca (eds.), *Acoustics and Vibration of Mechanical Structures – AVMS-2021*, Springer Proceedings in Physics 274,
https://doi.org/10.1007/978-3-030-96787-1_19

The angular momentum has two components: one that came from linear velocity of the segment and is computed respecting to an external point (orbital angular momentum) and the second one that is generated by the own rotation of the segment around an axis that passes to the Com (spin angular momentum) [10, 13, 14].

This study presents a three-segment model of the lower limb for computing the angular momentum during a gait cycle. The results can be used to understand the muscle torques and reaction torques that are generating and canceling the angular momentum, and how the forward progression is achieved.

2 Material and Methods

The method of kinematic assessment is based on an ultrasound transmission-reception system (Zebris CMS-HS, Zebris GmbH, Germany). For acquiring multiple gait cycles of the same subject, an electric treadmill was used. The speed was selected to ensure a comfortable walk of the subject. Prior to recording gait kinematics, the subject trained for 10 min at the same velocity as the measurements were conducted. A session of five minutes of treadmill walk was recorded. The recording includes spatial coordinates of anatomical landmarks: hip, knee and foot for both limbs. The data file was exported in asci format for further processing. At this point, we have the joint angles of the lower limbs according to all anatomical planes, and the spatial coordinates of the landmarks.

The second part of the study implies the construction of the three segments analytical model that simulates the thigh, shank and foot (Fig. 1). This was created for computing the angular momentum in the sagittal plane of the body, where this parameter is the most significant in value and biomechanical effect. The spin angular momentum (K_S) was computed taking into consideration the individual axes of rotations that pass through the Com of each segment and their moments of inertia (Eq. 1). The value of inertia was computed based on Dempster radius of gyration, as percentage of the segment length [15] while the angular velocity was determined by numerical differentiation of the joint angles relative to time. The orbital angular momentum (K_o) was computed relative to the hip joint point (joint center), using the linear momentum of each body segment and the position vector that runs from the reference point to each Com (Eq. 2). The total amount of angular momentum can be computed using the Eq. (3) [10, 11, 14].

$$\vec{k}_{s}^{seg} = J_{Com}^{seg} \cdot \vec{\omega}_{Com}^{seg} \quad (1)$$

$$\vec{k}_{o}^{JC/seg} = \vec{r}_{JC/seg} \times m_{seg} \cdot \vec{v}_{Com} \quad (2)$$

$$\vec{k}_{limb} = \vec{k}_{s}^{seg} + \vec{k}_{o}^{JC/seg} \quad (3)$$

The notations used in Eqs. (1)–(3) signify: $\vec{k}_{s}^{seg}$—spin angular momentum of a segment; $\vec{k}_{o}^{JC/seg}$—orbital angular momentum of a segment in respect to JC; k_{limb}—total angular momentum of the limb; J_{Com}^{seg}—mass moment of inertia of a segment about

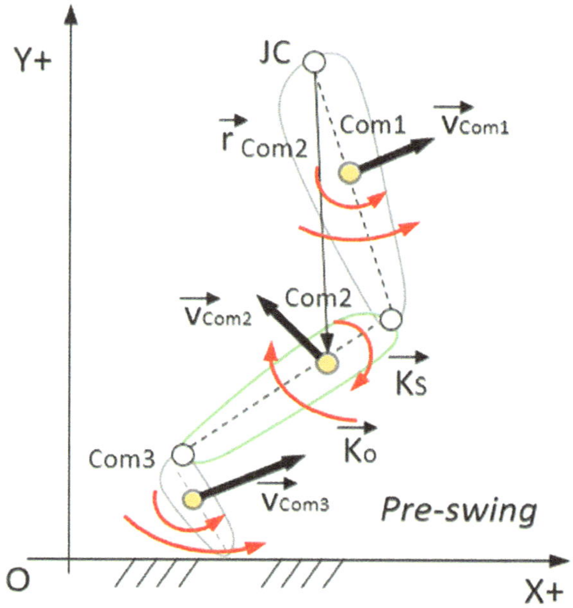

Fig. 1. Three-segment analytical model of lower limb

the transversal axis that passes through the Com; $\vec{\omega}_{Com}^{seg}$—angular velocity of a segment; m_{seg}—mass of the segment; $\vec{v}_{Com}$—linear velocity of the Com of a segment.

The mass of the segments was determined as percentages of total body mass while for inertia, the gyration radii were used. The gyration radii are computed as percentages of the total length of a segment and applied from the proximal end [15]. The position vector and the Com of every segment are determinable from the landmarks law of movements.

3 Results and Discussions

The K_S and K_o variations in one gait cycle are presented in Figs. 2 and 3. The body segments are represented as different lines in order to observe how the kinetic parameters are changing from one segment to another. The X axis represents one gait cycle, computed by linear normalization of the time data.

The highest spin angular momenta are recorded for thigh and shank for two reasons: one is the high value of the moment of inertia for these two segments and second reason is the high angular velocity associated with high angular motion. The spin momentum of the foot is very low and can be considered as participating in an insignificant quantity.

Results look different for orbital momentum. The larger distance from the point considered as reference (JC) to the Com of foot and shank comparing to thigh, makes these two segments to be significant. They possess high linear velocity during walking so a large linear momentum, which corroborated with large distance from the reference point led to high angular momentum.

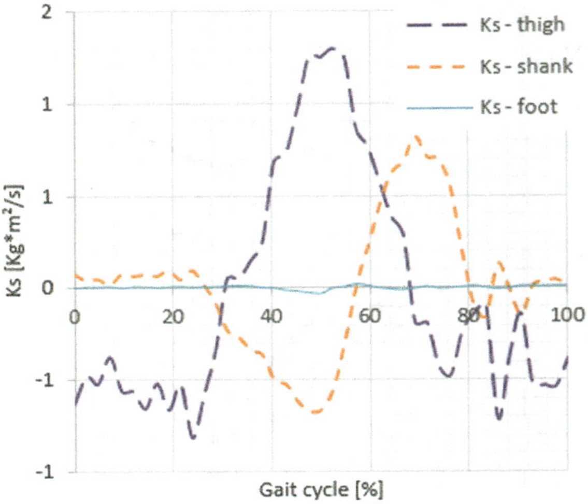

Fig. 2. Spin angular momentum (Ks) in a gait cycle

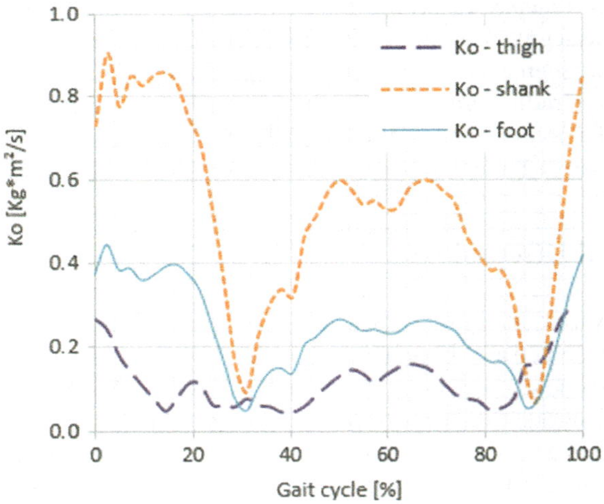

Fig. 3. Orbital angular momentum (Ko) in a gait cycle

The negative values of the spin components are present due to the changing direction of the angular velocity, which occurs when the flexion movement is replaced by the extension, at any segment level.

In Figs. 3 and 4, the K_S and K_o variations are presented for a recording of 10 s, in relation to the X position of the center of mass of each segment. The X position of Com was used in representation because it represents the progression of the body segments during treadmill walking. The cyclic aspect of graphs is caused by the progression of Com in a loop (treadmill consequence), the support foot being carried in the stance phase by the movement of the belt.

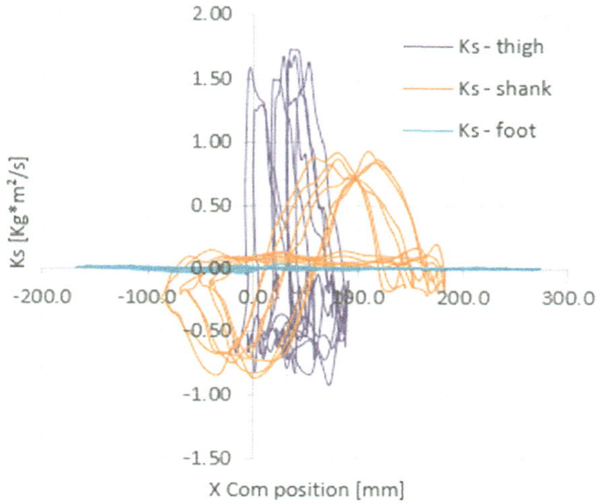

Fig. 4. Spin angular momentum (Ks) against Com progression (X direction)

The representations of angular momentum for 10 s walk (Figs. 4 and 5) also reveal the significance of each quantity. Here the relative constant values in angular momentum are created by the belt movement of the treadmill, leaving the wrong impression of conservation.

4 Conclusions

The paper presents an analytical three-segment model for computing the angular momentum of the lower limb when the subject is walking on a treadmill. The model relies on classical mechanics (Newton-Euler formulation for rotational motion) and the kinematical parameters obtained by motion analysis.

The results reveal the changing in angular momentum of each considered segment, both as spin and orbital components. The highest value belongs to the thigh spin momentum, while the less significant is of the foot. The orbital components are well represented by the shank and foot.

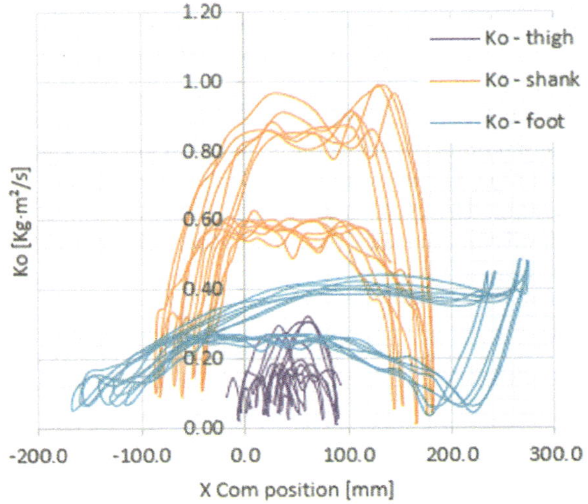

Fig. 5. Orbital angular momentum (Ko) against Com progression (X direction)

References

1. Sun, J., Wang, Y., Li, J., Wan,W., Cheng, D., Zhang, H.: View-invariant gait recognition based on kinect skeleton feature. Multimed. Tools Appl. **77**(295) (2018)
2. Luca, R., Bejinariu, S.-I.: Classification method for human locomotion. In: Proceedings of the 10th International Conference on Electronics, Computers and Artificial Intelligence, ECAI 2018, Iasi, Romania (2019)
3. Baritz, M.I., Cotoros, D.: Oscillatory movements analysis at knee level. Appl. Mech. Mater. **436**, 271–276 (2013)
4. Fukuchi, C.A.; Fukuchi, R.K.; Duarte, M.: A public dataset of overground and treadmill walking kinematics and kinetics in healthy individuals. Peer J. **6**, e4640 (2018)
5. Dragulescu, D., Perdereau, V., Drouin, M., Ungureanu, L., Menyhardt, K.: 3D active workspace of human hand anatomical model. Biomed. Eng. Online **6**, 15 (2007)
6. Tarnita, D., Georgescu, M., Tarnita, D.N.: Applications of nonlinear dynamics to gait analysis on plane & inclined treadmill. New Trends Med. Serv. Robots **39**, 59–73 (2016)
7. Tarnita, D., Catana, M., Tarnita, D.N.: Nonlinear analysis of normal human gait for different activities with application to bipedal locomotion. Rom. J. Techn. Sci. Appl. Mech. **58**(1–2), 177–192 (2013)
8. Dragulescu, D., Ungureanu, L., Menyhardt, K., Stanciu, A.: 3D active workspace of the human hand shaped end effector. In: Proceedings of the 13th IASTED International Conference on Robotics and Applications, Wurzburg, Germany (2007)
9. Nagy, R., Maruta, R.S., Menyhardt, K.: Active vibration control of test equipment through feedback algorithm. In: 14th International Conference on Acoustics and Vibration of Mechanical Structures (AVMS), vol. 198, pp. 247–255 (2017)
10. Gaffney, B.M.M., Christiansen, C.L., Murray, A.M., Silverman, A.K., Davidson, B.S.: Separation of rotational and translational segmental momentum to assess movement coordination during walking. Hum. Mov. Sci. **51**, 99–111 (2017)
11. Bennett, B.C., Russell, S.D., Sheth, P., Abel, M.F.: Angular momentum of walking at diferent speeds. Hum. Mov. Sci. **29**, 114–124 (2010)

12. Hamill, J., Knutzen, K.M.: Biomechanical Basics of Human Movement, 2nd edn. Lippincott Williams &Wilkins (2003)
13. Stoia, D.I, Vigaru, C., Nicoara, A., Herisanu, N.: Angular momentum about the total body center of mass computed at different speeds. In:15th Conference on Acoustics and Vibration of Mechanical Structures, AVMS. Springer Proc. Phys. **251**, 227–233 (2021)
14. Kasdin, N.J., Paley, D.A.: Engineering Dynamics: A Comprehensive Introduction. Princeton University Press (2011)
15. Dempster, W.T.: Space requirements of the seated operator: geometrical, kinematic, and mechanical aspects of the body, with special reference to the limbs (1955)

Influence of the Stent Geometry on Flow Parameters After Arterial Stent Placement

Maria-Cristina Ioncica[1], Alin-Florin Totorean[1], and Sandor Ianos Bernad[2](✉)

[1] Department of Mechanics and Strength of Materials, Politehnica University of Timisoara, Timisoara, Romania
[2] Romanian Academy, Centre of Fundamental and Advanced Research in Engineering Sciences, Timisoara Branch, Timisoara, Romania
sandor.bernad@upt.ro

Abstract. Atherosclerosis is a cardiovascular disease that hardens, narrows, and finally blocks the medium and large arteries. Depending on the stenosis severity, different treatments are applied. One of the most common approaches used in moderate stenosis is percutaneous transluminal coronary angioplasty and stent implantation. In this paper, computational fluid dynamics (CFD) methods were used to understand how the stent's design influences the hemodynamic parameters associated with the blood flow inside the stented arterial segment in unsteady-state conditions. This paper aims to characterize the blood flow inside a stent using pulsatile flow conditions related to the natural cardiac cycle. The analyzed parameters were the velocity field and pressure variation as global parameters, and wall shear stress (WSS) and recirculation areas as local parameters. The results may contribute to a better understanding of the blood flow near the arterial walls in the stent's presence and could predict critical regions prone to thrombosis or in-stent restenosis.

Keywords: Stent · Stenosis · Velocity field · Pressure · Wall shear stress · Recirculation regions

1 Introduction

Atherosclerosis is a pathology that affects the medium and large arteries, characterized mainly by vessel narrowing with effects on tissue perfusion. In the case of coronary stenosis, depending on the condition's severity, atherosclerosis may lead to ischemia and myocardial infarction. In moderate coronary stenosis, percutaneous transluminal coronary angioplasty is performed, characterized by stent implantation inside the affected coronary artery. The presence of the stent may cause local hemodynamic environment alteration, especially in the vicinity of the stent and coronary artery wall that could lead post-intervention to thrombosis, as a short term complication or to in-stent restenosis as a medium and long term condition. From the hydrodynamic point of view, the appearance of the in-stent restenosis is strongly related to regions characterized by low values of Wall Shear Stress and the presence of recirculation regions [1–4].

N. Herisanu and V. Marinca (eds.), *Acoustics and Vibration of Mechanical Structures – AVMS-2021*, Springer Proceedings in Physics 274,
https://doi.org/10.1007/978-3-030-96787-1_20

The aim of this paper is to assess the hemodynamic parameters associated to blood flow induced by the stent geometry in a simplified coronary stent configuration under the physiological pulsatile flow condition using the Computational Fluid Dynamics tools. The main hemodynamic parameters considered in the analysis are: the wall shear stress distribution on the artery wall, the velocity field in the vicinity of the stent and arterial wall region and pressure variation along the vessel's centerline.

2 The Simplified Stented Arterial Model

A bi-dimensional simplified stent model was reconstructed associated to a longitudinal section from a 3D stent configuration implanted inside a coronary artery with a diameter of 4 mm. The struts were considered circular shaped with a diameter of 0.4 mm. The geometrical configuration was modeled considering half of the struts being protruded in the artery wall after the balloon inflation and stent placement inside the artery. The geometry was extended upstream and downstream the stent so that the flow in the region of interest will not be affected, as described in Fig. 1 and Table 1.

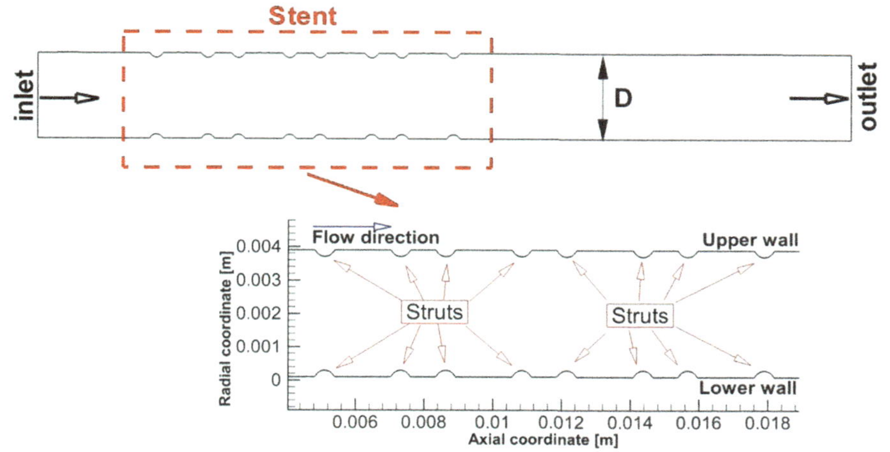

Fig. 1. Geometrical characteristics of stent configuration used in the numerical analysis

Table 1. Geometrical characteristics of the stented arterial 2D model

Geometrical characteristics	
Vessel diameter (D) (mm)	4
Stent length (mm) (mm)	15
Length of the flow domain upstream the stent (mm)	4
Length of the flow domain downstream the stent	16

An unstructured mesh (Fig. 2a) was generated in the ANSYS Workbench 19.2 Mesh tool with 69,658 triangular elements.

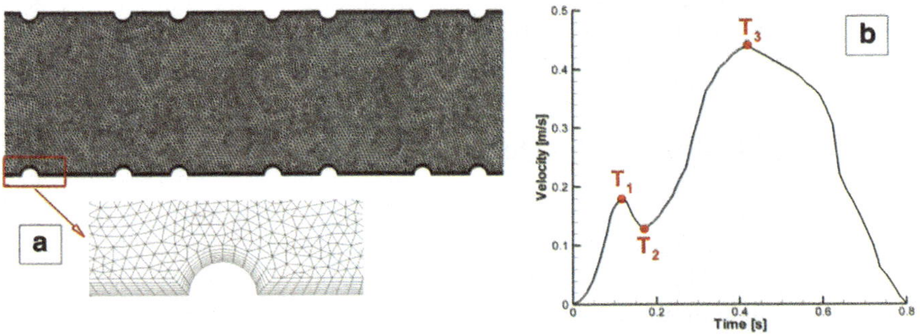

Fig. 2. **a** Mesh characteristics associated to the stented region; **b** physiological cardiac cycle used in the unsteady-state numerical analysis

For a better characterization of the blood flow in the vicinity of the stent's struts, a boundary layer was generated, as described in Table 2.

Table 2. Mesh characteristics of the stent arterial 2D model

Characteristics	
Number of nodes	39,464
Number of elements	69,658
Type of elements	Triangular
Boundary layer – Number of layers – Layer maximum thickness – Layer growth rate (this represents an increase of 20% in size from one layer to the next one)	7 0.1 mm 1.2

3 Numerical Setup

The boundary conditions used for the numerical analysis with respect to the hemodynamic particularities [4–6] are:

- Inlet: a velocity profile associated to a physiological cardiac cycle (Fig. 2b), as described in the literature [5] was used as a condition for the inlet section;
- Outlet: pressure was set to 0 Pa in the outlet section (to reduce the computational effort);

- Walls: are assumed to be no-slip, impermeable and rigid, considering that an artery affected by stenosis together with stent structure suffer low deformations under the blood flow condition during a cardiac cycle;
- Fluid was set to be homogeneous, incompressible, having the physical properties similar to blood, with density of 1050 kg/m^3 and dynamic viscosity of 0,004 Pa.s.

The numerical analysis was performed in the commercial software ANSYS Fluent 19.2, using the laminar model. The numerical simulation ran for two consecutive cardiac cycles. The results presented in this paper show the evolutions of the hemodynamics parameters corresponding to three representative time steps from the cardiac cycle: T1 (maximum systole peak), T2 (minimum velocity associated to the transition from systole to diastole) and T3 (maximum diastole peak) [5].

4 Results

Wall Shear Stress (WSS) distribution along the lower wall of the stented artery is represented in Fig. 3.

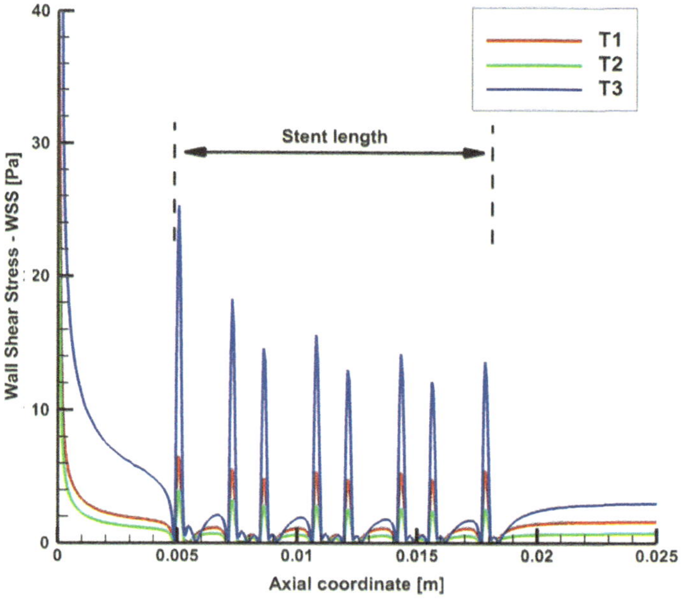

Fig. 3. Wall Shear Stress distribution along the stented arterial lower wall

High values of WSS are associated to the upper regions of the struts, whereas WSS significantly decreases downstream the struts. The highest value of WSS appears in the region of the first strut and decreases as blood flows inside the stent.

The highest values for WSS are associated to the diastole peak time step, T3, and WSS has low values in the case of T2 time step associated to the transition from systole to diastole. It can be observed that the variation of WSS is strongly related to the stent geometry. In the regions where the distance between two consecutive struts is high (e.g. strut 3–strut 4 as presented in Fig. 4), the WSS tends to increase on the upper zone of the second strut compared to the previous strut, due to the fact that the higher gap leads to flow resistance decrease, whereas in the case of two close consecutive struts (e.g. strut 2–strut 3) WSS significantly decrease, as blood flow encounters a higher resistance.

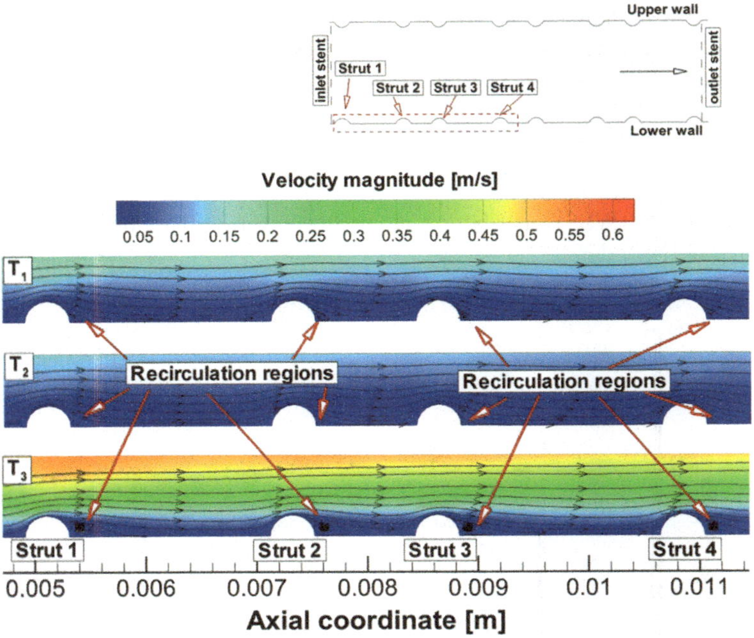

Fig. 4. Velocity field characteristics in the vicinity of the stented arterial lower wall associated to the first four struts associated to three different time steps of the cardiac cycle. Representation of streamlines put in evidence the presence of the recirculation regions downstream each strut

Velocity field associated to the blood flow in the vicinity of the strut and near the arterial lower wall is represented in Fig. 4. The presence of the recirculation regions is put in evidence through the representation of streamlines. It can be observed there for all the investigated time steps there are developed recirculation regions downstream each strut, which are more significant as the inlet velocity increases. The largest recirculation regions are associated to the diastole peak of the cardiac cycle (T3) whereas in the case of the T2 time step specific for the transition from systole to diastole, due to the deceleration associated to the systole decrease curve, the recirculation regions are significantly reduced.

By correlating Figs. 3 and 4 it can be noticed that regions with low values of WSS are associated to the presence of the recirculation regions.

Figure 5 describes the pressure variation along the centerline of the stented arterial region. It can be observed that the presence of the struts induces pressure drop. This variation is more significant in the case of two close consecutive struts (e.g. region delimited by strut 2 and 3 as described in Fig. 4). The highest pressure variation appears at the T3 time step of the cardiac cycle, whereas the transition from systole to diastole time step is characterized by low pressure drop.

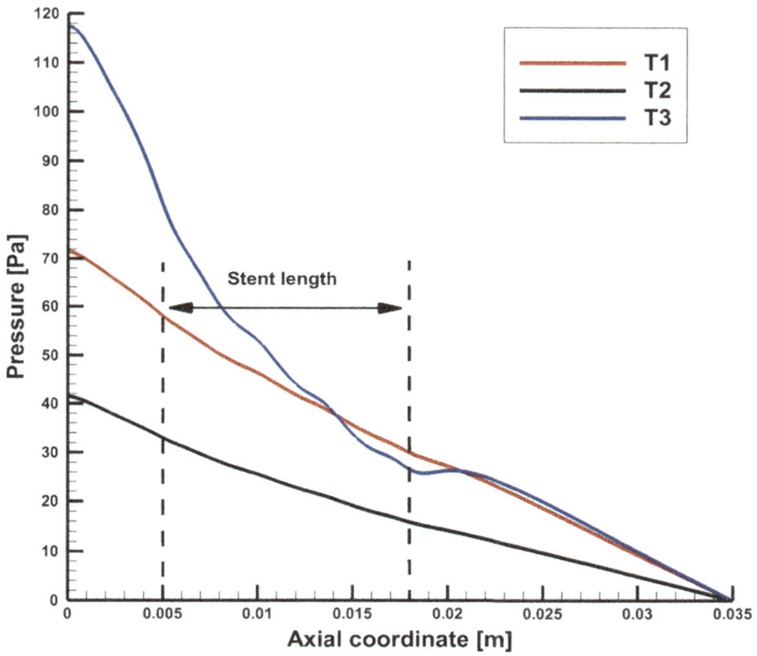

Fig. 5. Pressure variation along the centerline of the vessel associated to the stented region

5 Conclusions

The presence of the stent alters the blood flow environment in the vicinity of the arterial wall, creating regions with low Wall Shear Stress downstream the struts and development of recirculation regions, together with pressure variation. By understanding the hemodynamic characteristics associated to the blood flow through a stent, this could contribute to therapeutic procedure optimization, stent design improvement, or drug targeting optimization to reduce the post-interventional complications.

The present study has its limitations regarding the geometrical model which is a simplified bi-dimensional configuration that do not completely mimic the real patient condition. However, the results obtained in this paper could offer a first step to understand the blood flow phenomena inside a stented arterial model that could help further studies in the direction of drug targeting, which can be of real interest in reducing the post-interventional complications.

References

1. DePaola, M.A.G., Jr, P.F., Davies, P.F.: Dewey Jr, Vascular endothelium responds to fluid shear stress gradients. Arterioscler. Thromb. **12**(11), 1254–1257 (1992)
2. Ng, J., Bourantas, C., Torii, R., et al.: Local hemodynamic forces after stenting implications on restenosis and thrombosis. Arterioscler. Thromb. Vasc. Biol. **37**, 2231–2242 (2017)
3. Kolandaivelu, K., Swaminathan, R., Gibson, W.J., Kolachalama, V.B., Nguyen-Ehrenreich, K.L., Giddings, V.L., Coleman, L., Wong, G.K., Edelman, E.R.: Stent thrombogenicity early in high-risk interventional settings is driven by stent design and deployment and protected by polymer-drug coatings. Circulation **123**, 1400–1409 (2011)
4. Hudrea, I.C., Totorean, A.F., Gaita, D.: Computational fluid dynamics analysis of coronary stent malapposition. IFMBE Proc **68**(1) (2019)
5. Banerjee, R.K., Back, L.H., Back, M.R., Cho, Y.I.: Physiological flow simulation in residual human stenoses after coronary angioplasty. ASME J. Biomech. Eng. **122**(4), 310–320 (2000)
6. Totorean, A., Totorean, I.-C., Bernad, S., Gaita, D.: Numerical investigation of an idealized overlapping coronary stents configuration. Springer Proc. Phys. **251**, 251–258 (2021)

Numerical Analysis of a Total Hip Prosthesis Under Static Loading Conditions

Corneliu Drugă(✉), Ionel Şerban, Irina Ioniţă, and Anca Stanciu

Transylvania University of Brasov, Eroilor 29, Brasov, Romania
druga@unitbv.ro

Abstract. In this paper a numerical investigation of replacement implant for Total Hip Arthroplasty (THA) is presented. The long-term stability of hip implants depends, among other things, on the loads acting across the joint. Forces occurring in vivo can be much greater than the recommended test values, because a typical gait cycle generates forces up to 6–7 times the body weight in the hip joint. A Finite Element Analysis (FEA) was performed using 3D models to examine the mechanical behavior of the femoral component at forces ranging from 2 to 3 kN. This implant design was chosen for numerical analysis because stress concentration in femoral component leads to implant fracture. Results show that the force magnitudes acting on the implant are of interest, and that they can cause implant stress field changes and implant stability problems, which can lead to implant failure.

Keywords: Total hip arthroplasty · Finite element analysis · SolidWorks

1 Introduction

The hip joint is the largest and most stressed synovial joint in the human body along with the knee joint, and gives stability to the entire human structure. The human joint is a self-acting and dynamic load bearing structure, which uses a porous and elastic biomaterial as well as highly non-Newtonian lubricants for its functioning [1]. The loading cycle to which an articular joint is subjected is complex and the demands this places on the joint are numerous [2]. As additional complications, the forces that operate within the joint are three dimensional, time dependent and vary with speed and length of stride and, of course, vary from person to person. The function of the cartilage is threefold: to distribute the load on the joint, to reduce friction and eliminate wear and to absorb energy during dynamic loading [1]. From an anatomical point of view, the hip joint consists of a femoral head that articulates with the acetabulum. The destruction of the articular cartilage surface, due to the tribological phenomena (wear, friction and lubrication) that take place between its components, the appearance of fractures or other causes (pathological, traumatic) eventually lead to the inclusion of natural joints with artificial ones (total or partial endo-prosthesis or partial). Currently, there is a wide variety of hip stents endo-prosthesis available on the market, which are differentiated by: rod

N. Herisanu and V. Marinca (eds.), *Acoustics and Vibration of Mechanical Structures – AVMS-2021*, Springer Proceedings in Physics 274,
https://doi.org/10.1007/978-3-030-96787-1_21

length, design, outer diameter of the femoral head, implantation method, combinations of materials used in construction and so on [3].

Total hip arthroplasty (THA) is a surgery that replaces the damaged coxo-femoral joint with an artificial implant called a hip stent and is one of the most effective and successful orthopedic interventions for many decades because it restores reproductive function and reduces pain in the hip joints was pathological. After that, the pain disappears, and the person can regain his mobility of the hip, resuming his daily activities. Hip arthroplasty (THA) is recommended especially in cases where damage to the coxo-femoral joint causes pain and affects the functional capacity of the person intervening in the normal course of normal activities. Most commonly, joint damage is caused by hip osteoarthritis, but also rheumatoid arthritis, post-traumatic osteoarthritis, osteonecrosis can also be among the causes. Currently, there are two methodological options for THA, cement and cementless implant. Cemented THA has a higher 10-year survival rate than uncemented THA, i.e. a lower risk of revision (including aseptic loosening). Therefore, in summary, the surgery to which the patients are subjected compensates functional deficiencies afferent bones (femur and acetabulum), related to genetic or trauma consequences, using a wide variety of endo-prosthesis. Briefly, two main options are available: the bone cement used to fix the stem of the new femur causes reduction in bone density as a result of removal of normal stress from the bone [4]. In the THA case are frequently used two types of stems namely modular and non-modular. Both stems are identical except the modular neck stem has a separate neck piece that allows for 60 different options for head center and the non-modular neck has only 10 options. The most used biomaterials used in the construction of hip implants are: cobalt-chromium- alloys, titanium alloys, stainless steels, UHMWPE (Ultra High Molecular Weight Polyethylene), aluminum oxide (Al_2O_3), zirconium oxide (ZrO_2) are popular due to their biocompatibility and biofunctionality. The basic requirement of all these biomaterials is to be biocompatible with the host tissue and to produce a wear rate as low as possible (the coefficient of friction should be as low as possible), so that the survival rate (life of the implant) is as higher as possible [5]. Finite Element Analysis (FEA) is a very popular tool for analyzing the behavior of any developed model (the Ti–6Al–4V alloy was considered in this study). FEA allows static and dynamic mechanical characterization but in the case of this study, it was approached only numerical analysis of a total hip prosthesis under static loading conditions.

2 Materials and Methods

2.1 The Computer-Aided Design (CAD) Model of the Hip Prosthesis

The CAD model of the hip prosthesis presented in this paper was done using the SolidWorks 2016 program (Figs. 1 and 2). SolidWorks is part of the most advanced CAD-CAM systems currently used in designing. Compared to other similar systems, a number of facilities can be used working with 3D spatial shapes.

2.2 Metallic Biomaterials Used in the FEA

Among bio-inert metallic biomaterials, 316L stainless steel, cobalt-chromium-based alloys and titanium-based alloys are the most commonly used biomaterials for fracture

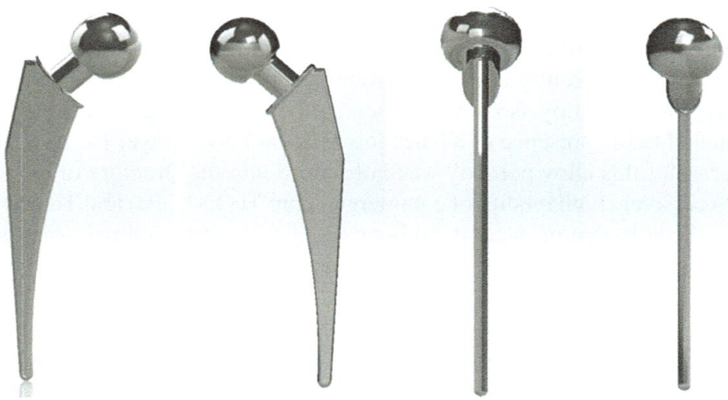

Fig. 1. The CAD model of the analyzed hip implant (different views).

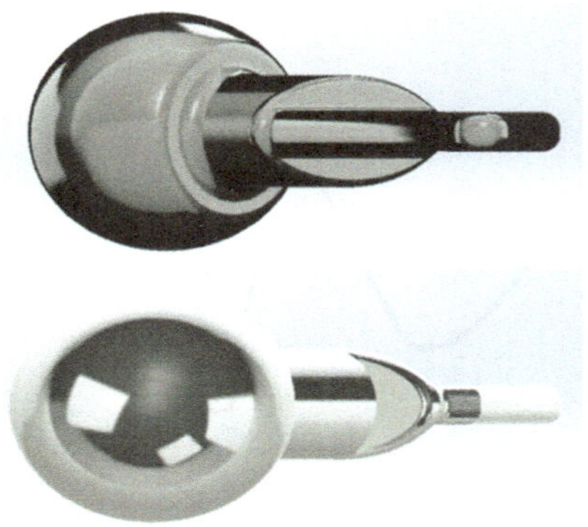

Fig. 2. Bottom and top view of the total hip prosthesis.

fixation, angioplasty and bone remodeling [6]. This is primarily due to their long-term chemical stability in acidic/reactive environments (high corrosion resistance) in vivo and excellent mechanical properties [6]. To obtain a mechanical strength and corrosion resistance, the percentage of chromium is between 10 and 30%. Due to the presence of chromium, a layer of passive oxide is formed, which determines the above-mentioned parameters. Also in order to obtain good parameters for corrosion and resistance, iron, molybdenum, nickel are added to these alloys. Although these materials are considered to have low corrosion, wear, friction and extremely aggressive environments can lead to biomaterial degradation and the associated release of unwanted metal ions, which can induce local tissue damage and inflammatory reactions, such as osteolysis gradual attachment of adjacent tissues, as well as systemic lesions such as hypersensitivity to metal [6].

Titanium and Ti-Based Alloys. The Ti–6Al–4V alloy containing 5.5–6.5% by weight aluminum and 3.5–4.5% by weight vanadium is frequently used to make a wide range of implants because it provides mechanical resistance to fatigue much better compared to non-alloy titanium in trade. Addition of aluminum increases the hardness of Ti by 32% without significantly affecting its other properties [6]. Addition of niobium to Ti, as in $Ti_{15}Nb_4TaZr$ alloy, is known to increase both strength and wear resistance, the latter attributed to the presence of a hard, low-friction Nb_2O_5 layer [6, 7].

In the case of this alloy porosity was introduced into the structure of the implant to bring the excessively high moduli of commercial pure Ti (112 GPa) and Ti–6Al–4V (115 GPa) alloy closer to that of the cortical bone (7–30 GPa), to minimize stress shielding, rather than to promote tissue regeneration or vascularization [8, 9].

2.3 Biomechanics of Hip Joint

In the case of the coxo-femoral joint, the loads acting on the femoral head vary depending on the variation of body weight (B.W), body position and externally applied forces (Fig. 3). When a person is in an upright position, body weight (B.W) is transmitted through the lumbar vertebra L5 to the sacral base.

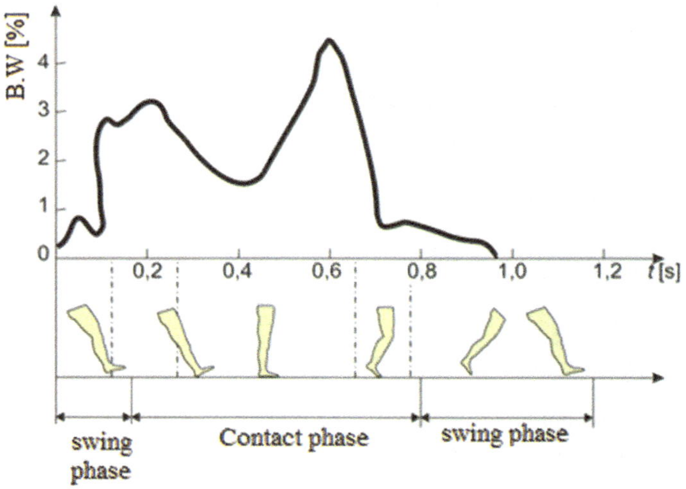

Fig. 3. The variation of the resultant of the forces acting in the hip joint during a walking cycle.

During normal gait, on heel-strike, the hip moves into 3 degrees of flexion and at toe-off (when the foot is finally off the ground) about 10 degrees of extension. The range of abduction to adduction is about 11°, and for internal–external rotation, the range is about 8. During different phases of gait cycle, different forces act on femoral head [4]. The amplitude of the resultant of the forces acting on the femoral head (and implicitly on the total hip prosthesis) was calculated for people with different body weights according to references [11, 12]. Thus, for a person with a body weight of 52 kg, the resultant of the forces acting on the femoral head is 2044 N (Figs. 5 and 6), and for a person of 70 kg the resultant of the forces is 2747 N (Figs. 7 and 8). As in FEA application, only static analysis was considered, the highest force acting on the femoral neck is developed in the position standing on a single led. Biomechanics studies [12] mention that this force is four times greater than the subject's weight, wherefrom the above-mentioned values of the resultant forces.

2.4 Finite Element Analysis (FEA)

See Fig. 4.

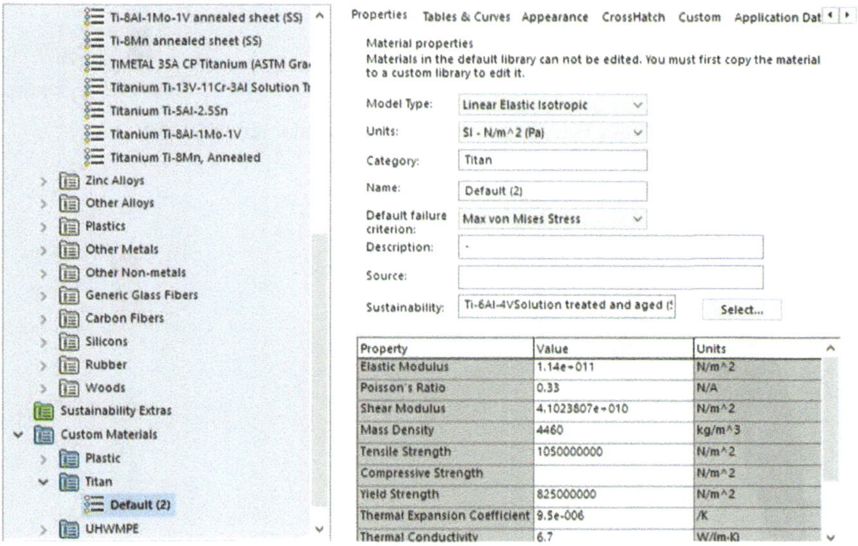

Fig. 4. Choosing the material used in the FEA.

Case I: person with body weight: 52 kg

See Figs. 5 and 6.

Case II: person with body weight: 70 kg

See Figs. 7 and 8.

Finite Element Analysis (FEA) is becoming one of the most important tools in orthopedic biomechanics, being used more and more to evaluate the stress and displacements, the biomechanical behavior of healthy bones and joints, of implant-bone assembly or human joint–prosthesis/orthosis assemblies [10]. In this study, we are interested in the field of von Mises Stress and URES (Figs. 5, 6, 7 and 8) and the used biomaterial is Ti–6Al–4V (Fig. 4). Hexahedral elements of Solid 186 type and tetrahedral elements of Solid 187 type are both solid elements with middle nodes which are necessary for better approximation of results and for better accuracy; they were both used in this analysis, resulting in number of 160,298 elements and of 334,911 nodes.

Fig. 5. Stress Von Misses yield for an applied force of 2044 N.

Fig. 6. Visualization of displacements (URES) when applying a force of 2044 N.

3 Results and Conclusions

To perform the numerical simulations of prosthetic hip joint, the following components that make up joint-prosthesis assembly have been considered the two components of the prosthesis: femoral component and acetabulum component. Finite Element Analysis, in this case, was performed in static mode. Although FEA is currently the most widely used simulation method computer-implemented in all areas of engineering, it will not replace in vitro or in vivo tests, but may supplement them. In addition to its numerous advantages, FEM has also a number of disadvantages, such as:

Fig. 7. Stress Von Misses yield when applying a force of 2747 N.

Fig. 8. Visualization of displacements (URES) when applying a force of 2747 N.

- The analysis with MEF is not done for the real structure but for a (calculation) model of it and, the obtained results represent an approximation of the states of displacements, tensions, temperatures and so on of the real structure being analyzed.
- it is not possible to estimate—in the vast majority of real situations—with a quantifiable level of confidence, how well the FEA approximates the exact (unknown) solution of the analyzed problem.

In other words, it is very difficult—sometimes even impossible—to estimate the deviations of the values (displacements, tensions, efforts, frequencies, etc.) calculated with MEF compared to the real, unknown ones.

The meshing and the solution for the analysis stages, the correct positioning of components have been achieved in the simulation environment of SolidWorks 2016 application, which allows advanced modeling and discretization using FEA. The centralized results from the simulation are presented in Table 1.

Table 1. Results of the FEA simulation.

Resultant force	Von Mises stress yield		Resultant displacements (URES)	
	Max. [N/m^2]	Min .[N/m^2]	Max. [mm]	Min. [mm]
2044 N	2.752e + 08	5.255e-02	1.592e-01	1.000e-30
2747 N	3.699e + 08	7.062e-02	2.140e-01	1.000e-30

The least maximum von Mises stress can reduce the long-term problem of the implant failure [11]. The maximum von Mises stress under static and dynamic conditions did not reach the yield strength of hip prosthesis. Yield strength of titanium alloy is 85e + 07(N/m^2) [11], which shows that the hip prosthesis is safe for static and dynamic condition of hip contact. The maximum von Mises stress on the hip prosthesis under static condition was higher than dynamic condition 0.14% (19e + 04 N/m^2) under walking and 0.03% (4e + 04 N/m^2) under climbing stairs condition (values mentioned in literature).

References

1. Szeri, A.: Fluid Film Lubrication, 2nd edn. Cambridge University Press (2011)
2. Paul, J.P.: Forces transmitted by joints in the human body. In: Proceedings of Institution Mechanical Engineers Conference Proceedings, vol. 181, pp. 8–15 (1967)
3. Chethan, K.N., et al.: Finite element analysis of different hip implant designs along with femur under static loading conditions. J. Biomed Phys. Eng. **9**(5), 507–517 (2019)
4. Izzo, M.G.: Support for total hip replacement surgery: structures modeling, gait data analysis and report system. Europ. J. Transl. Myol. Basic Appl. Myol. **22**(1&2), 69–121 (2012)
5. Chethan, K.N., et al.: Static structural analysis of different stem designs used in total hip arthroplasty using finite element method. Heliyon **5**, 1–8, e01767 (2019)
6. Prasad, K., et al.: Metallic biomaterials: current challenges and opportunities. J. Mater. **10**, 884 (2017). https://doi.org/10.3390/ma10080884
7. Nakada, H., et al.: Assessment of the quality of newly formed bone around titanium alloy implants by using X-ray photoelectron spectroscopy. Int. J. Biomater. 615018 (2012)
8. Niinomi, M.: Recent metallic materials for biomedical applications. Metall. Mater. Trans. A **33**, 477–486 (2002)
9. Aherwar, A., Singh, A.K., Patnaik, A.: Cobalt based alloy: a better choice biomaterial for hip implants. Trends Biomater. Artif. Organs **30**, 50–55 (2016)
10. Finite element method.: https://en.wikipedia.org/wiki/Finite_element_method#:~:text=The%20finite%20element%20method%20. Accessed 13 May 2021

11. Roşca, I., Şerban, I.: Biomecanică-Îndrumar de laborator, 2nd edn Transilvania University of Brasov Publisher, Braşov (2017)
12. Roşca, I., Şerban, I.: Fundamente de Biomecanică, 1st edn. Transilvania University of Brasov Publisher, Braşov (2013)
13. Chalernpon, K., Aroonjarattham, P.: Static and dynamic load on hip contact of hip prosthesis and thai femoral bones. World academy of science. Eng. Technol. Int. J. Mech. Mechatron. Eng. **9**(3), 251–255 (2015)

Designing and FEA Analyzing an Intramedullary Nail for Tibial Fractures

Irina Ionita, Corneliu Druga(✉), Ileana Rosca, and Ionel Şerban

Faculty of Product Design and Environment, Transylvania University of Brasov, Eroilor 29, Brasov, Romania
druga@unitbv.ro

Abstract. The paper presents a series of aspects regarding the design and FEA analysis of an intramedullary nail used for tibial fractures. The theoretical section consists of the analysis of bone fractures and orthopedical implants. The practical section contains the CAD model and Finite Element Analysis (FEA) of the intramedullary nail most commonly used during the tibial prosthesis intervention. The last part of the work is dedicated to the results and conclusions in order to determine if the movement of the nail inserted in the intramedullary canal is due to its design or due to design of the external guidance system.

Keywords: Intramedullary nail · FEA analysis · External guidance system for locked nails

1 Introduction

People have been trying to replace or support bones and joints for thousands of years. At first, they addressed the consequences of trauma and fractures. Later they became concerned about abnormalities in bone structure. For this reason, some of the oldest implants were not designed to accurately reproduce bone anatomy, but to support or reshape it. Orthopedics uses a variety of implants in the treatment of injuries, diseases, and musculoskeletal deformities. The orthopedic implant can be grouped into the following categories: fracture fixation, joint restoration or replacement, soft tissue replacement, and spinal implants.

It is important that orthopedic implants be designed so that they have mechanical and chemical properties appropriate to the anticipated clinical requirements. This requires understanding the properties of the materials and biomechanics. It also requires mechanical testing of implants to ensure that they meet performance specifications.

N. Herisanu and V. Marinca (eds.), *Acoustics and Vibration of Mechanical Structures – AVMS-2021*, Springer Proceedings in Physics 274,
https://doi.org/10.1007/978-3-030-96787-1_22

2 Tibial Fractures and Instruments

2.1 Tibial Fractures

The diaphysis of the long bones corresponds to the middle portion, located between the two epiphyseal extremities (Fig. 1). A cross section of the diaphyseal axis allows us to observe that the diameter of the diaphysis is not constant, the medullary canal progressively widening toward the extremities. The diaphyseal medullary canal does not contain hematopoietic bone marrow (located at the level of the epiphyses), but fatty bone marrow and vessels of medullary origin.

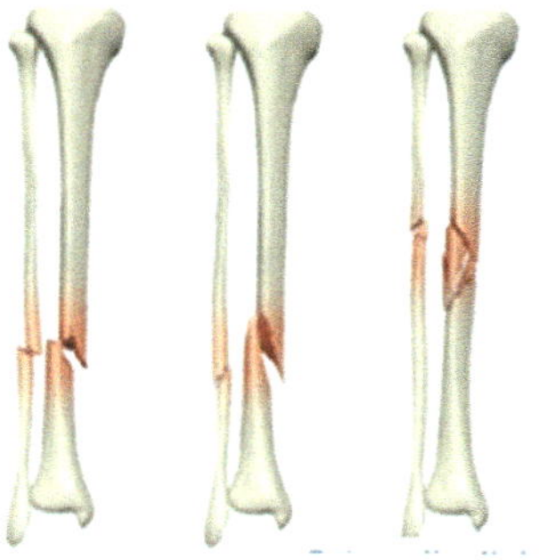

Fig. 1. Distal tibial fractures [1].

The cortical bone follows a different evolution during the life of the individual. Thus, aging leads to a progressive physiological resorption of the cortex. This decrease in cortical thickness also occurs in pathological cases, such as during prolonged immobilizations that lead to the appearance of immobilizing osteoporosis, much more pronounced in the case of spongy bone than in cortical bone.

The vascularization of the diaphyseal bone has a double origin, namely endosteal vascularization and periosteal vascularization [2].

Fractures can be found in practice in a wide variety of types but, for practical reasons, they can be divided into two well-defined groups, namely complete fractures and incomplete fractures. In complete fractures, there is a complete disruption of bone continuity and the bone is completely fractured into two or more fragments. Displacement in a complete fracture occurs, partly due to the force of the trauma, partly due to gravity, and partly due to the muscles that are inserted on the fractured bone fragments (Fig. 2).

Incomplete fractures are fractures in which the anatomical continuity of the skeletal part is preserved. The bone is incompletely fractured and the periosteum retains its continuity. In adults, this type of fracture occurs through cracking or compression, when the spongy bone is compacted [3].

Thus, simple diaphyseal fractures of the tibia include spiroid fractures, oblique fractures that are composed of two types of fractures (oblique fractures below 30° and oblique fractures with an angle of more than 30°) and transverse fractures. Complex fractures are fractures where a third bone fragment appears and they can be bifocal, oblique or transverse.

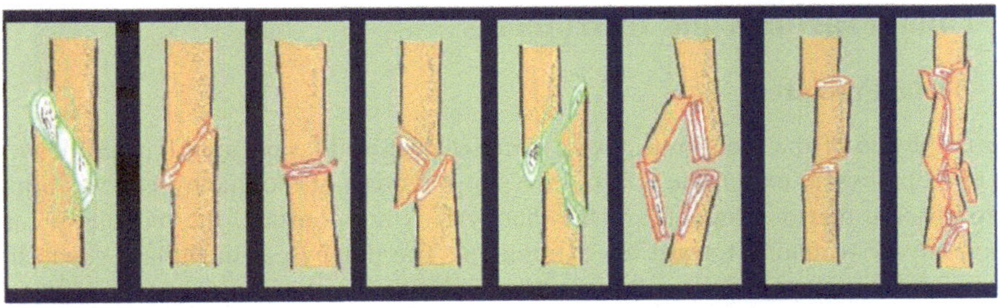

Fig. 2. Different types of fracture [3].

2.2 Intramedullary Locked Nails

The principle of the intramedullary nail is to use the cavitary medullary canal to insert it and ensure the rigidity of the fractured bone until it is strengthened. Cominutive fractures and fractures located at the extremities of the diaphysis do not allow to obtain a satisfactory mechanical restraint.

The locked intramedullary shaft is perforated at both ends and 4–6 screws are inserted transversely into the bone and into the holes in the shaft. This locking method can also be applied to unstable fractures, complex metaphysics. Dynamization involves removing the lower or upper screws, which block (depending on the location of the fracture), which allows the compression of the fracture focus and promotes the corticalization of the callus.

The insertion of the intramedullary stem destroys the bone marrow and therefore part of the cortical vascularization. After a few days, a cortical neovascularization occurs, which covers this initial loss. The absence of a direct approach to the fracture site plays an important role because it avoids deperiostation and therefore devitalization of the fragments, as well as the risk of infection [4].

At present, two concepts of treatment of diaphyseal fractures of long bones are confronted: the concept of rigid fixation, with open focus and the concept of elastic fixation, of the intramedullary beginning, in which the fracture area is respected by approach with closed focus; the introduction of a nail into the neutral axis of the diaphysis, mechanically as effective as possible, authorizes a certain elasticity, and is, indisputably, a benefit in relation to open focus osteosynthesis [4] (Fig. 3).

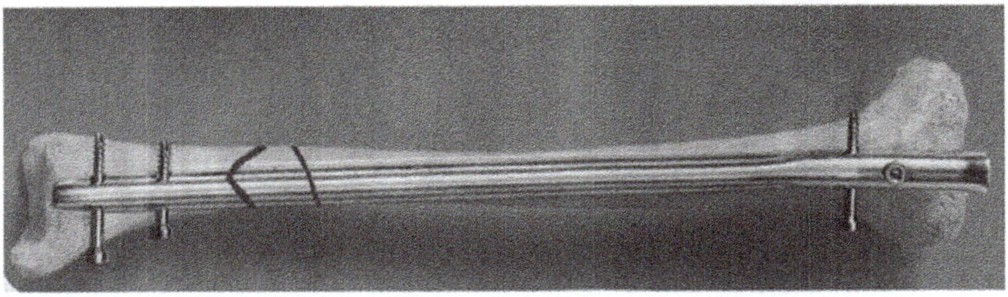

Fig. 3. Static intramedullary locked nail [4].

2.3 The External Guidance System

To perform the intramedullary blockage with the nail by surgical method we need the surgery kit composed of the nail guide and anxillary instruments as well as a kit for opening and entering the intramedullary canal of the tibia. The nail will be chosen according to the anatomical structure of the tibia of each patient (Fig. 4).

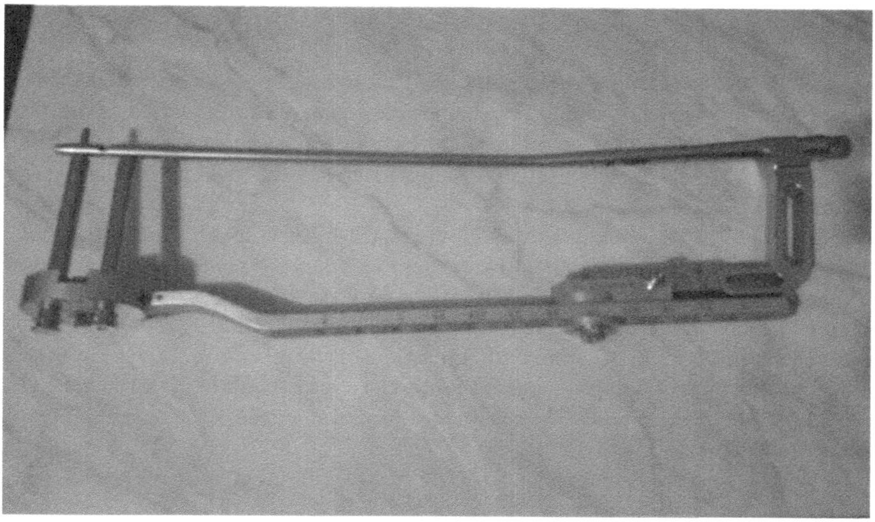

Fig. 4. The nail fixed in the external guidance system.

3 Designing and FEA Analyzing an Intramedullary Nail

3.1 The Computer-Aided Design Model of the Intramedullary Nail

In this paper, the CAD (Computer-Aided Design) model of the nail was made using the CATIA v5R21 program. Since its fifth version, CATIA introduces the notion of virtual prototype. This term is used to describe the whole data that allows handling CAD objects as in real life (Fig. 5).

Fig. 5. The Kuntscher nail after which the CAD model was made [5].

The modeling of the geometric domain was made starting from the dimensions of the nail existing in the catalogs offered by the manufacturer. The next step was to describe the plane profiles that formed the basis for generating 3D elements in "Sketcher" mode. Another step was to describe the 3D elements, using the "Pad" mode. The description of the assemblies and subassemblies was made using the "Assembly Design" mode by using specific options.

3.2 FEA Analysis

The Finite Element Analysis (FEA) provides solutions to problems in identifying the fields of variation in stresses and displacements, mainly. Discretization is the fundamental approach required by the finite element method and consists in moving from the continuous structure (with an infinity of points) to a discrete model with a finite number of points [6, 7].

To generate the finite element model, the CATIA Analysis & Simulation package is launched, from which the Generative Structural Analysis mode is selected and then the Static Analysis option from the New Analysis Case window, which involves static structure analysis under imposed constraints and time-independent loads.

The generation of the structure with finite elements was done in 2 stages: discretization and introduction of finite element properties. Discretization was performed by selecting the three-dimensional domain using the Octree Tetrahedron Mesh option. Constraint modeling was done by:

- modeling the connections with the base using the Clamp option;
- modeling the connections with the elements adjacent to Rigid Connection.

After discretizing and modeling the constraints, was established the load to which the assembly is subjected.

Depending on the weight of the subject, the force acting on the bone has different values. For the present situation, a subject weighing 60 kg was chosen, so the force was loaded vertically, on the tibial plateau, about 250 MPa, evenly distributed.

The figures (Figs. 6, 7) represents the Von Mises stress resulting in case if the assembly is subjected to the load.

During the surgery of distal fractures of the tibia, on the intramedullary nail is applied a load similar to the one used in the FEM analysis.

Commonly intramedullary nails used for tibial fractures have important disadvantages, such as complicated orientation, handling, and positioning in the bone, and at the same time difficulty in positioning the distal screws using the classical nail guide. Also, all these aspects can lead to errors or additional holes in the tibia, causing a decrease in bone strength and at the same time increasing the duration of surgery with unpredictable effects on bone recovery.

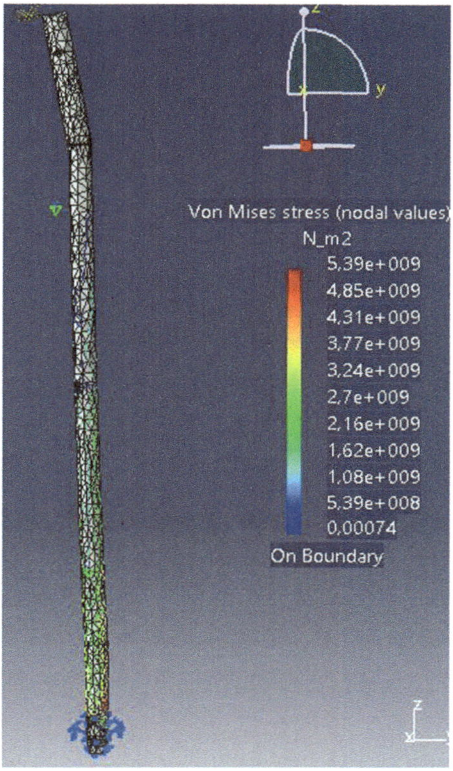

Fig. 6. The 250 MPa tension applied on the nail.

4 Results and Conclusions

Following the aspects related to the fixation system of the diaphyseal fractures of the tibia and the problems encountered during the operations, it was wanted to make a study to determine the stresses of the nail in the distal area, to see if the main problem is in the design of the nail or in the design of the guidance system used to insert the nail.

Several factors are important to determine the stability of the nail, including the size of it, the stresses applied, and where the number of locking screws are positioned.

Due to the fact that during the insertion of the nails, they often bend under the influence of the displacement of the fractured bone fragments and the muscular traction, the position of the distal guide holes of the screws in the guide level no longer corresponds to the position of the distal holes, following this virtual simulation, it can be seen that the most requested area is the one where the distal screws are located, more precisely near the proximal hole of the distal block.

In this case, the surgeon is being forced to fix the distal screws without guide, through numerous successive attempts to trigger the distal holes under the radiological-TV screen, which leads to a significant additional irradiation of the medical staff and the patient (sometimes more greater than the irradiation required to fix the nail).

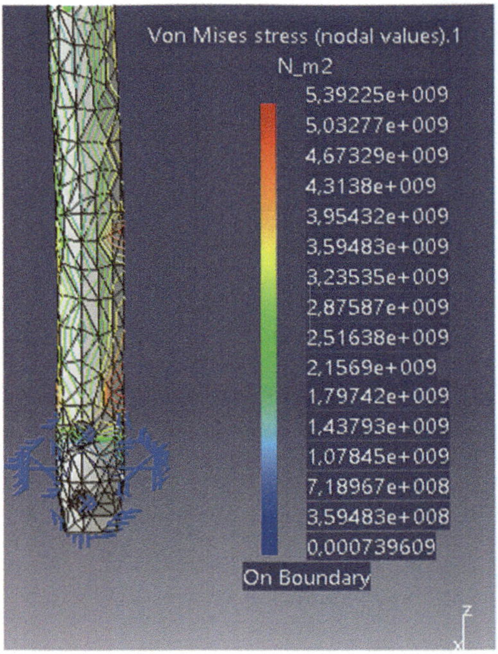

Fig. 7. Tensions arising in the designed assembly.

In conclusion, this paper was based on the study of the importance of implants such as intramedullary nails in the healing of bone skeletal injuries, but especially the importance of good implant functionality, subjected to the demands of the biological environment, the importance generated by the appropriate modeling of the mechanical elements through the FEM analysis.

It can be stated that, in order to obtain a short healing time of the affected bone structure, as well as to avoid a failure of the implant, it is important to observe some major conditions:

- choosing the nail following a detailed study on the fractured bone, reported to the patient;
- choosing the right material of the nail so it can face external forces;
- improving the guidance system so it can show more precisely the distal holes position;

References

1. Cristoiu, F.C.: Diaphyseal fractures of the leg. University of Medicine and Pharmacy Dr, Carol Davila-Bucharest (2014)
2. Tomoaia, G.: Osteoarticular traumatology. In: Anatomical and physiological bases of the bone system, Cluj (2004)

3. Botez, P.: Orthopedics, caduceus collection. Publishing House. ISBN: 978-973-756-075-9, Iaşi (2008)
4. Nicolescu, M., Ravcica, G., Ojoga, F.: Orthopedic-surgical treatment, and physiotherapy in musculoskeletal disorders, part I, Alma Mater Bacau Publishing House (2008)
5. www.zimmerbiomet.com. Accesed 21 June 2021
6. Lates, M.T.: Metoda elementelor finite, Ed. Universităţii Transilvania din Brasov (2008)
7. Cotoros, D.: Analysis of Skeletal prosthesis component elements at structural level. Metal. Int. **15**(7):116–120 (2010)

Smart Posture Corrector

Ramona Nagy(✉), Bianca Popa, and Karoly Menyhardt

Politehnica University Timisoara, Timişoara, Romania
{ramona.nagy,bianca.popa,karoly.menyhardt}@upt.ro

Abstract. Home working or computer activities have serious effects on the backbone. These afflictions range from neck to shoulder pain and even to permanent deformation of the spinal column. In this paper we present a smart device that can monitor and alert the user of prolonged incorrect postures in order to self-teach and take actions.

Keywords: Accelerometer · Spinal column · Correction

1 Introduction

Modern day workers (programmers, designers, even accountants) have long been tormented to use computers for lengthy periods of time. With the advent of Covid-19 threats, even more people have been forced to take home jobs in front of computer desks sitting on chairs in unnatural positions (Fig. 1).

Setting up the workstation in the correct way in order to keep ourselves comfortable, working hard all day long does not guarantee a pain-free experience.

While ensuring one has the correct chair height, good posture and the right equipment, over time one may relax too much and end up fatigued, injured or even with chronic illnesses from wrong postures. Having a bad posture has been known to have a very poor effect on mood. Those who sit for extended periods of time without a correct workstation setup show higher levels of depression. Thus it is important to passively and actively maintain a good posture [1, 2]. Wearable posture correction sensors are useful tools to help office workers maintain good postures in the office environment and avoid future consequences [3, 4]. Besides from using harnesses [5] to maintain or correct excessive deviation, it aids to have someone or something to constantly remind these slips.

In this paper, we present such a smart electronic device that analyses current orientation and in case of prolonged deviations it informs the users.

2 Posture Corrector Development

With the aid of microcontrollers, one can correct posture deviations throughout the day without the help of specialized or supplementary personnel [6]. The microcontroller's job is to monitor the current orientation and to warn the user if needed in case of exceedance of a given angular range.

N. Herisanu and V. Marinca (eds.), *Acoustics and Vibration of Mechanical Structures – AVMS-2021*, Springer Proceedings in Physics 274,
https://doi.org/10.1007/978-3-030-96787-1_23

Fig. 1. Person working in unhealthy posture

Microelectromechanical systems (MEMS) have many applications, among which is measuring either linear acceleration along one or several axis, or angular motion about one or several axis as an input to control a system (Fig. 2).

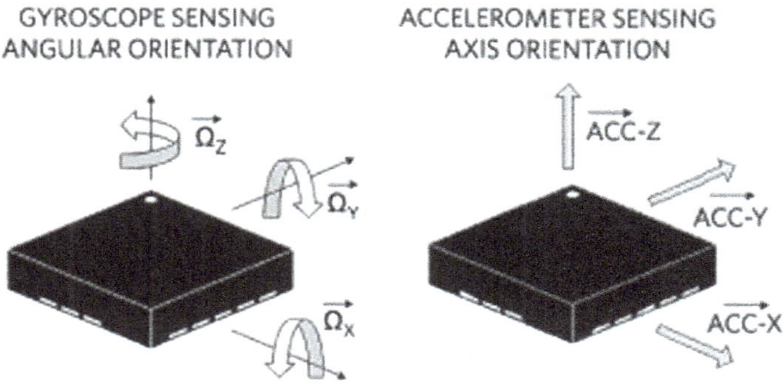

Fig. 2. Maxim integrated [7]

Gyroscopes, measure both the displacement of the resonating mass and its frame because of the Coriolis acceleration (a_c).

$$\overline{a_c} = 2\left(\overline{\Omega} \times \overline{v_r}\right) \quad (1)$$

where $\overline{\Omega} = \Omega_x\bar{i} + \Omega_y\bar{j} + \Omega_z\bar{k}$ is the angular transport velocity and $v_r = e\omega\cos(\omega t)$ is the relative velocity of the excited mass (m) at an amplitude of displacement e with a frequency ω, considering the plane position being $e\sin(\omega t)$.

Thus, a piezoelectric material can be forced to vibrate, and the lateral motion due to the Coriolis force ($\overline{F_c} = -m \cdot \overline{a_c}$) can be measured:

$$z = \frac{2m\Omega e\omega\cos(\omega \cdot t)}{k} \tag{2}$$

where k is an elastic coefficient in the out of plane motion.

The initial and current orientation can be determined with the aid of MEMS gyroscopes [8]; in particular, we used MPU6050 (Fig. 3).

Fig. 3. MPU6050 MEMS gyroscope

The MPU6050 has programmable full-scale range of ±250 up to ±2000°/s (dps), thus it can be used for either fast or slow applications.

From a logical point of view (software), oversimplified as explanation, the microcontroller keeps polling data from the inertial measurement unit (IMU) and if the processed values lay outside a given range then an alarm is activated (Fig. 4).

The size of the device, having 87 × 36 × 14 mm, weighing 52 g (batteries included) can further be shrunk by using Li or LiPo batteries and using a dedicated controller with built in gyroscope (Fig. 5).

Besides actively warning any slips, the microcontroller also logs sensorial data throughout the day in order to offer a daily activity report. This report can be compared with previous ones and a trend can be determined.

3 Device Alerting

In order to make real-time posture corrections the correcting device must alert the user every time he/she/it has an incorrect posture. Flashing signals or shrieking sounds are not desired because the first one may be disregarded and second one may be disrupting. Thus the obvious choice is vibrating alarms. For this, coreless motor (motor without iron core) has been chosen as alerting element.

To have an efficient warning system, the vibration given by the motor was measured in various signal control modes. Three scenarios were considered having a fill factor of 50% (blue), 75% (red) and 100% (green) along 3 axes: X, Y and Z (Figs. 6, 7 and 8).

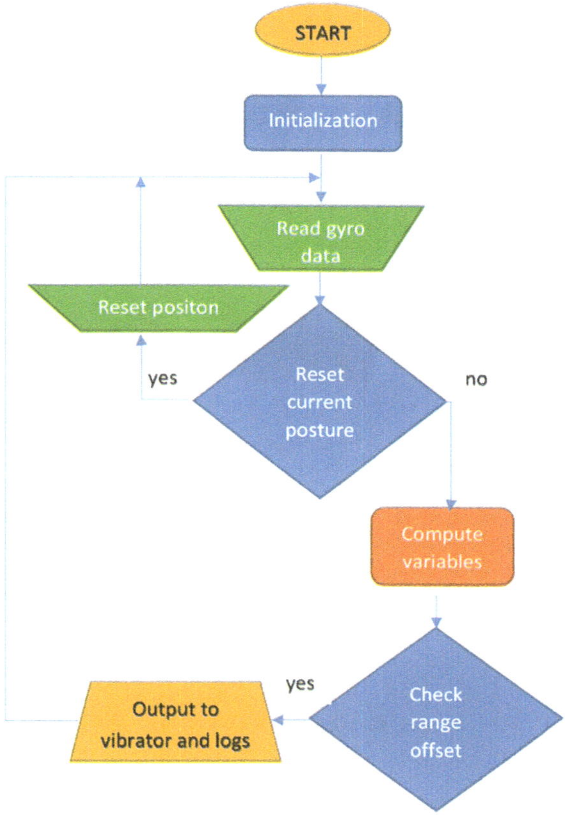

Fig. 4. Posture controller flow chart

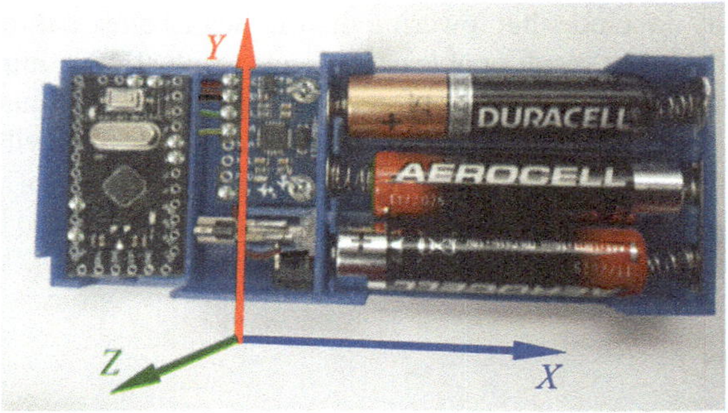

Fig. 5. Smart posture correction device (and measurement axes)

The measurements were made with a mobile device having a sampling frequency of 500 Hz, maximum sensor acceleration of 78 m/s^2 and a resolution of 0,0012 m/s^2.

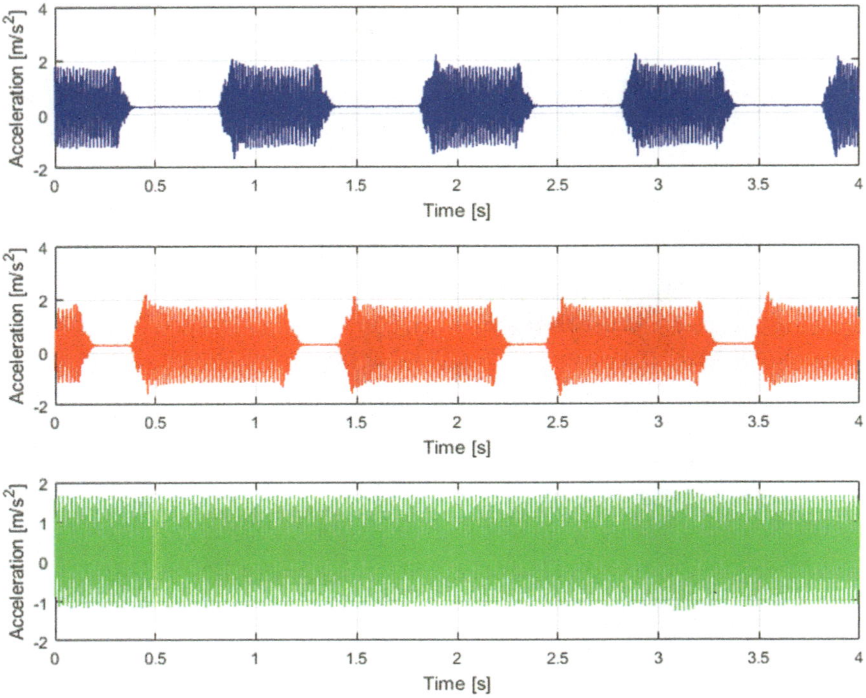

Fig. 6. Acceleration on X axis

With the acquired data, the power spectral density (PSD) was computed to choose the optimal alert control algorithm. As it can be observed in Figs. 9, 10 and 11, intermittent vibration offers a higher energy output than continuous ones at 100 Hz.

There is no noticeable difference between 50% fill factor or 75% fill factor. On this reasoning, one may conclude that lowering the fill factor below a threshold of 50% fill factor would yield a better result or at least the same result. Taking into account the fact that higher fill factors use more electric energy this reasoning is sane. But, there is a problem with the damping factors of the human flesh that attenuate vibrations and there need to be a significant amount of active duration for a vibration of this amplitude to be felt. The resonance frequency of standing humans is considered to be between 9 and 16 Hz and independent of mass, height and body mass index [9].

4 Discussions

Studies [10, 11] show that more and more people need some sort of postural corrective devices. The hypothesis is that the wearable sensor would assist a person in having a more upright posture than without the wearable sensor. Data regarding differences in performance with and without the wearable sensor are not part of this study.

The objective of this study was to investigate the susceptibility or body awareness to vibrations as effect of a wearable posture correction device. The novelty of this correction

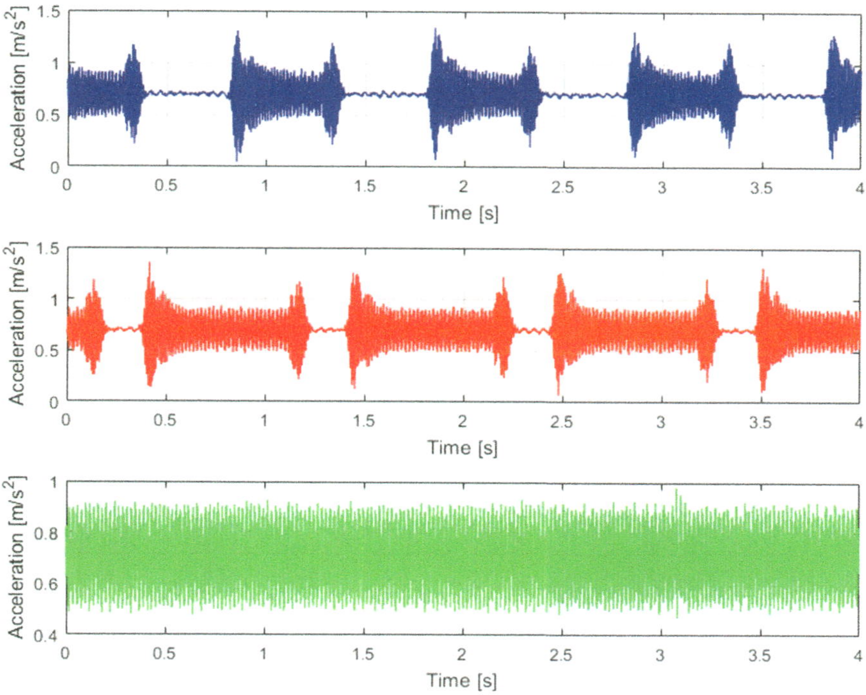

Fig. 7. Acceleration on Y axis

device consists in the usage of MEMS to detect angular position and alert the wearer being in bad posture for prolonged periods of time through vibrations. Other alerting signals, such as light or sounds were neglected for considerations of privacy.

The position of the sensor was chosen to be on the back of the wearer on the upper torso (thoracic vertebrae), but could also be positioned on lumbar vertebrae. Due to the frequent neck flexion, the cervical vertebrae were ruled out from adequate usage.

The postural corrector device is designed to function after a necessary angular reset after orthostatic positioning. Starting from this position a working range of $\pm 5^{\circ}$ is allowed and if this range is exceeded for a period of more than 5 s (programmable) an alarm is activated. The range of freedom can be modified in accordance with the type of activity and physician recommendation.

Fat tissue absorbs more vibration, and an exact correlation between such damping factor and body mass index is hard to determine [9, 12]. Taking into account the desire to have a small as possible postural corrector, a coreless motors with a mass of 2 g was used. This motor has a working frequency between 175 and 295 Hz. Having a voltage-dependent vibration amplitude, at 3 V the acceleration amplitude is around 0.86 g, the rotational frequency240 Hz and the absorption current is 250 mA. The undertaken control method was through pulse width modulation. Acceleration was measured along all 3 axes and for 50, 75 and 100% duty cycle in order to determine the power spectral

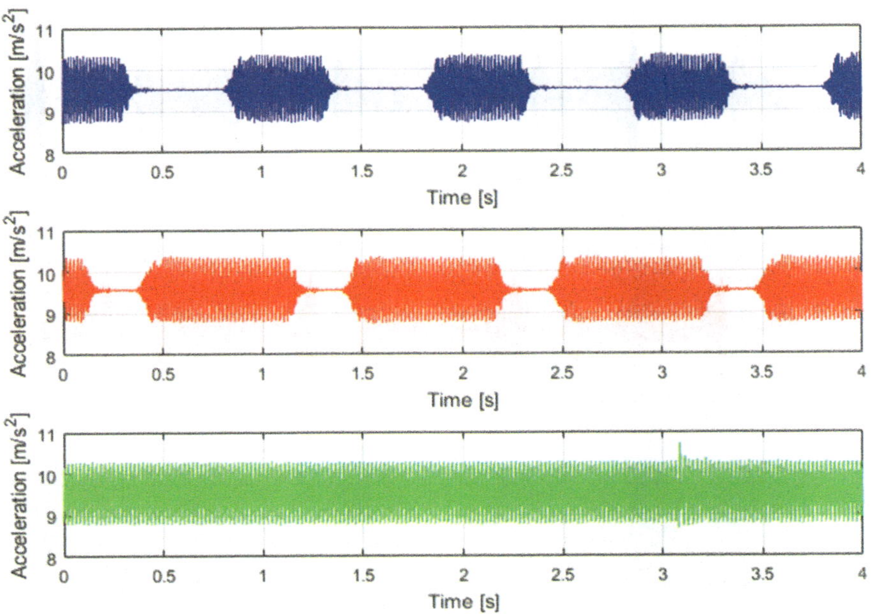

Fig. 8. Acceleration on Z axis

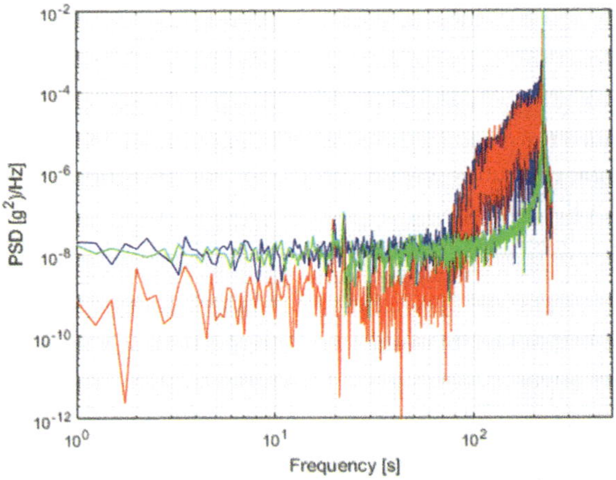

Fig. 9. Power spectral density on X axis

density. The best fit was a 100% duty cycle. Although the least energy efficient, it has the frequency spectra nearest to the human resonance [9].

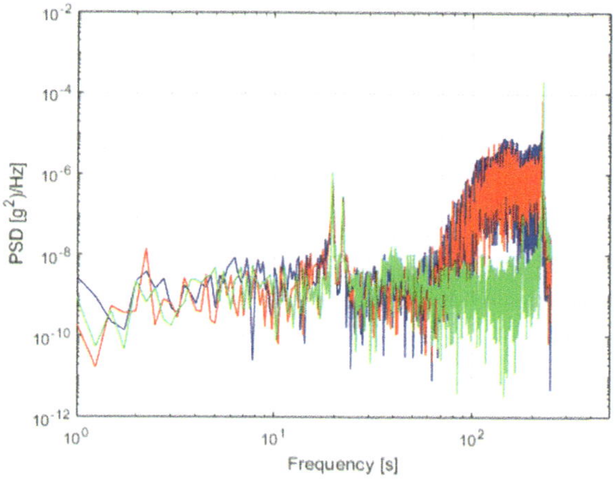

Fig. 10. Power spectral density on Y axis

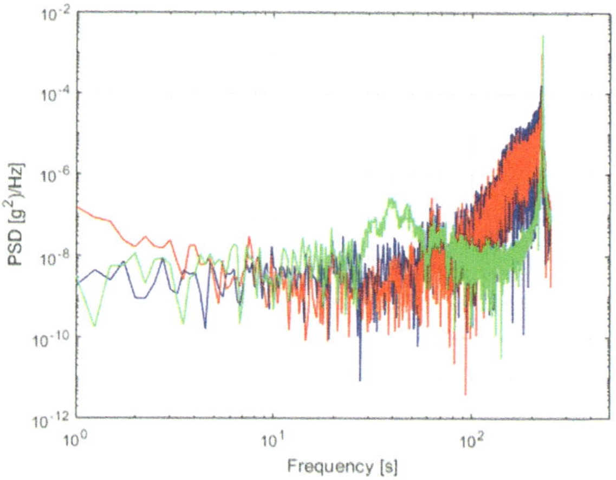

Fig. 11. Power spectral density on Z axis

5 Conclusions

In this paper, a vibration analysis was presented in order to choose the best silent alert method for a smart posture corrector device.

Although battery life is inversely proportional to the fill factor of the vibrator signal, lowering the fill factor too much will lead to a lower PSD in the domain of the human standing resonant frequencies, thus a higher fill factor for the alerting signal is desired.

Being unable to quantify the human flesh vibration damping, the highest PSD was chosen with a fill factor of 100% over the lower ones.

This project started as an alternative to posture analysis and corrective control methods. Similar ideas have risen in the last year(s) throughout the world, this shows the importance of such corrective devices and the struggle to design and develop smart spinal tools which will be necessary in the coming years.

References

1. Matuska, S., Paralic, M., Hudec, R.: A smart system for sitting posture detection based on force sensors and mobile application. Mob. Inf. Syst. (2020)
2. Stoia, D.I., Vigaru, C., Rusu, L.: Relative and absolute angles computed from pathologic gait data. In: 14th International Conference on Acoustics and Vibration of Mechanical Structures (AVMS) (2017)
3. Ailneni, R.C., Syamala, K.R., Kim, I.S., Hwang, J.J.: Influence of the wearable posture correction sensor on head and neck posture: sitting and standing workstations. Work-A J Prev Assess Rehabil **62**, 27–35 (2019). https://doi.org/10.3233/WOR-162839
4. Lakshmi, S., Kumar, A., Aravindan, C.: Posturector-the posture corrector. Int. J. Innov. Sci. Res. Technol. **2**(3) (2017)
5. Flutur, G., Movileanu, B., et al.: Smart chair system for posture correction. In: 2019 22nd Euromicro conference on digital system design (DSD), pp. 436–441 (2019). https://doi.org/10.1109/DSD.2019.00069
6. Chung, H., Chung, Y., Liang, C.: Design and implementation of a novel system for correcting posture through the use of a wearable necklace sensor. JMIR Mhealth Uhealth, **7**(5), e12293. https://doi.org/10.2196/12293. PMID: 31140439, PMCID: 6660123
7. https://www.maximintegrated.com/en/design/technical-documents/app-notes/5/5830.html
8. Karanth, G., Pentapati, N., Gupta, S., Ravish, R.: IoT based smart posture detector. In: Mandal, J., Mukhopadhyay, S., (eds.), Proceedings of the global AI congress 2019. Advances in intelligent systems and computing, vol. 1112. Springer, Singapore. https://doi.org/10.1007/978-981-15-2188-1_16
9. Randall, J.M., Matthews, R.T., Stiles, M.A.: Resonant frequency of standing humans. Ergonomics **40**(9), 879–886 (1997)
10. Balkó, S., Balkó, I., Valter3 L., Jelínek, M.: Influence of physical activities on the posture in 10–11 year old schoolchildren. J. Phys. Educ. Sport (JPES) **17**(1), 101–106, Art 16 (2017). https://doi.org/10.7752/jpes.2017.s1016
11. Silva, S., Queirós, S., Moreira, A.H., Oliveira, E., Rodrigues, N.F., Vilaça, J.L.: Classification algorithms for body posture. In: 2017 IEEE 5th International Conference on Serious Games and Applications for Health (SeGAH), pp. 1–6 (2017). https://doi.org/10.1109/SeGAH.2017.7939299
12. Lofvenberg, J., Johansson, R.S.: Regional differences and interindividual variability in sensitivity to vibration in the glabrous skin of the human hand. Brain Res. **301**(1), 65–72 (1984)

Vibratory Analysis of a Robotic Hand

Ştefan Dumitru, Andrei Craifaleanu(✉), and Roxana Alexandra Petre

University "Politehnica" of Bucharest, Bucharest, Romania

Abstract. The paper presents an anthropomorphic robotic hand which is to become part of a robot manipulator. The real human hand is a complex mechanism, made of 27 bones, out of which 19 can move relative to each other, therefore creating a mechanical system with 25 degrees of freedom. The practical implementation of a mechanical hand with such a high number of degrees of freedom causes major technological difficulties, hence robotic hands are often made in simpler configurations than those of a human hand. The studied device presented in the paper is such a simplified model. The functionality of the device imposes certain requirements on the accuracy of motions that can be affected by vibrations induced by various mechanical interactions. By considering first the fingers as independent systems, and, afterwards, the entire hand mechanism, with five fingers, a number of eigenfrequencies and eigenmodes of the device are determined. The paper presents also the variation curves showing the influence of the system configuration on the eigenfrequencies.

Keywords: Anthropomorphic robotic hand · Eigenfrequency · Eigenmode · Vibration

1 Introduction

The current trend of industrial process automation requires extensive use of robots with different functional characteristics. Among them, the anthropomorphic robots and, in particular, the robotic hands, stand out due to their versatility.

The term "robot" was created in 1921 by Karel Čapek, a Czechoslovakian playwriter for his play "Rossum's Universal Robots", where humans create mechanical devices to serve as workers. The term "robotics" was introduced by science fiction author Isaac Asimov in his 1942 short story "Runaround", where he formulated his concept of "The Three Laws of Robotics" [1–3].

By the year 1950 engineers were developing machines to handle difficult or dangerous repetitive tasks for both defense and consumer manufacturing. Since robots were meant to replicate the pattern of motions that a human would make while lifting, pulling, pressing, or pushing, some of the designs were based on the anatomical structure of a human arm [1–3].

The functionality of these devices is still conditioned by the precision of the mechanical motions, which can be severely affected by the vibrations induced by various mechanical interactions [4–7].

N. Herisanu and V. Marinca (eds.), *Acoustics and Vibration of Mechanical Structures – AVMS-2021*, Springer Proceedings in Physics 274,
https://doi.org/10.1007/978-3-030-96787-1_24

This paper studies the vibrations of an anthropomorphic robotic hand that is to become part of a robot manipulator, intended for repetitive mechanical operation, for current use (Fig. 1).

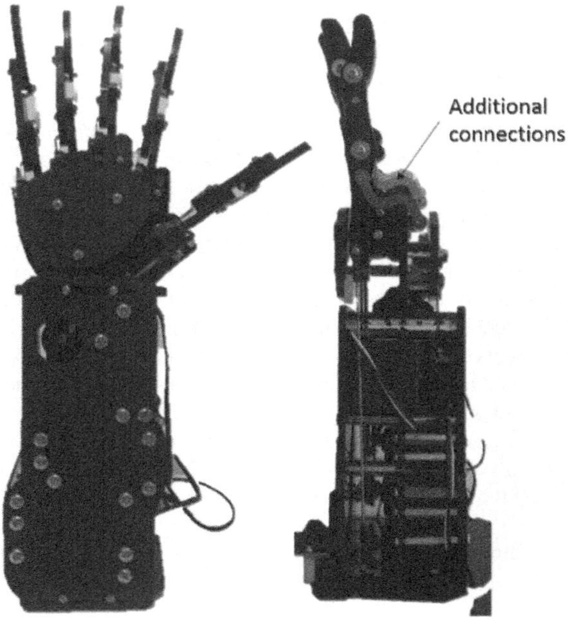

Fig. 1. Robotic hand.

The robotic hand is modeled as a system of elastic bodies with various geometric configurations and is analyzed from the vibratory point of view, using the CATIA V5 software.

2 Configuration of the Robotic Hand

The real human hand is a complex mechanism, made of 27 bones, out of which 19 can move relative to each other, therefore creating a mechanical system with 25 degrees of freedom [8, 9].

The practical implementation of a mechanical hand with a high number of degrees of freedom leads to major technological difficulties, such as: increasing the complexity of the system, using a high number of actuators and electronic control components, increasing the system mass, and need of large amounts of energy in order to operate.

As a result, robotic hands are often made in simpler configurations than those of a human hand, with a smaller number of mobilities, the motion of certain elements being correlated, by using additional connections. Such devices cannot perform motions as complex as those of a real hand, but they can provide a functional configuration close to the real one, enough for a large number of practical applications.

The studied device is such a simplified model, with only five mobilities (Fig. 2). The phalanges are modeled as elastic bodies connected by hinge joints.

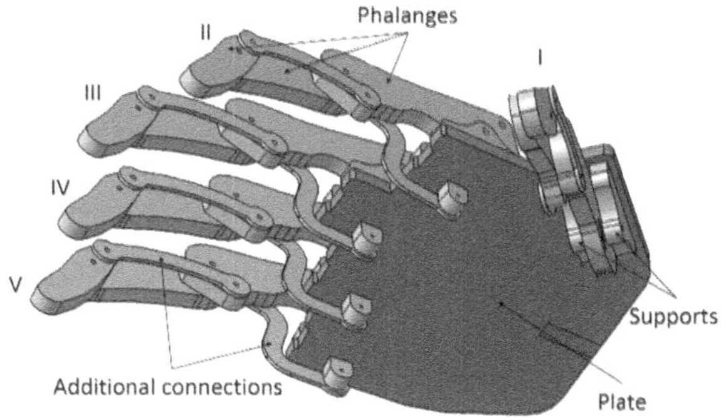

Fig. 2. Robotic hand 3D model.

The constructive characteristics of the studied robotic hand are presented in Fig. 3, while the material characteristics are given in Table 1.

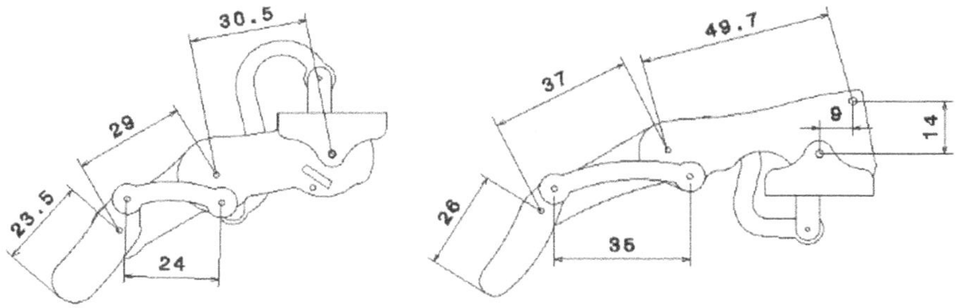

Fig. 3. Constructive characteristics [mm] of finger I (left) and fingers II–V (right).

Table 1. Material characteristics of the robotic hand elements

Element	Material	Density (kg/m^3)	Young's modulus (N/m^2)	Poisson ratio
Phalanges, supports, plate	Wood	610	1.4×10^{10}	0.45
Additional connections	Duralumin	2710	7×10^{10}	0.346

3 Numerical Study and Results

The free vibrations of the robotic hand were analyzed using the CATIA V5 software, separately for the following case studies:

1. finger I (thumb), considered as an independent system (Fig. 4—left);

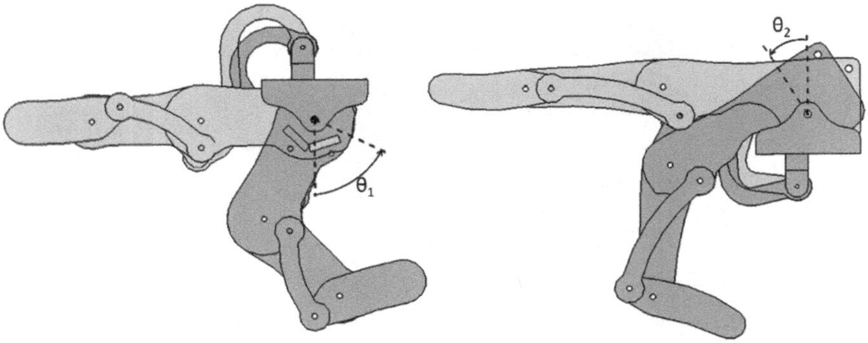

Fig. 4. Finger I (left) and fingers II–V (right), considered as independent systems.

2. one of the fingers II–V, which are presumed identical, considered as an independent system (Fig. 4—right);
3. the whole hand, considering the fingers fixed on the plate that models the carpals and metacarpals (Fig. 2).

   All three cases were analyzed for two configurations (Fig. 4), expressed by angle θ_1 for the thumb and θ_2 for the other finger:

- extended phalanges (light gray)—$\theta_1 = 0°$, $\theta_2 = 0°$.
- flexed phalanges (dark gray)—$\theta_1 = 64°$, $\theta_2 = 32°$.

Supplementary, the fingers as independent systems, were also considered in seven intermediate configurations between the above two, with the steps $\Delta\theta_1 = 8°$ and $\Delta\theta_2 = 4°$, respectively.

Table 2 presents the first five eigenfrequencies[10] of the extended and flexed configurations.

Figure 5 illustrates the variation of the first two eigenfrequencies (f_1 and f_2) of the thumb and of the other finger, considered as independent systems, with respect to angle θ_1 and θ_2, respectively.

Figures 6, 7, 8, 9, 10, 11, 12 and 13 illustrate some of the determined eigenmodes.

Table 2. First five eigenfrequencies (Hz)

Eigenfrequency	f_1	f_2	f_3	f_4	f_5
Extended finger I	433	607	1291	2208	2351
Extended finger II–V	344	559	1864	2548	2631
Extended hand	298	298	298	298	371
Flexed finger I	557	634	1259	1384	1891
Flexed finger II–V	396	641	1203	2100	2343
Flexed hand	338	338	338	338	414

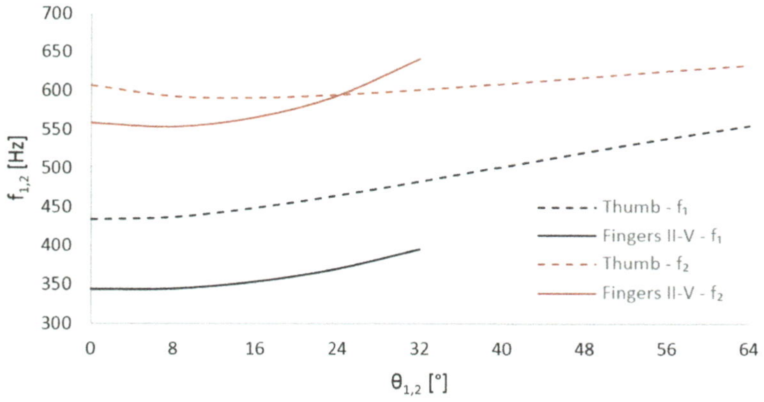

Fig. 5. Variation of the first two eigenfrequencies with respect to angles θ_1 and θ_2.

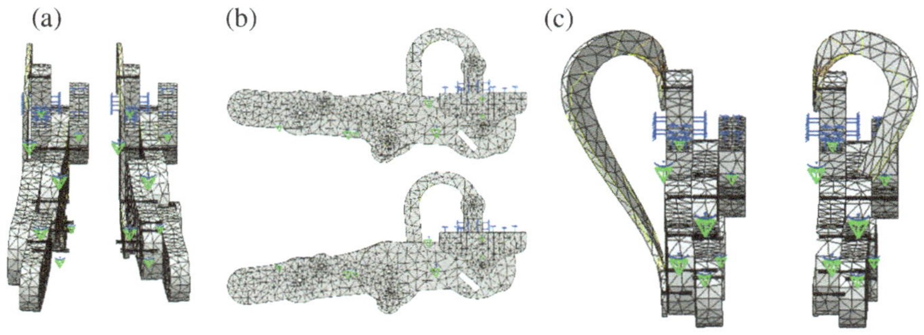

Fig. 6. Extended finger I (thumb)—eigenmodes 1, 2, 3.

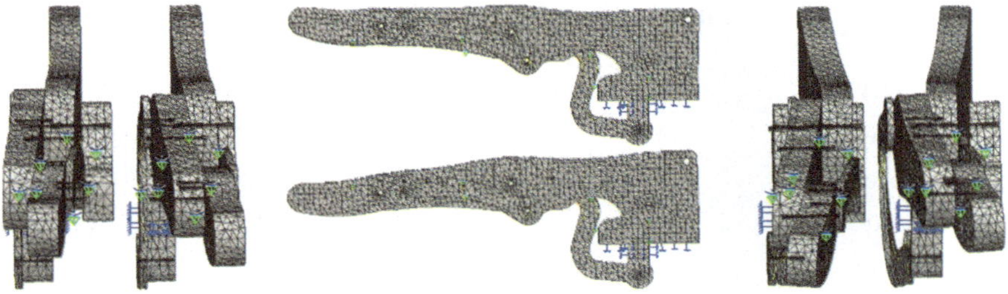

Fig. 7. Extended fingers II–V—eigenmodes 1, 2, 3.

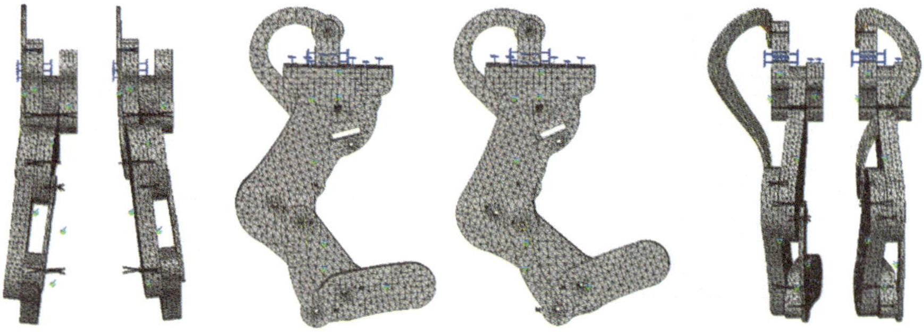

Fig. 8. Flexed finger I (thumb)—eigenmodes 1, 2, 3.

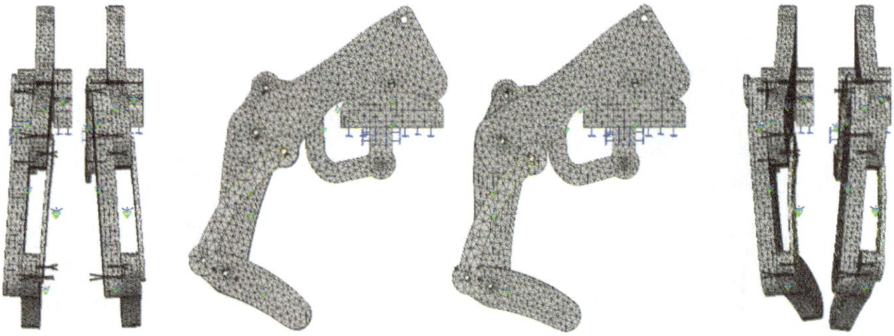

Fig. 9. Flexed fingers II–V—eigenmodes 1, 2, 3.

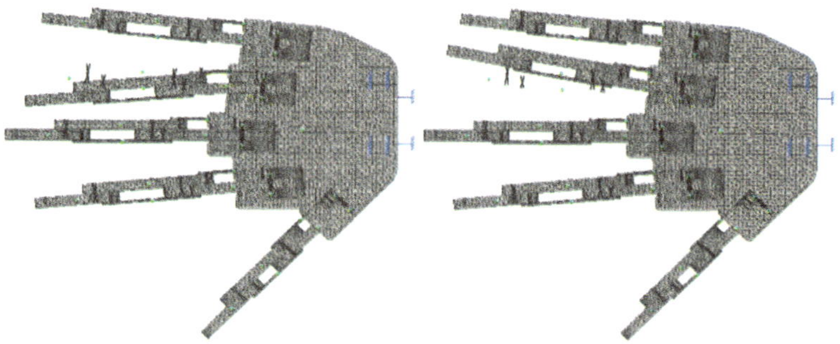

Fig. 10. Extended hand—eigenmode 1.

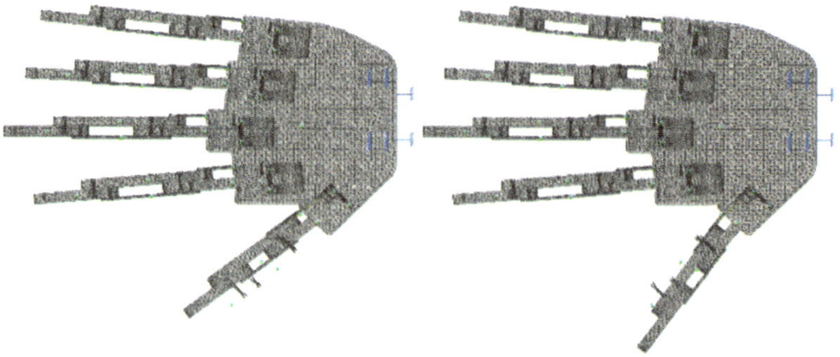

Fig. 11. Extended hand—eigenmode 5.

Fig. 12. Flexed hand—eigenmode 1.

Fig. 13. Flexed hand—eigenmode 5.

4 Conclusions

The vibratory analysis of the anthropomorphic robotic hand using the CATIA V5 software leads to the following conclusions:

1. the first eigenmode of all five fingers, considered as independent systems, is a bending one, in a plane parallel to that of the plate;
2. the second eigenmode of all five fingers, considered independent systems, is a bending one in a plane perpendicular to that of the plate;
3. the higher eigenmodes show coupled deformations and different aspects from one configuration to another; in some cases, the oscillations of the additional connections are prevailing;
4. the first two eigenfrequencies are higher for finger I (thumb), than for fingers II–V, which can be explained by the size difference between the first finger, which is smaller, and the other four. This is not valid for the higher modes, which can be explained by their different shape, from one configuration to another;
5. the first two eigenfrequencies, generally, tend to increase with the flexing degree (expressed by parameters θ_1 and θ_2, respectively), which is explained by the smaller general size of the system. This property is also not valid for higher eigenmodes;
6. for the first four eigenmodes of the hand, the bending of fingers II–V in a plane parallel to that of the plate is predominant, and the corresponding eigenfrequencies are very close in terms of values (coinciding at the displayed precision), due to the identical characteristics of the respective fingers; these modes correspond to the first vibration mode of fingers II–V, considered as independent systems;
7. for the fifth eigenmode of the hand, the bending of finger I (thumb) in a plane parallel to that of the plate is predominant; this mode corresponds to the first eigenmode of the finger I.

The above conclusions can be used for:

- the design and the optimization of a robotic hand;
- determining the optimal working parameters of such a device.

References

1. Matarić, M.J.: The robotics primer. The MIT Press, Cambridge (2007)
2. Murray, R.M., Li, Z., Sastry, S.: A mathematical introduction to robotic manipulation. CRC Press Inc., Boca Raton (1994)
3. Conrad, J.M., Mills, J.W.: Stiquito for beginners: an introduction to robotics, 1st edn. Wiley-IEEE Computer Society Pr (1992)
4. Masuda, T., Futakawa, A., Koga, E., Inoue, K.: A parametric study of robot arm vibration by simulation analysis. Adv. Robot. **1**(4), 325–342 (1986)
5. Tsetserukou, D., Kawakami, N., Tachi, S.: Vibration damping control of robot arm intended for service application in human environment. In: 2008 8th IEEE-RAS International Conference on Humanoid Robots, pp. 441–446 (2008)
6. Thomsen, D.K., Søe-Knudsen, R., Balling, O., Zhang, X.: Vibration control of industrial robot arms by multi-mode time-varying input shaping. Mech. Mach. Theory **155** (2021)
7. Wang, P.K.C., Wei, J.-D.: Vibrations in a moving flexible robot arm. J. Sound Vib. **116**(1), 149–160 (1987)
8. Abdel-Malek, K., Yang, J., Marler, T., Beck, S., Mathai, A., Zhou, X., Patrik, A., Arora, J.: Towards a new generation of virtual humans. Int. J. Hum. Factors Model. Simul. **1**(1), 2–39 (2006)
9. Săvescu, A.V., Cheze, L., Xuguang, W., Beurier, G., Verriest, J.P.: A 25 degrees of freedom hand geometrical model for better hand attitude simulation. Trans. J. Aerosp. **113**, 270–275, section 1 (2004)
10. Meirovitch, L.: Fundamentals of vibrations. McGraw-Hill, New York (2001)

Vibration Problems in Industrial Processes

Experimental Active Vibration Control of a Highly Flexible Composite Manipulator with Acceleration Feedback

Mehmet Uyar[1,2](✉), Levent Malgaca[2], Ş.İpek Lök[2], and Seda Vatan Can[2,3]

[1] Bayburt University, Bayburt, Turkey
muyar@bayburt.edu.tr
[2] Dokuz Eylül University, İzmir, Turkey
[3] Manisa Celal Bayar University, Manisa, Turkey

Abstract. In this study, vibration control of a single-link flexible smart composite manipulator (FSCM) is experimentally studied. The experimental system includes a real-time data acquisition control card, accelerometer, amplifier, servo motor, and driver. The FSCM is an epoxy-glass composite manipulator with the piezoelectric actuator and has the orientation of [0/90] lay-up. Modal analysis is experimentally conducted to find the natural frequencies by applying a chirp signal to the piezoelectric actuator. Natural frequencies are found by taking the Fast Fourier Transform (FFT) in MATLAB. Triangular motion profiles are used to drive the FSCM. Motion parameters are found by utilizing the natural frequencies of the system for different deceleration times (t_{dec}). with different deceleration times (t_{dec}). Residual vibrations are suppressed by using PID control with acceleration feedback. The acceleration signals at the tip point are measured by utilizing the accelerometer. The controller gains are determined with the trial and error method. Experimental residual vibration results are presented for the active control. It is observed that the suppression of vibration amplitudes further increases with active control.

Keywords: Smart composite manipulator · Active vibration control · Acceleration feedback

1 Introduction

Flexible manipulators are widely used in many fields such as manufacturing, automation, robotics due to their lightweight, low energy consumption, and easy transportation [1, 2]. Flexible manipulators can be produced from aluminum, steel or composite structures. Composite manipulators have advantages such as lighter weight and higher payload/weight ratio. However, besides advantages, they have some disadvantages such as residual vibrations at the tip point of the manipulators because of their flexibilities. Unwanted residual vibrations of the manipulators must be eliminated or reduced to increase repeatability and performance at high speeds.

N. Herisanu and V. Marinca (eds.), *Acoustics and Vibration of Mechanical Structures – AVMS-2021*, Springer Proceedings in Physics 274,
https://doi.org/10.1007/978-3-030-96787-1_25

Vibration control of flexible manipulators is widely studied in the literature. Active control method is used for vibration suppression of the flexible manipulators. Suitable sensors and actuators are used in active control. Vibration control of composite beam and manipulators is a new research topic. Literature survey of vibration control of composite beam and manipulators is described below.

Mitra et al. investigated vibration control of a composite box beam using piezoelectric patches as an actuator [3]. They applied a PI controller using acceleration feedback via simulation and experiments. Active vibration control of laminated composite box structure containing piezoelectric actuators and sensors was studied via simulation in [4]. Passive vibration control of a flexible composite manipulator with [45/−45] and [0/90] lay-ups was studied with triangular and trapezoidal velocity profiles in [5]. A novel method was presented in [6] to suppress residual vibrations of a single-link flexible glass fabric reinforced epoxy-glass composite manipulator. An exponential harmonic velocity excitation was added to the end of the driving signal of the manipulator which is trapezoidal velocity profile and residual vibrations were reduced in simulation and experiment. Passive and active control of a composite box manipulator with [0/90] lay-up was investigated with displacement and strain feedback control in [7]. Residual vibrations of a flexible active composite manipulator were reduced with piezoelectric patch actuators and sensors in [8].

In this paper, active vibration control (AVC) of a single-link flexible smart composite manipulator (FSCM) is studied experimentally. PID controller with acceleration feedback is used to suppress residual vibrations of the FSCM. Control signals are applied to the piezoelectric actuator. Residual vibrations of the FSCM are suppressed successfully in the experiment.

2 Experimental Analysis of Active Vibration Control

The experimental system of a single-link flexible smart composite manipulator (FSCM) is established as shown in Fig. 1.

The experimental system consists of a FSCM, a servo motor and driver, a multifunction I/O data acquisition (DAQ) control card, a wireless accelerometer sensor (WAS) and base station (WBS), and a personal computer (PC). The multifunctional DAQ control card of PCI-e 6363 model of National Instruments [9] and G-Link WAS and WBS of Microstrain Company [10] are utilized in the experiment. The servo motor is a Mitsubishi-Electric servo motor with 200 W, Model HC-KFS23B/ MR-J2S-20A, and its driver is employed. The FSCM occurs from a composite manipulator with the piezoelectric (PZT) material patched on the surface of it. The FSCM used in the experiment is produced from epoxy-glass material and has the orientation of [0/90] lay-ups. The FSCM has the material and geometric properties given in Table 1.

The PZT material is utilized as an actuator to reduce the residual vibrations in the AVC. The PZT actuator is located at 5 mm distance from the root. The lower surface of the PZT actuator on the manipulator is connected to the ground of the control voltage while the upper surface of the PZT actuator is defined as the actuation voltage. The actuation voltage is transmitted to the PZT actuator by using a high voltage amplifier.

In this experiment system, the WAS is attached to the tip point of the manipulator. The acceleration signals at the endpoint of the FSCM are measured with the WAS. The

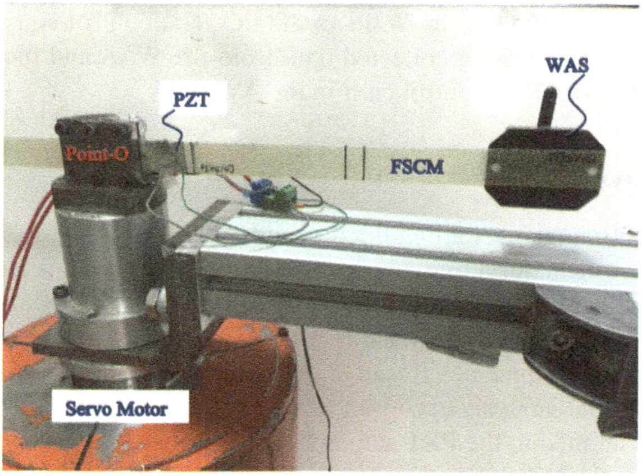

Fig. 1. Flexible smart composite manipulator

Table 1. Dimension and material properties of the FSCM

Epoxy-glass composite structure	Piezoelectric material, PZT-5H
$\rho = 1600$ kg/m^3	$\rho = 7500$ kg/m^3
Young modulus (GPa)	Elastic stiffness matrix (N/m^2)
E1 = 40.51	C11 = 126.0×10^9
E2 = 13.96	C12 = 79.50×10^9
E3 = 13.96	C13 = 84.10×10^9
Poisson ratio	C33 = 117.0×10^9
$\nu1 = 0.22$	C44 = 23.30×10^9
$\nu2 = 0.22$	Piezoelectric strain matrix (C/m^2)
$\nu3 = 0.11$	E31 = –6.5
Shear modulus (GPa)	E33 = 23.3
G1 = 3.1	E15 = 17.0
G2 = 3.1	Dielectric matrix (F/m)
G3 = 1.5	$\varepsilon11 = 15.03 \times 10^{-9}$
	$\varepsilon22 = 15.03 \times 10^{-9}$
	$\varepsilon33 = 13.00 \times 10^{-9}$
Numerical values of system	
Length × Height × Thickness of composite link	270 mm × 20 mm × 0.85 mm
Length × Height × Thickness of piezoelectric material	25 mm × 20 mm × 1 mm
Distance of receiving point from the center of sensor	30 mm

acquired data is transferred via the WBS to the PC. The control signal is produced in MATLAB/Simulink concerning collected data from the WAS and then, transmitted to the FSCM by using the DAQ control card in the AVC.

2.1 Modal Analysis

The triangular motion profiles are defined to move the manipulator. To determine the motion parameters, the natural frequencies of the system are used. Therefore, the experimental modal analysis is performed to find the natural frequencies and motion parameters. Modal analysis is realized by using a chirp signal applied to the PZT actuator. The initial and target frequencies of the chirp signal are selected as 0.01 and 100 Hz. The duration of modal analysis is chosen as 70 s. The generated chirp signal in MATLAB/Simulink is applied to the PZT actuator. The endpoint vibrations are measured by using the WAS via the WBS. Fast Fourier Transform (FFT) is enforced to the acceleration signal to find the natural frequencies. The frequency and acceleration responses in the experimental modal analysis are obtained as given in Fig. 2.

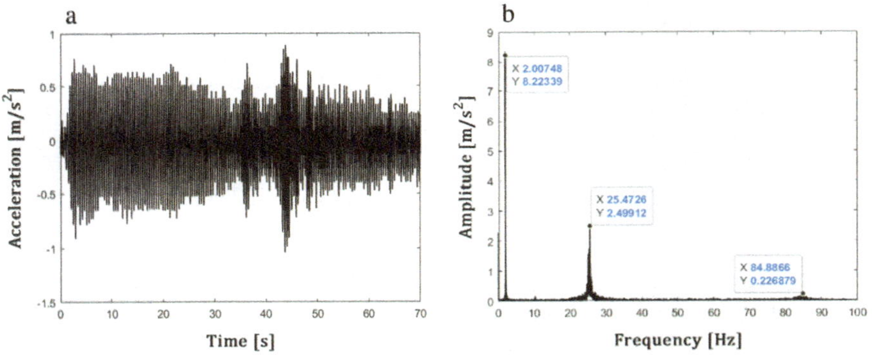

Fig. 2. **a** Vibration and **b** FFT responses.

As shown in Fig. 2, the experimental natural frequencies are found as 2.007, 25.47, and 84.89 for the first three vibration modes. The time parameters of motion profiles are calculated from the modal analysis results.

2.2 Motion

The triangular motion profile given in Fig. 3 is described to drive the FSCM with the servo motor at Point-O.

Figure 3 shows the triangular velocity profile with the time and velocity parameters. In the motion profile, time variables t_{acc}, t_{dec}, and t_m are described as the acceleration, constant, deceleration, and motion times, respectively. Also, ω_0 is the maximum angular velocity. The triangular velocity parameters are determined according to the first natural frequency obtained from the modal analysis. The relationship between the velocity

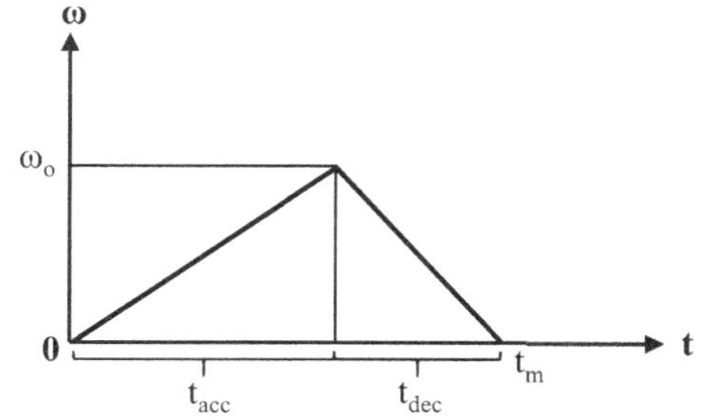

Fig. 3. Triangular velocity profile

profile and motion parameters is given with the following equations.

$$\omega_o = \frac{\varphi_m}{(0.5t_{\text{acc}} + 0.5t_{\text{dec}})} \tag{1}$$

$$\omega_m = \frac{\omega_o t}{t_{\text{acc}}} \text{for } 0 < t < t_{\text{acc}} \tag{2}$$

$$\omega_m = \frac{\omega_o(t_m - t)}{(t_m - t_{\text{acc}})} \text{ for } t_{\text{acc}} < t < t_m, \ \omega_m = 0 \text{ for } t \geq t_m \tag{3}$$

where φ_m is the angular position of the servo motor and ω_{m} is the angular velocity. The motion parameters for angular displacement 60° and motion time $t_m = 1.25$ s are listed in Table 2.

Table 2. Motion parameters for triangular velocity profile

Motion Profiles	θ (°)	*t*acc (s)	*t*dec (s)	*tm* (s)	ωmax (rad/s)
MP1	60	1.0009	0.2491	1.25	1.6755
MP2		0.7517	0.4983		
MP3		0.5026	0.7474		
MP4		0.2535	0.9965		

As seen from Table 2, different motion profiles are created with the deceleration time and velocity variables. The motion profiles that occurred from different velocity parameters are applied to the FSCM via the servo motor. The FSCM moves from a rest position at $t = 0$ to a stopping position φ_m at $t = t_m$. Then, the AVC with the acceleration feedback PD control is started to reduce the residual vibrations after the manipulator has completed the movement of the FSCM at $t = t_m$.

2.3 Active Control

For the AVC, the closed-loop block diagram is shown in Fig. 4.

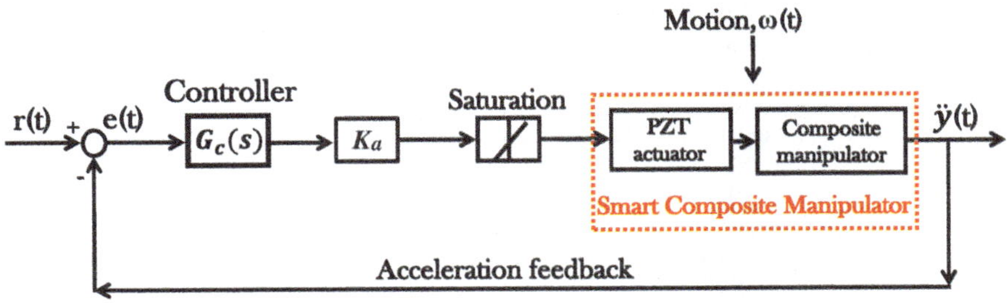

Fig. 4. Closed-loop block diagram

The FSCM is driven with the given motion profiles and then, the acceleration signal $\ddot{y}(t)$ at the endpoint is obtained by using the WAS. $\ddot{y}(t)$ is used as feedback. Closed-loop control is performed by using $\ddot{y}(t)$ in the AVC. The error signal $e(t)$ can be calculated as $e(t) = r(t)\text{-}\ddot{y}(t)$ where $r(t)$ is selected as reference input that is zero. $e(t)$ is first multiplied by Kp, and its derivative is multiplied by K_d. Then, the control signal is found by summing these two signals. The actuation signal $V_a(t)$ is obtained by multiplying the power amplification gain K_a with the control signal. The value $V_a(t)$ is applied to the PZT actuator to suppress the residual vibration of the FSCM.

Because of the limitation of the experimental control system, saturation is considered for $V_a(t)$. The limit of analog output of DAQ card is ±10 V. Therefore, the set value of saturation in the experiment is chosen as ±9 V.

3 Results and Discussion

The experimental results for the AVC are obtained with the defined motion profiles. For the motion profiles, the uncontrolled and controlled vibration responses with the control signals are shown in Fig. 5.

As seen from Fig. 5, the experimental results are presented for the AVC. The acceleration and actuation responses are obtained with the various PD gains for MP_1, MP_2, MP_3, and MP_4. The gains of the PD controller are determined with the trial and error method. To evaluate the performance of the AVC, the percent reduction ratios and root means square (RMS) values of the residual vibrations are calculated to indicate the reduction ratio over the RMS values of the non-controlled and controlled responses. The time of residual vibrations is chosen as 10 s. The RMS values and reduction ratios for the corresponding motion profiles and the controller gains are given in Table 3.

Table 3 gives the RMS values and reduction ratios of residual vibrations obtained from experiments. The reduction ratios are determined by using the RMS values of the non-controlled and controlled residual vibrations. The RMS values of the control signal are calculated by using the actuation signal. The RMS values of non-controlled responses are considered as reference values for comparison. As seen from Table 3, maximum reduction ratios are found as 35.76%, 37.73%, 36.73%, and 27.52% for MP_1, MP_2, MP_3, and MP_4, respectively. When the reduction ratios are compared, the best result for the AVC is obtained as 37.73% reduction ratio for MP_2.

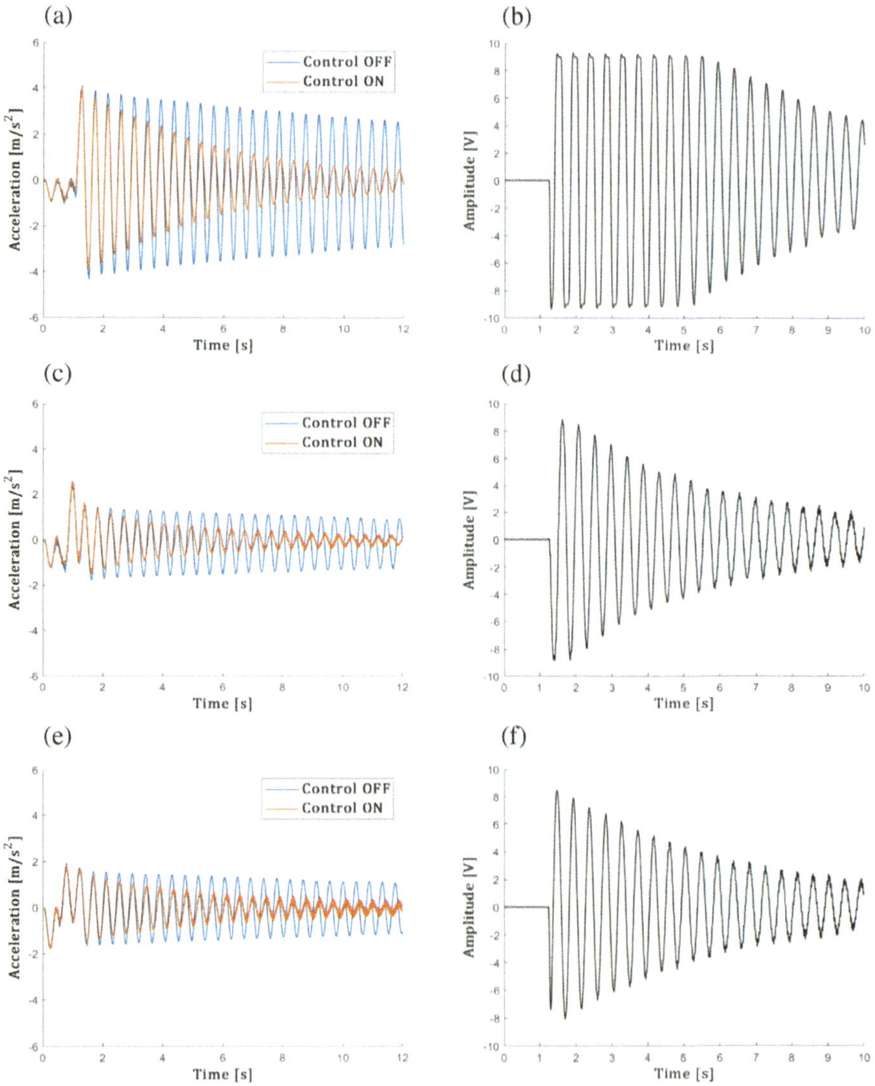

Fig. 5. Example controlled and uncontrolled **a**–**c**–**e** vibration responses and **b**–**d**–**f** control signals, MP1 for (**a**) and (**b**); MP2 for (**c**) and (**d**); MP3 for (**e**) and (**f**)

Table 3. RMS and reduction values for the AVC

Motion profiles	Controller gains	AVC		Control signal (VRMS)
	Kp	RMS	Reduction (%)	
MP1	0	2.3258	–	–
	2	1.8257	21.50	3.5801
	5	1.4942	35.76	5.6830
MP2	0	0.9663	–	–
	5	0.6229	35.54	2.5810
	7	0.6017	37.73	3.2763
MP3	0	0.9875	–	–
	5	0.6875	30.38	2.9024
	6	0.6248	36.73	3.0872
MP4	0	2.1971	–	–
	3	1.5925	27.52	3.9787

As indicated in Table 3, the maximum RMS values of actuation responses are increased from 0 V to 5.68 V, 3.27 V, 3.08 V, to 3.97 V for MP_1, MP_2, MP_3, and MP_4, respectively. It is observed that the maximum reduction ratio is calculated for MP_2 while the maximum RMS value of control voltage is obtained as 5.68 V for MP_1. The minimum RMS value of actuation signals is calculated for MP_3. If the PD controller gains can be optimized considering the stability and maximum operating voltage, it may be possible to increase the efficiency of the actuator.

4 Conclusion

In this study, the suppression of residual vibrations of an FSCM is experimentally studied with the active control using the PD controller. The manipulator is moved with a triangular motion profile. The parameters of motion profiles generated for motion time of 1.25 s and stopping position of 60° are created with the experimental modal analysis. Then, AVC is conducted with the acceleration feedback. Experimental results are presented with various PD gains. The effect of the motion parameters and various PD controller gains are investigated. The residual vibrations of the FSCM are successfully suppressed by using the PD controller.

Acknowledgements. This study is supported by Dokuz Eylül University—Department of Scientific Research Projects with the research project with project number 2020.KB.FEN.009. The authors offer their special thanks to Dokuz Eylül University Research Fund for giving support.

References

1. Tokhi, M.O., Azad, A.K.M.: Flexible robot manipulators modelling, simulation and control, vol. 86. The Institution of Engineering and Technology (2017)
2. Rigatos, G., Busawon, K.: Robotic manipulators and vehicles: control, estimation and filtering. Springer International Publishing AG (2018)
3. Mitra, M., Gopalakrishnan, S., Bhat, M.S.: Vibration control in a composite box beam with piezoelectric actuators. Smart Mater. Struct. **13**(4), 676–690 (2004)
4. Ramkumar, K., Chandran, S., Ganesan, N.: Finite element analysis of composite box structure containing piezoelectric actuators and sensors for active vibration control. In: SAE Technical Paper, pp. 1–6 (2009)
5. Yavuz, S., Malgaca, L., Karagülle, H.: Vibration control of a single-link flexible composite manipulator. Compos. Struct. **140**, 684–691 (2016)
6. Yavuz, S.: An enhanced method to control the residual vibrations of a single-link flexible glass fabric reinforced epoxy-glass composite manipulator. Compos. B Eng. **159**, 405–417 (2019)
7. Malgaca, L., Uyar, M.: Hybrid vibration control of a flexible composite box cross-sectional manipulator with piezoelectric actuators. Compos. Part B: Eng. **176**(2) (2019)
8. Ma, K., Ghasemi-Nejhad, M.N.: Adaptive control of flexible active composite manipulators driven by piezoelectric patches and active struts with dead zones. IEEE Trans. Control Syst. Technol. **16**(5), 897–907 (2008)
9. National Intruments Inc.: https://www.ni.com. Accessed 01 Apr 2021
10. MicroStrain Inc.: http://www.microstrain.com/wireless/sensors. Accessed 01 Apr 2021

Artificial Neural Network Approach to Extension of the Frequency Range for Experimental Determination of Dispersion Relationship Using Sparse Spatial Data

Jelena Tomić, Vladimir Sinđelić, Snežana Ćirić Kostić, Nebojša Bogojević, and Zlatan Šoškić(✉)

Faculty of Mechanical and Civil Engineering in Kraljevo, University of Kragujevac, Dositejeva 19, 36000 Kraljevo, Serbia
soskic.z@mfkv.kg.ac.rs

Abstract. In the case of sparse spatial data, the experimental determination of a dispersion relationship of mechanical waves using correlation method meets the challenges caused both by the discrete nature of the measurement data and by the discrete nature of wavenumber sets used by the method. This paper presents an effort aimed to overcome the challenges using an artificial neural network which includes knowledge about previously determined points of dispersion relationship into the process of determination of its new points. The results show that the application of the neural network contributes to extension of the frequency range for experimental determination of dispersion relationship of flexural waves in beams, but that it also may be used for other mechanical structures.

Keywords: Mechanical waves · Dispersion relationship · Frequency range · Neural network

1 Introduction

Theoretical prediction and experimental measurement of dispersion relationship (abbreviated as "DR" in the following text) within a structure represent a key step in modern studies of wave propagation through 3D printed periodic and aperiodic structures [1]. The common approach to experimental determination of the DR comprises measurement of the response to an excitation in selected points of a structure. In the majority of practical applications, the excitation is impulse force, and the response is acceleration. Using spectral decomposition, the measurement results are used to calculate the accelerations that correspond to a particular frequency f in each of the measurement points, thus obtaining the acceleration wavefield with that frequency. The experimentally obtained wavefield is then systematically compared to theoretical forms of wavefields constructed using different wavenumbers. The wavenumber k that provides minimal difference between the experimental and constructed wavefields is considered to be the wavenumber that

N. Herisanu and V. Marinca (eds.), *Acoustics and Vibration of Mechanical Structures – AVMS-2021*, Springer Proceedings in Physics 274,
https://doi.org/10.1007/978-3-030-96787-1_26

corresponds to the frequency f in the DR. The methods for construction of the wavefields may be divided in two groups, the methods based on decomposition of the response into a series of damped sinusoids, and the methods based on systematic repetitive testing of candidate wavefields. The first group of methods [2–4] is computationally efficient, but they request equidistant measurement points and require a large number of measurement points, which are two requests that cannot be satisfied in many cases. Conversely, the second group of methods is slower, but the methods may be applied to sparse spatial measurements and to measurements taken in non-equidistant points. Alongside the methods that rely on least-squares-method and different algorithms for non-linear optimization [5, 6], to the second group of methods belongs also the correlation method [7, 8], which is the subject of the study in this paper.

The correlation method [7] compares the experimental wavefields at frequency f to the wavefields of harmonic progressive waves with the same frequency f (i.e., angular frequency $\omega = 2\pi f$) and different test wavenumbers k_t. In the one-dimensional case, the direction of the wave propagation is adopted to be the x-axis. If the experimental wavefield would be a progressive harmonic wave with wavenumber k, then the correlation between that wavefield and a test wavefield

$$\underline{W}_{\omega}(k_t) = \int_{-\infty}^{+\infty} \underline{w}(x, \omega)\underline{w}_t^*(x, \omega)dx \tag{1}$$

would have the highest modulus when $k_t = k$. In the expression (1), $w(x, \omega)$ and $w_t(x, \omega)$ represent complex amplitudes of the progressive harmonic wave with wavenumber k, and the test wavefield with the wavenumber k_t, respectively. The complex values are indicated by underlined symbols. The correlation method determines the DR of a wave in a frequency range $f_{\text{min}} < f < f_{\text{max}}$ by calculating the correlations between the wave and progressive harmonic waves with wavenumbers in the range $k_{\text{min}} < k_t < k_{\text{max}}$ for each of the frequencies f, assuming that the value of k_t that leads to highest correlation between the wave and a test wavefield is the wavenumber $k(f)$ that corresponds to the frequency f in the DR.

A numerical realization of correlation method requires resolution of three aspects of the Eq. (1): the infinite limits of the integral, the continuity of the spatial coordinate x, and the continuity of the range $k_{\text{min}} < k_t < k_{\text{max}}$. In order to resolve the first two aspects, the implementations of the correlation method use a sufficiently large, but still finite number of spatial points x_l ($l = 1, 2, \ldots, L$) to describe the wavefield of the propagating wave by measurements of complex amplitude of the response function (most often accelerances) $\underline{w}(x_l,\omega) = \underline{w}_{l\omega}$, which reduces the correlation (1) to the *correlation function* Y_{ω},

$$Y_{\omega}(k_t) = \left| \sum_{l=1}^{L} \underline{w}_{l\omega} \exp(-ik_t x_l)\Delta x_l \right|. \tag{2}$$

To resolve the second aspect, the calculations of the correlations (1) in the range $k_{\text{min}} < k_t < k_{\text{max}}$ are reduced to calculations of the correlation function (2) in a sufficiently large, but still finite number of points within the wavenumber range $k_t^{(j)}$ ($j = 1, 2, \ldots, J$), which, as a consequence, reduces the correlation function (2) to the *correlation array*

$Y_{\omega}^{(j)} = Y_{\omega}\left(k_t^{(j)}\right)$ ($j = 1,2, \ldots, J$). With such an implementation, the test wavenumber k_t that corresponds to the largest member of the correlation array is assumed to be the wavenumber that corresponds to the frequency f.

As it may be seen, the correlation method does not make any assumption, nor imposes any limitation, on the number and position of data measurement points x_l, which, together with its simplicity, represents the fundamental advantage of the method. The correlation method was used to determine DR in steel and composite plates [7]. A further extension of the correlation method [8] was applied for determination of DR in materials with high material losses.

A more detailed description of the correlation method may be found in [9], where it was shown that the described principles of practical realization of the correlation method lead to two conceptual problems due to (1) the finite number of measurement points L and (2) the finite number of wavenumber test points J [9].

Due to the finite number of measurement points, the correlation function $Y_{\omega}(k_t)$ may have multiple maxima with the same value within a wavenumber range $k_{\text{min}} < k_t < k_{\text{max}}$. In fact, in the case of equidistant measurement points, the correlation function is periodic with period $2\pi/d$, where d stands for the distance between the measurement points (the basic period, also called the first Brillouin zone, is denoted by IBZ).

On the other hand, due to the finite number of wavenumber test points, the maximum of the correlation function may not be included in the correlation array in the process of discretization. Therefore, any straightforward implementation of the correlation method by using the correlation function in the form given by Eq. (2) is limited to wavenumber ranges where the correlation function has a unique maximum, which in turn limits the frequency range for experimental determination of the DR. In the case of equidistant measurement points, the existence of only one maximum of correlation function limits the experimental determination of the DR to the IBZ, i.e., to the frequency range $f < f_{\text{BZ}}$, where f_{BZ} is the frequency that corresponds to the border of the IBZ. Even in the case of non-equidistant measurement points [10], the frequency range decreases with increase of the distance between the measurement points, which makes experimental determination of DR using sparse spatial data a particularly challenging task.

In the paper [9] three algorithms are studied for extension of the frequency range for experimental determination of dispersion relationship of flexural waves in beams using sparse spatial data, and this paper presents the efforts to use machine learning to the same goal.

2 Method

One of the advantages of the methods based on systematic repetitive testing of candidate wavefields is that it is possible to include previous knowledge about the vibrating structure or propagating waves in the process of determination of DR. For example, the algorithms developed in the paper [9] used the fact that the DR of flexural waves in a beam, given in the form [11]

$$\omega = \sqrt{\frac{EI}{\rho A}} k^2 \Rightarrow k = \pm \left(\frac{\rho A}{EI}\right)^{1/4} \sqrt{\omega}, \tag{3}$$

is a continuous and monotonous function. In the Eq. (3) ρ stands for density and E for Young's modulus of the material of the beam, while I stands for the area moment of inertia, and A for the area of the cross-section, of the beam.

The theoretical expression for the DR given by Eq. (3) shows that the experimental determination of DR of flexural waves in beams by straightforward application of correlation method is, at best, limited to the frequency range

$$f < f_{BZ} = \frac{1}{2\pi}\sqrt{\frac{EI}{\rho A}}\left(\frac{\pi}{d}\right)^2 = \frac{\pi}{2d^2}\sqrt{\frac{EI}{\rho A}}. \tag{4}$$

while the results obtained in the paper [9] show that the applicability of the algorithms that rely on monotonicity and continuity of the DR is limited by the fact that the experimentally obtained DR is expected to be neither continuous (due to the application of discrete Fourier transformation) nor monotonous (due to the measurement errors). While the frequency range for experimental determination of DR using the algorithms may be extended in comparison to straightforward application of correlation method, the simultaneous requests for monotonicity and continuity lead to failure of the algorithms to resolve all of the situations when the two requests, due to experimental measurements or round-off numeric errors, contradict each other. In some cases, the algorithms are able to recover from the misidentification of corresponding wavenumbers at a certain frequency, but, sooner or later, such errors ultimately lead to incorrect determination of wavenumbers in wide frequency ranges.

On the other hand, "manual" resolution of the conflicting requests, when a software user makes an educated guess about the wavenumber corresponding to a frequency f under these circumstances, turned out to be remarkably effective, since such an "assistance" led the same algorithms to successful determination of DR in the whole measurement range. While the human participants in the study were able to lead the process of resolution of the conflicts between the requests of continuity and monotonicity, they were not able to define the rules for their decision, describing it more as a "feeling" about the possible deviations of experimental results from the behaviour predicted by theoretical models.

Such circumstances, when humans are able to solve some problems, but are not able to define the rules of decision-making, are usually suitable for solution by artificial neural networks, and that was the reason to try to overcome the problems of algorithms based on monotonicity and continuity of DR by use of that machine learning tool. An artificial neural network (ANN) is a computational tool inspired by biological neural network of the human nervous system. It is a massively parallel distributed processor with the capability to learn, memorize and generalize to novel data. Therefore, ANNs are used for modelling complex functional relationships or finding patterns in data.

An experimental DR represents a parallel array structure (f_n, k_n, $n = 1,2,\ldots,N$) with the array of frequencies f_n and the array of corresponding wavenumbers k_n being the subarrays of the parallel array. Each of the elements of the parallel array structure represents a point of the experimental DR. In the implementation of the ANN, for the purposes of extension of the frequency range for experimental determination of the DR, the array of corresponding wavenumbers is considered as a sequential data, and the recurrent neural network is developed to serve as a sequence prediction tool for estimation of the wavenumber corresponding to a frequency using the wavenumbers of the previously determined points of the DR. The ANN, presented in Fig. 1 implements the sliding window technique, using the wavenumber values of the five previous DR points as input data. The output of the ANN is the difference Δk between the current and the previous wavenumber values, and that output serves as the target value at the network output layer. Developed neural network contains one hidden layer which includes five nodes. Bipolar sigmoid function and linear function are used as transfer functions of the hidden layer and the output layer, respectively. During the neural network training process, Adam optimization algorithm [12] is used for updating weight and bias values according to the training data set.

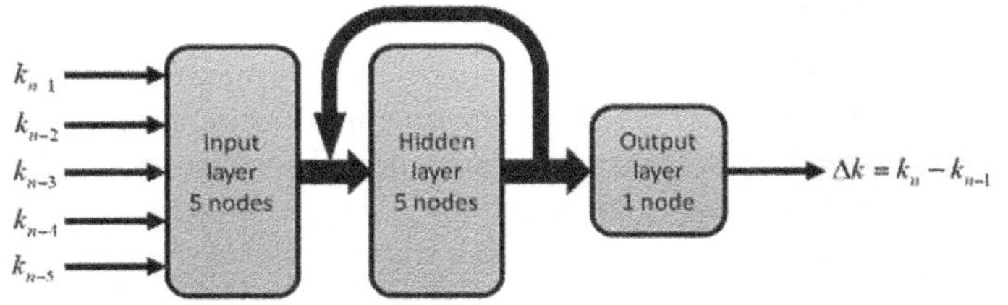

Fig. 1. The structure of the ANN network developed for extension of the frequency range

Neural network model is trained on the sequence of the wavenumbers that correspond to frequencies lower than $f_{BZ}/2$, where the straightforward application of correlation method is justified and successful.

The algorithm for determination of the DR of flexural waves in beams is based on the algorithm developed in [9], but it relies on the developed ANN for the selection of the wavenumbers corresponding to frequencies higher than $f_{BZ}/2$. Therefore, here will be presented only the outline of the algorithm, and the readers may find the complete algorithm in [9].

Among the other inputs, the algorithm uses the following data:

- the ordered array of frequencies f_n ($n = 1, 2, \ldots, N$) that represents the frequency range $f_{min} < f < f_{max}$ where DR is determined;
- the ordered array of positions x_l ($l = 1, 2, \ldots, L$) of the points where measurements were taken; the array should
- the calculated complex amplitudes of frequency response functions represented by an array of arrays w_{ln} ($l = 1, 2, \ldots, L$ and $n = 1, 2, \ldots, N$);
- the ordered array of wavenumbers, $k_t^{(j)}$ ($j = 1, 2, \ldots, J$) that represents the wavenumber range $k_{min} < k_t < k_{max}$ where correlation function is calculated;
- the selected proximity limit ε_k ($0 < \varepsilon_k < 1$).

The algorithm, which determines the DR in the form of a parallel array structure (f_n, k_n, $n = 1, 2, \ldots, N$), consists of a preparatory phase (phase 1) and the construction loop phase (phase 2). The preparatory phase comprises calculation of variables needed in the construction loop phase, such as the Brillouin zone frequency f_{BZ}. During each of the cycles of the construction loop is determined one pair of the DR values (f_n, k_n), (n = 1, 2, ..., N). The first step of the n-th cycle is calculation of the correlation array $Y_\omega^{(j)}$ (j = 1, 2, ..., J) using the Eq. (2) and $\omega = 2\pi \cdot f_n$ for all test wavenumber points $k_t^{(j)}$ (j = 1, 2, ..., J). In the next step is derived the candidate array k_m (m = 1, 2, ..., M), which contains the wavenumbers that correspond the local maxima of the correlation array. By the algorithm, the wavenumber k_1 that corresponds to the first point ($n = 1$) of the DR is the member of the candidate array for the lowest frequency f_1 with the smallest absolute value. The selection of the wavenumbers that correspond to the other points of the DR ($n > 1$) depends on the already calculated wavenumbers corresponding to the previous points of the DR, and this procedure represents the difference between various methods for implementation of the correlation method presented here and in [9]. In order to facilitate the comparisons between the methods, the steps of the procedure are numbered as in [9], with 2.5.1 representing the step that corresponds to the first point of DR ($n = 1$). Having that in mind, the procedure for selection of the wavenumbers that correspond to the other points of the DR ($n > 1$) in this paper is defined as follows:

2.5.2 If ($f_n < f_{BZ}/2$), then the wavenumber corresponding to the frequency f_n is the member of the candidate array closest to the wavenumber of the previous point and the algorithm loops to the next DR point; otherwise, it proceeds with the step 2.5.3;

2.5.3 Once algorithm comes here, then it holds ($f_n \geq f_{BZ}/2$) and the following procedure applies:

2.5.3.1 Calculate the estimation of the wavenumber corresponding to the current frequency based on previous DR points $k^{(est)}$ by feeding the five previous points of the DR, k_{n-p} (p = 1, 2, ..., 5), as inputs to the developed ANN, and adding the output of the ANN to the wavenumber of previous DR point:

$$k^{(est)} = k_{n-1} + \Delta k \tag{5}$$

2.5.3.2 Find the smallest element of candidate array higher than $k^{(est)}$—referred further as the "Hi" candidate $k^{(Hi)}$ and determine its proximity rate $r^{(Hi)}$:

$$r\left(k^{(Hi)}\right) = \frac{k^{(Hi)} - k_e}{k_e} \tag{6}$$

2.5.3.3 Find the largest element of the candidate array lower than $k^{(est)}$—referred further as the "Lo" candidate $k^{(Lo)}$ and determine its proximity rate $r^{(Lo)}$ by applying following equation:

$$r\left(k^{(Lo)}\right) = \frac{k_e - k^{(Lo)}}{k_e} \tag{7}$$

2.5.3.4 If one of the elements $k^{(Hi)}$ or $k^{(Lo)}$ does not exist, then the existing one is selected as the final candidate $k^{(final)}$ and its proximity ratio as the proximity ratio of the final candidate $r^{(\mathrm{final})}$; the algorithm proceeds with the step 2.5.3.6;

2.5.3.5 If both $k^{(\mathrm{Hi})}$ and $k^{(\mathrm{Lo})}$ exist, then the one closer to $k^{(\mathrm{est})}$ is selected as the final candidate $k^{(\mathrm{final})}$ and its proximity ratio is set as the proximity ratio of the final candidate $r^{(\mathrm{final})}$ (therefore, if $r^{(\mathrm{Lo})} < r^{(\mathrm{Hi})}$ then $k^{(\mathrm{final})} = k^{(\mathrm{Lo})}$ and $r^{(\mathrm{final})} = r^{(\mathrm{Lo})}$, otherwise $k^{(\mathrm{final})} = k^{(\mathrm{Hi})}$ and $r^{(\mathrm{final})} = r^{(\mathrm{Lo})}$);

2.3.5.6 If the proximity ratio of the final candidate point is lower than the selected proximity limit, $r^{(\mathrm{final})} < \varepsilon_k$, then the final candidate wavenumber $k^{(\mathrm{final})}$ is accepted as the wavenumber corresponding to the frequency f_n, $k_n = k^{(\mathrm{final})}$; However, if the proximity ratio of the final candidate point is higher than the selected proximity limit, then the candidate array is considered ambiguous, and the value $k^{(\mathrm{est})}$ is accepted as the wavenumber corresponding to the frequency f_n, $k_n = k^{(\mathrm{est})}$;

Previous research [9] resulted in three algorithms for extension of the frequency range that are remarkably similar to the algorithm presented in the previous section, since the only difference is the most elaborate step 2.5.

The algorithms in [9] are based on properties of DR predicted by theory, given by the Eq. (3), and the properties that experimentally determined DR has to have as a discretized form of the theoretical DR. Since theoretical DR is monotonous and continuous, the candidate array member that corresponds to a frequency should be as close as possible to the previous point of DR (due to the continuity), and larger than it (due to the monotonicity). If a candidate array does not contain any member that satisfies both requests, then a conflict between the two requests arises, and the candidate array is ambiguous. The differences between the method proposed in this paper and the methods proposed in [9] are the definition of the ambiguous candidate arrays and handling of the conflicting requests.

In the paper [9], the ambiguity is detected by comparison of the candidate array members to the wavenumber of the previous DR point, while in this paper the ambiguity is detected by comparison of the candidate array members to the estimated value of the wavenumber $k^{(\mathrm{est})}$. That difference is expressed by the algorithm steps 2.5.2–2.5.3. Due to the new definition of the ambiguity, the candidates are actually compared *to the trend* established by previous points of DR, instead to be compared *to the previous point* of the DR.

The other difference with respect to algorithms developed in [9] is the handling of the ambiguous arrays, expressed by the algorithm steps 2.5.3.4–2.5.3.6. Three algorithms proposed in [9] differ by handling the conflicts, i.e., by selection of wavenumber of the next point of DR under conflicting requests, since the "Close" method favours the continuity of DR, the "CloseHi" method favours the monotonicity of DR while "AutoSkip" method skips to the next point of DR;

The change of the definition of ambiguity of the candidate array reflects in the resolution of the conflicting requests. In such cases, the ANN method sets the estimated wavenumber value $k^{(\mathrm{est})}$ to be the wavenumber corresponding to the next point of the DR, which then becomes a part of the further determination of the DR. In comparison to the three described methods from [9], it may be said that the ANN method favours continuity, but only if the closest member of the candidate array follows the trend established by previous points of DR.

However, the most essential difference between methods presented in [9] and the ANN method is that the former implement only monotonicity and continuity of DR of flexural waves in beams, while the latter uses ANN training process to develop any knowledge about the DR, which means that the ANN method may be used for determination of DR in arbitrary structures.

3 Experiment

The proposed ANN method for implementation of correlation method was studied by application to experimental results. The experiment consisted of measurement of time variation of transversal accelerations $a_x(t)$ in points x selected along a beam that was excited by a hit in transversal directions. The excitation was performed by an instrumented impact hammer that measured variation of the force during the impact $F(t)$. which further enabled calculation of the accelerances $w_x(f)$. The ends of the beam were supported by thick and soft foam, so that the response to the excitation could be interpreted as propagation of a flexural wave in the beam with free ends.

The experiment is described in [9], and the authors will repeat here only the basic data relevant for further reference. The beam was made of steel, and had length $D \approx 1{,}65\ m$, and square cross-section with side $b \approx 1\ cm$. Ten measurement points were equidistant, with distances $d \approx 15\ cm$. Due to the spatial distribution of the measurement points, the limit of the I BZ was $k_{BZ} = \pi/d \approx 21{,}0\ \text{m}^{-1}$ and $f_{BZ} \approx 1{,}01$ kHz. The frequency range of the measurement equipment used in the experiment was 10 Hz $< f <$ 6 kHz.

4 Results and Discussion

This section presents both the results of the application of the algorithms proposed in [9] and the ANN method proposed in this paper. Since the analysed algorithms are developed to extend the applicability of the correlation method beyond the IBZ, the range of interest for further analysis and comparison of the algorithms comprises wavenumbers $k > \pi/d$, i.e. for frequencies above f_{BZ}.

For the sake of comparison, the following figures also include graphical representation of theoretical predictions obtained using the Eq. (3), with Young modulus E = 200 GPa and density of steel $\rho = 7800\ \text{kg/m}^3$. That theoretical prediction for DR is represented by the solid line that passes through the origin in Fig. 2. The other solid lines in Fig. 2 represent the branches of the equivalent maxima that arise due to the periodicity of the correlation function caused by equidistant distribution of the measurement points, as explained in the introduction. The increasing branches describe the transfer of the energy in the direction of the incident wave, and the decreasing branches describe the transfer of the energy in the direction of the reflected wave. The increasing and decreasing branches intersect and become indistinguishable at the borders of the Brillouin zones, when the equidistant measurement points lead to false detection of standing waves. The points in Fig. 2 represent the results obtained by the straightforward application of the correlation method, and they demonstrate the inability of the approach to distinguish the wavenumbers corresponding to the DR from the wavenumbers corresponding to the branches of equivalent maxima.

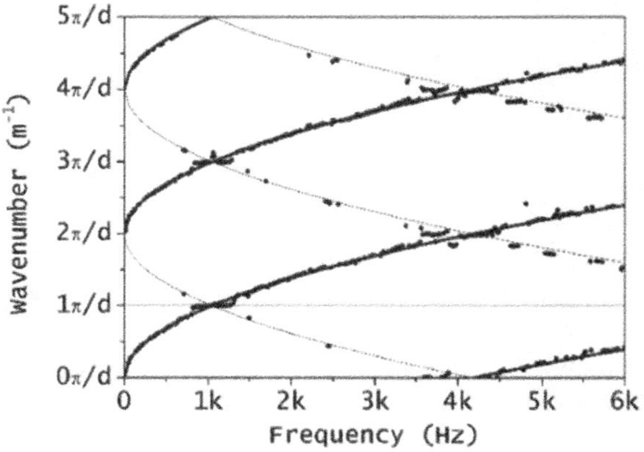

Fig. 2. Wavenumber branches and the values detected by the straightforward implementation of correlation method [9]

The results of the application of the "CloseHi" method (that favours monotonicity of the DR over its continuity) and the "Close" method [9] (that favours continuity of the DR over its monotonicity) are presented in Fig. 3 left. The obtained results show that these methods may be used for the frequencies higher than f_{BZ} to some extent, but also clearly demonstrate the weaknesses of both methods.

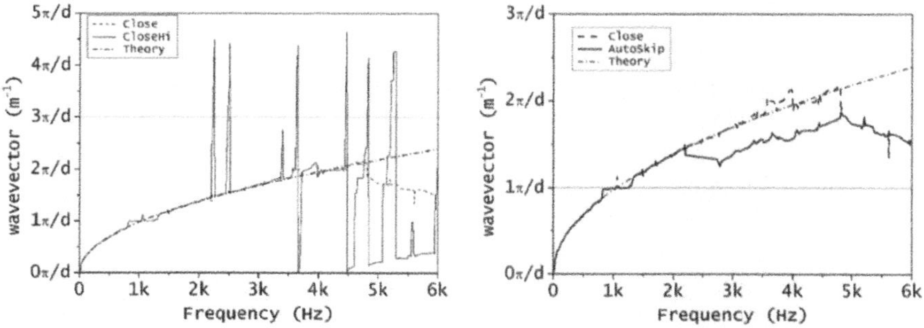

Fig. 3. Comparison of the DRs obtained by "Close", "CloseHi" and "AutoSkip" method" with theoretical predictions [9]

The failure of the "CloseHi" method arises when round-off errors decrease the value of the wavenumber of the local maximum of the correlation function that corresponds to the DR. If the value is smaller than the wavenumber corresponding to the previous point k_{n-1}, the algorithm selects that wavenumber as the candidate point $k^{(Lo)}$. Consequently, a value of the wavenumber corresponding to the next higher branch of the equivalent maxima is selected as the candidate point $k^{(\mathrm{Hi})}$. Since the "CloseHi" method selects $k^{(\mathrm{Hi})}$ as the corresponding wavenumber, under such circumstances it selects a wrong candidate and may recover only when the equivalent branch of the maxima increases above the range of tested wavenumbers. Such misidentifications cause the occasional "glitches" that may be observed in the range 2–4 kHz, as well as the "oscillations" between the increasing and decreasing branches of equivalent maxima in the range 4.5–6.0 kHz.

The "Close" method fails near the high-frequency borders of Brillouin zones, when the slope of DR becomes sufficiently low and DR intersects with some of the decreasing branches of equivalent maxima. In that case, the wavenumber of a candidate point belonging to the decreasing branch may have closer value to the wavenumber of the previous point of DR k_{n-1} than the wavenumber of the candidate point belonging to the DR. The "Close" method, which selects the closest candidate point, under these circumstances selects the candidate point belonging to the decreasing branch. Once the "Close" algorithm "switches" to monotonously decreasing branch of maxima, its tendency to favour continuity results in further incorrect selection of the wavenumbers that belong to that branch of maxima (observable in the range above 5 kHz).

Figure 3 right compares the results of the application of the "AutoSkip" and the "Close" methods [9], clearly demonstrating that the "AutoSkip" method is even more sensitive to the high-frequency low slopes in the DR, because "AutoSkip" method keeps the last selected point as the previous point after skipping over a sequence of data points. Since the DR is monotonously increasing, the value of the point that was selected before skipping is low in comparison with the following points, which results in "switching" to a lower branch of local maxima, which is observable around 2 kHz.

Figure 4 presents the results of application of the ANN method, showing that its application leads to extension of the applicability of the correlation method for determination of dispersion relationship of flexural waves beyond the IBZ, extending it to whole frequency range measurable by the experimental equipment, up to frequencies of 6 kHz. It may be concluded that the proposed method is superior to the straightforward application of correlation method, as well as the other analysed methods.

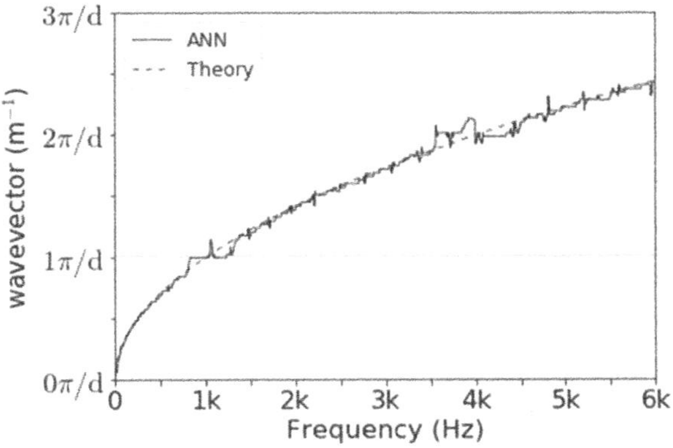

Fig. 4. Comparison of the DR obtained by application of ANN method and theoretical predictions

The superiority of the ANN method is a consequence of inclusion of knowledge about previous DR points in the process of prediction of the next DR point, which makes it less prone to select candidates which are members of other branches of equivalent maxima, but also more capable to recover after the events of misidentification. The previous conclusion is supported by the following considerations: (1) the rejection of the candidates that are too distant from the estimated next point of DR prevents "glitches" and "oscillations" that appear in the "CloseHi" method, as well as the "switches" to monotonously decreasing branches of maxima with lower values of candidate array members, which appear in the "Close" method; (2) the inclusion of the estimated point in the DR keeps the established trend of DR in sufficiently wide frequency ranges to avoid "falling" to monotonously increasing branches of maxima with lower values of candidate array members, which appear in the "Auto Skip" method;

The only significant deviation of the experimentally determined DR with respect to the theoretical predictions occurs in the ranges around frequencies 1.0 and 4.1 kHz, which represent the borders of the BZ. Within these frequency ranges, the method leads to constant wavenumber values. However, such behaviour is observed for all the methods, and as explained at the beginning of this chapter, is a consequence of the equidistant positioning of the measurement points. Therefore, the inability of the algorithms to detect the monotonously increasing DR near the BZ borders is only a consequence of the inadequate input, which is, in turn, a consequence of the experimental setup. Such an explanation is confirmed by the results of the experiments performed using non-equidistant measurement points [9].

5 Conclusions

The paper presents an artificial neural network approach for the determination of the dispersion relationship of flexural waves in beams using the correlation method. The aim of the proposed method is to contribute to extension of the frequency range for determination of dispersion relationship using correlation method by overcoming limitations due to spatial aliasing caused by discrete nature of the measurement data, which restricts applicability of correlation method to the first Brillouin zone of wavenumber space (which is defined by positions of the measurement points).

In the paper were compared four methods for determination of dispersion relationship for the case of equidistant measurement points. The three of the methods were previously described [9], and these are the "CloseHi" method, which favours monotonicity of DR in cases of conflicting requests, the "Close" method, which favours continuity of DR in cases of conflicting requests, and the "AutoSkip" method, which skips the data points of DR in cases of conflicting requests. The "ANN" method, presented in this paper, uses the knowledge about previous DR points to resolve the cases of conflicting requests.

Due to the smaller sensitivity to round-off and measurement errors, the "Close" method outperforms both "CloseHi" method and "AutoSkip" method. The "Close" method enabled calculation of the dispersion relationship in a frequency range up to 5 kHz, except in frequency ranges around BZ borders. On the other hand, the application of the ANN provided the best results, extending the frequency range for determination of the dispersion relationship to the whole frequency range used in the experiment. Due to its ability to learn and include information about previously determined points of DR into the process of estimation of new points, the ANN is far less likely to misidentify a wavenumber corresponding to a certain frequency, as well as to return to the process of correct identification of the following points of DR after misidentification.

Acknowledgements. The authors wish to express their gratitude to prof. Neil Fergusson from Institute of Sound and Vibration of University of Southampton for inspiration and support to this research effort.

The authors also wish to acknowledge the support of Ministry of Education, Science and Technology Development of Republic of Serbia through Grant 451-03-9/2021-14/200108.

References

1. Lucklum, F., Vellekoop, M.J.: Bandgap engineering of three-dimensional phononic crystals in a simple cubic lattice. Appl. Phys. Lett. **113**(20), 201902 (2018)
2. Steedly, W.M., Ying, C.-H.J., Moses, R.L.: Statistical analysis of TLS-based Prony techniques. Automatica **30**(1), 115–129 (1994)
3. Tufts, D.W., Kumaresan, R.: Estimation of frequencies of multiple sinusoids: making linear prediction perform like maximum likelihood. Proc. IEEE **70**(9), 975–989 (1982)
4. Grosh, K., Williams, E.G.: Complex wave-number decomposition of structural vibrations. J. Acoust. Soc. Am. **93**(2), 836–848 (1993)
5. McDaniel, J.G., Dupont, P., Salvino, L.: A wave approach to estimating frequency-dependent damping under transient loading. J. Sound Vib. **231**(2), 433–449 (2000)
6. McDaniel, J.G., Shepard, W.S., Jr.: Estimation of structural wave numbers from spatially sparse response measurements. J. Acoust. Soc. Am. **108**(4), 1674–1682 (2000)
7. Ferguson, N.S., Halkyard, C.R., Mace, B.R., Heron, K.H.: The estimation of wavenumbers in two-dimensional structures. In: Proceedings of ISMA2002: international conference on noise and vibration engineering, Leuven (Belgium), pp. 799–806 (2002)
8. Bareille, O., Ichchou, M., Berthaut, J., Jezequel L.: Numerical extraction of dispersion curves from experimental data. In: The Second Conference on Noise and Vibration Emerging Methods—NOVEM 2005, St. Raphael, France (2005)
9. Sinđelić, V., Ćirić-Kostić, S., Nikolić, A., Šoškić, Z.: Extension of the frequency range for experimental determination of dispersion relationship of flexural waves in beams by correlation method. IMK-14—Res. Dev. Heavy Mach. **26**(4), 95–107 (2020). https://doi.org/10.5937/IMK2004095S
10. Maciejewski, M.W., Qui, H.Z., Rujan, I., Mobli, M., Hoch, J.C.: Nonuniform sampling and spectral aliasing. J. Magn. Reson. **199**(1), 88–93 (2009)
11. Beard, C.F.: Structural vibrations—Analysis and damping. Elsevier, Oxford, (UK) (1996)
12. Kingma, D.P., Ba, J.: Adam: a method for stochastic optimization (2014). arXiv:1412.6980

Parametric Assessment of the Elastomeric Anti-vibration Devices According to the Dynamic Test Regime

Polidor Bratu(✉)

Research Institute for Construction Equipment and Technology, ICECON SA, Sos. Pantelimon 266, Bucharest, Romania
icecon@icecon.ro

Abstract. It is presented the dynamic test method on special stands for the anti-vibration elastomer devices according to the dynamic loading regimes, namely: ante-resonance, resonance and post-resonance. The results obtained, both on the basis of numerical modeling and experimental testing, confirm that the rigidity and damping parameters as well as durability at repeated cycles can be achieved on specially designed specialized dynamic stands. Thus, for this article, an innovative inertial vibrator stand was used, whose frequency may be modified by a control system within the range of 2–60 Hz. The transmitted force and the kinematic parameters were measured in real time, and the excitation given by the inertial centrifugal force of the dynamic vibratory imbalance masses was determined based on the frequency measured over time. Thus, there were highlighted the characteristics of the three dynamic regimes of the elastomeric vibration isolation and damping devices for supporting the equipment mounted and installed in the buildings (pumping groups, ventilation systems, motors, etc.). The established calculation relations and the parametric curves highlight the modality of variation of the energy dissipation capacity depending on the dynamic operating regime of the dynamic equipment supported on the anti-vibration devices of elastomers.

Keywords: Elastomeric · Anti-vibration devices · Dynamic test regime · Hysteretic loop · Dissipated energy

1 Introduction

The research made were aimed at optimizing the vibration isolation and damping systems for several categories of dynamic equipment for public buildings, process industries and energy groups.

At ICECON Bucharest, dynamic test benches were designed and constructed where significant dynamic regimes were generated consistent with the operating parameters of the dynamic equipment for which the antivibration devices were made.

In this context, it is specified that the dynamic test bench used is equipped with three vertical groups for mounting the antivibration devices connected in parallel with the

N. Herisanu and V. Marinca (eds.), *Acoustics and Vibration of Mechanical Structures – AVMS-2021*, Springer Proceedings in Physics 274,
https://doi.org/10.1007/978-3-030-96787-1_27

inertial vibrator placed at the top. The three vertical groups are attached to the bottom of the stand's frame.

Each group is equipped with an installation-mounting system of the antivibration device that can be rotated and fixed at angled index positions at 15° from 0 to 90°.

The angular position corresponds to the positioning angle and it is defined between the vertical direction of the direction of the unidirectional perturbator force of the vibrator and the axis of compression of the elastomeric anti-vibration device. In this manner, there were measured for various angles of positioning for compression, shearing and compression-shearing.

For each case, there was raised the curve of the dissipated energy and the hysteretic loop represented in the system $F - x$ as well as in the $Q - x$ system where $F = F(t)$ is the disturbing excitatory force and $Q = Q(t)$ is the force transmitted to the foundation (the fixed part of the stand).

2 Dissipated Energy

For a dynamic regime with the relative pulse $\boldsymbol{\Omega} = \frac{\omega}{\omega_n}$, where ω is the excitation pulse and is own pulsation, the dissipated energy W_d may be expressed as

$$W_d(\boldsymbol{\Omega}, \zeta) = 2\pi \left(\frac{m_0 r}{m}\right)^2 k\zeta \frac{\boldsymbol{\Omega}^5}{\left(1 - \boldsymbol{\Omega}^2\right)^2 + (2\zeta\boldsymbol{\Omega})^2} \tag{1}$$

where ζ the fraction of the critical damping.

k—rigidity of the system, N/m.

$m_0 r$—static momentum of the exciter with rotating motion eccentric bodies, kg·m.

m—mobile mass of the stand, kg.

The maximum dissipated energy in ante-resonance regime $W_{d.a}^{max}$ corresponds to a fraction of the critical damping $\zeta_a^0 = \frac{1-\boldsymbol{\Omega}_a^2}{2\boldsymbol{\Omega}_a}$, that is we have

$$W_{d.a}^{max} = 2\pi \left(\frac{m_0 r}{m}\right)^2 k \frac{\boldsymbol{\Omega}_a^4}{4\left(1 - \boldsymbol{\Omega}_a^2\right)} \tag{2}$$

where $\boldsymbol{\Omega}_a < 1$ is the relative pulsation of ante-resonance.

The maximum dissipated energy in post-resonance regime $W_{d.p}^{max}$ corresponds to the pulsation $\boldsymbol{\Omega}_p > 1$ and to the fraction of the critical damping $\zeta_p^0 = \frac{\boldsymbol{\Omega}_p^2-1}{2\boldsymbol{\Omega}_p}$, so that we have

$$W_{d.p}^{max} = 2\pi \left(\frac{m_0 r}{m}\right)^2 k \frac{\boldsymbol{\Omega}_p^4}{4\left(\boldsymbol{\Omega}_p^2 - 1\right)} \tag{3}$$

The dissipated energy in resonance regime W_d^{rez} corresponds to the pulsation $\boldsymbol{\Omega} = 1$ and to the fraction of the critical damping $\zeta_n = \frac{c}{2m\omega_n}$, so that we have

$$W_d^{rez} = 2\pi \left(\frac{m_0 r}{m}\right)^2 k \frac{1}{4\zeta_n} \tag{4}$$

From relations (2), (4) then from relations (3) and (4), the following correlations emerge

$$\mathbf{W}_{d,a}^{max} = \mathbf{W}_{d}^{rez} \frac{\boldsymbol{\Omega}_a^4}{1 - \boldsymbol{\Omega}_a^2} \zeta_n \tag{5}$$

$$\mathbf{W}_{d,p}^{max} = \mathbf{W}_{d}^{rez} \frac{\boldsymbol{\Omega}_p^4}{1 - \boldsymbol{\Omega}_p^2} \zeta_n \tag{6}$$

Correlation between $\mathbf{W}_{d,a}^{max}$ and $\mathbf{W}_{d,p}^{max}$ emerged from relations (5) and (6) as follows

$$\frac{\mathbf{W}_{d,p}^{max}}{\mathbf{W}_{d,a}^{max}} = \left(\frac{\boldsymbol{\Omega}_p}{\mathbf{a}}\right)^2 \frac{1 - \boldsymbol{\Omega}_a^2}{\boldsymbol{\Omega}_p^2 - 1} \tag{7}$$

2.1 Variation of the Dissipated Energy According on Damping

For the current variable ζ with $\Omega \neq 1$ and constant we have the following situations, namely [4–8]:

(a) ante-resonance with $\Omega_a < 1$ for which we have

$$W_{d,a}^{ciclu}(\zeta) = 2\pi \left(\frac{m_0 r}{m}\right)^2 k\zeta \frac{\Omega_a^5}{\left(1 - \Omega^2\right)^2 + (2\zeta\Omega)^2} \tag{8}$$

with graphical representation in Fig. 1.

(b) post-resonance with $\Omega_p > 1$ for which it is valid the relation

$$W_{d,p}^{ciclu}(\zeta) = 2\pi \left(\frac{m_0 r}{m}\right)^2 k\zeta \frac{\Omega_p^5}{\left(1 - \Omega^2\right)^2 + (2\zeta\Omega)^2} \tag{9}$$

with graphic representation in Fig. 2.

The three curves of Figs. 1 and 2 correspond to the cases where the elastomeric antivibration devices required at compression with k_c, combined requirement of compression and shear $k\alpha$ and shear stress k_f for the ante-resonance regime $\Omega_a = 0.8$, $\zeta_a^0 = 0.225$ and, respectively, post-resonance $\Omega_p = 1.5$, $\zeta_p^0 = 0.41$ [9, 10].

2.2 Variation of the Dissipation Energy According to the Relative Excitation Pulsation

The energy dissipated on cycle $W_d{}^{ciclu}(\Omega)$ according to the relative excitation pulsation $\Omega = \frac{\omega}{\omega_n}$ pentru for the given values of rigidity k and the fraction of the critical damping ζ is given by the relation

$$W_d^{cycle}(\Omega) = 2\pi \left(\frac{m_0 r}{m}\right)^2 k_i\zeta_i \frac{\Omega^5}{\left(1 - \Omega^2\right)^2 + (2\zeta_i\Omega)^2} \tag{10}$$

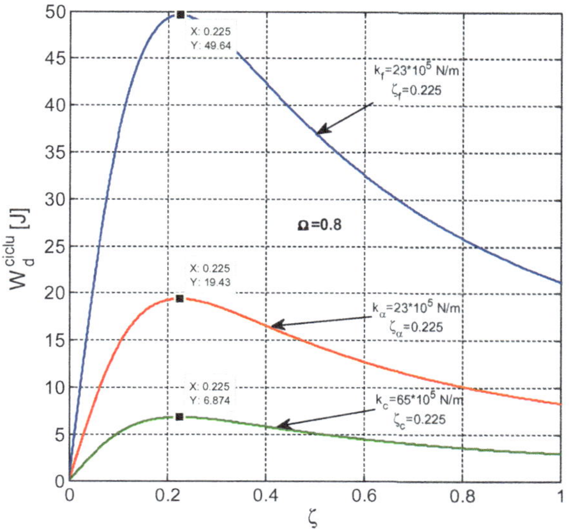

Fig. 1. Variation of the dissipated energy per cycle, in ante-resonance regime, in relation with the damping ratio ζ and rigidity *k*

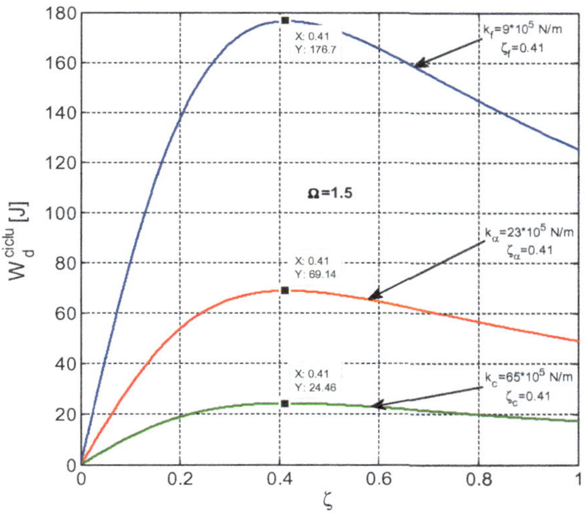

Fig. 2. Variation of the dissipated energy per cycle, in post-resonance regime, according to the damping ratio ζ and rigidity *k*

where $i = 1, 2, \ldots, n$ represents the index of the parametric order k_i, ζ_i which describes a curve of the curve family with the current variable Ω [11–13].

Figure 3 shows the family of curves on discrete values k_i, ζ_i and the continuous variation of Ω.

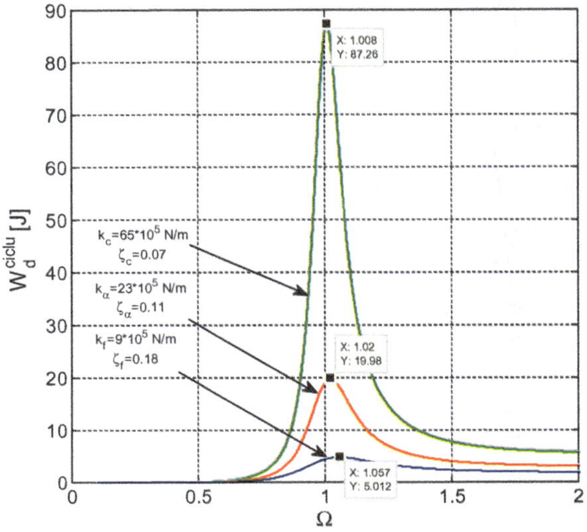

Fig. 3. Variation of the dissipated energy per cycle, in post-resonance regime, according to relative pulsation Ω and the discrete variation of rigidity k

3 Hysteretic Loops

For the viscoelastic linear dynamical system, there may be represented the ***elliptical*** hysteretic loops for the dynamic excitatory force $F(t)$ in relation to the deformation $x = x(t)$ in $F - x$ coordinates as well as for the dynamic force transmitted $Q = Q(t)$ depending on the deformation $x = x(t)$ in $Q - x$ coordinates.

(a) Hysteretic loops in $F - x$ coordinates

The equation of the elliptical hysteretic loops in the coordinate system $F - x$ is given by the parameterized relation by k and ζ discrete variable and x continuous variable so that $x \in [-A, +A]$.

Thus, we have

$$F(x, \Omega) = k\left[\left(1 - \Omega^2\right)x \pm 2\zeta\Omega\sqrt{A^2 - x^2}\right] \tag{11}$$

where $A = A(\Omega)$ is the amplitude of instantons displacement (deformation) $x = x(t)$.

$$A = A(\Omega) = \frac{m_0 r}{m}\Omega^2 \frac{1}{\sqrt{\left(1 - \Omega^2\right)^2 + (2\zeta_i\Omega)^2}} \tag{12}$$

In Fig. 4, for $\Omega_a = 0.8$ in ***ante-resonance regime*** it is presented the family of elliptic loops for the three significant cases of the viscoelastic system $\alpha_c = 0°$, $\alpha = 60°$ and $\alpha_f = 90°$ that is in situation k_c, ζ_c, k_α, ζ_α and k_f, ζ_f. It is mentioned that all ellipses are inclined in quadrants I and III.

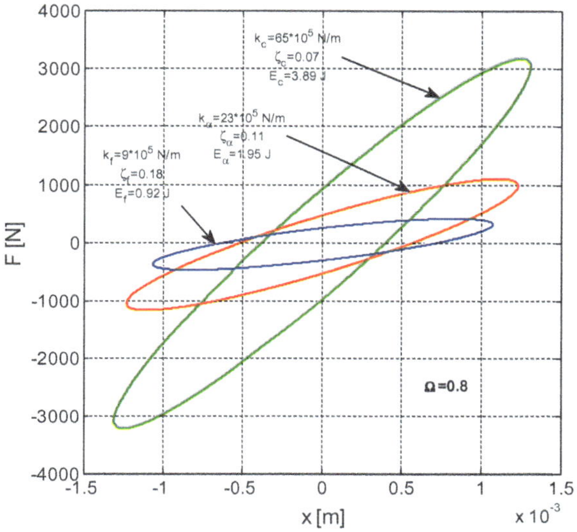

Fig. 4. Hysteretic loops $F - x$ in ante-resonance regime for $\Omega = 0.8$

In Fig. 5, for $\Omega_p = 1.5$ in ***post-resonance regime*** it is presented the family of elliptic loops for the three significant cases with the specification that all ellipses are inclined in quadrant III and IV as an effect of the post-resonance influence regime [14, 15].

In Fig. 6, there are presented the elliptic hysteretic loops in *resonance regime* for the three significant cases.

4 Conclusions

Based on the analysis of the analytical relations established both for the dissipated energy as well as for the representation of the hysteretic loops specific to the linear viscoelastic system that models a dynamic stand for elastomeric antivibration devices, the following conclusions can be drawn:

(a) the analytical expression of the dissipative energy provides the possibility of evaluation for two significant cases, namely:

- the variation of the dissipated energy according to the modification of the damping for the three dynamic regimes: ante-resonance, post-resonance and resonance;

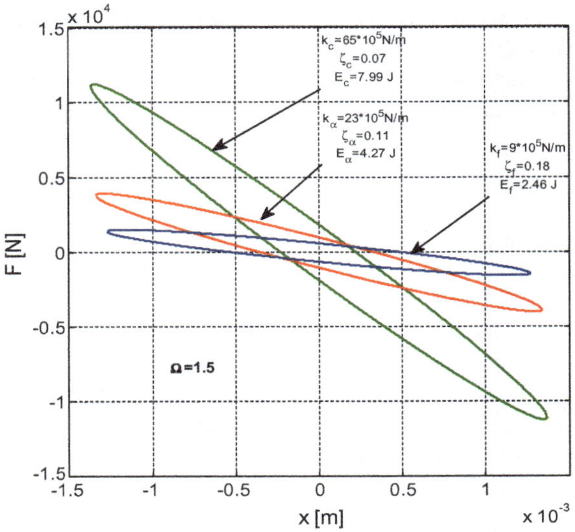

Fig. 5. Hysteretic loops $F - x$ in post-resonance regime for $\Omega = 1.5$

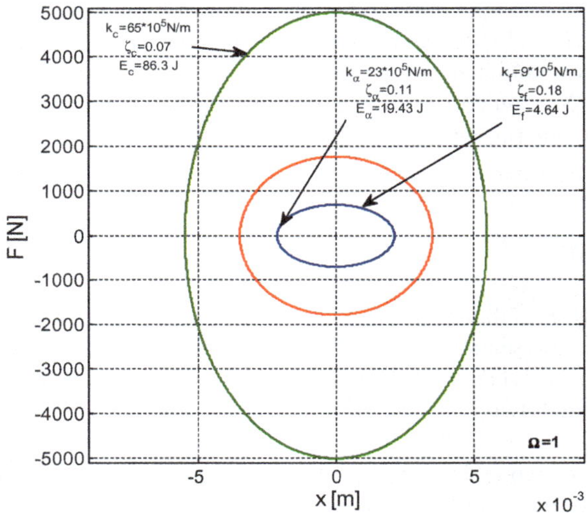

Fig. 6. Centered hysteretic loops $F - x$ in resonance regime for $\Omega = 1$

- variation of the dissipated energy according to the variation of Ω for sets of discrete variable values of k and ζ;

(b) the representation of elliptical hysteretic loops in the $F - x$ coordinate system for the three cases of dynamic regimes, namely: ante-resonance, post-resonance and resonance. It is found that in the post-resonance the tilting of the ellipse axes changes versus the ante-resonance regime changes as a result of the inertial mass effect, and in resonance the ellipses are symmetrically centered relative to the $F - x$ axes system.
(c) the elliptical hysteretic loops in the $Q - x$ system are tilted only in quadrants I and III, regardless of the dynamic regime.
(d) the areas of the ellipses represent the dissipated energy. The numerical results were verified by experimental lifts of hysteretic loops on the dynamic stand.

References

1. Dobrescu, C.: The rheological behavior of stabilized bioactive soils during the vibration compaction process for road structures. In: Proceedings of the 22th International Congress on Sound and Vibration, Florence, Italy, 12–16 July 2015
2. Chen, B.J., Lin, S.B., Tsai, C.S.: Theoretical and experimental study of high damping rubber bearings. In: Russel, G. (eds.): Proceedings of the Seismic Engineering, vol. 2. The 2000 ASME Pressure Vessels and Piping Conference, Seattle, WA, USA, 23–27 July 2001
3. Bratu, P.: Vibration transmissivity in mechanical systems with rubber elements using viscoelastic models. In: Proceedings of the 5th European Rheology Conference, Ljubljana, Slovenia, 6–11 Sept 1998
4. Delfosse, C.G.: Étude des vibrations linéaires d'un systeme mecanique complexe par méthode des mates normaux. Ann. l'ITBTP 122–136 (1976)
5. Giacchetti, R.: Fondamente di Dinamica delle Strutture e di Ingineria Sismica. EPC Libri, Roma, Italy (2004)
6. Rao, M.: Mechanical Vibrations. Addison-Wesley Pub. Co., Boston, MA, USA
7. Sireteanu, T., Giuclea, M., Mitu, A.M.: An analytical approach for approximation of experimental hysteretic by Bouc.-Wen model. Procc. Rom. Acad. Ser. A **10**(1), 43–54 (2009)
8. Le Tallec, P.: Introduction à la Dynamique des Structures. Cépaduès, Toulouse, France; Ellipses (2000)
9. Bratu, P.: Analyze insulator rubber elements subjected to actual dynamic regime. In: Proceedings of the 9th International Congress Sound and Vibration. University Orlando, FL, USA, July 8–11 2002
10. Bratu, P.: Rheological model of the neoprene elements used for base isolation against seismic actions. Mater. Plast. **46**, 288–294 (2009)
11. Trigili, G.: Introduzione alla Dinamica delle Strutture e Spettri di Progeto. In: Dario, F. (ed.). Laccorio, Palermo, Italy (2010)
12. Radeş, M.: Mechanical Vibrationes. Ed. Printech, Bucureşti, Romania (2006)
13. Stanescu, N.D.: Vibrations of a shell with clearances, neo-Hookean stiffness, and harmonic excitations. Rom. J. Acoust. Vib. **13**, 104–111 (2016)
14. Vasile, O.: Active vibration control for viscoelastic damping systems under the action of inertial forces. Rom. J. Acoust. Vib. **14**, 54–58 (2017)
15. Inman, D.: Vibration with Control. John Wiley & Sons Ltd., London, UK (2007)

The Dynamic Compaction Regime of the Discretely Variable Viscoelastic Soil Parameters During the Work Process

Cornelia Florentina Dobrescu(✉)

INCD URBAN—INCERC, Sos. Pantelimon, 266, Bucharest, Romania

Abstract. The dynamic compaction of soil for road works is conducted with vibrating rollers that generate harmonic vibrations on the soil layers during constant-speed movements. The categories of the road foundation land on natural or filling soil are physically and mechanically distinguished by granularity, cohesiveness, clay content, stabilizing additives, water, etc. In this context, the rheological viscoelastic model differs according to the nature of the soil so that the elasticity and viscosity constitute the determining synthesis parameters in the schematization of the dynamic behavior. The research conducted on 15 categories of soil consisting of soils stabilized with ecological substances showed a linear viscoelastic behavior with discreetly variable elastic and viscous parameters after each passage of the vibrating roller. The experiments on the Romanian construction sites emphasize the influence of the discrete variation of the rigidity and the amortization of the stabilized soil layer in the post-resonance regime that corresponds to the optimal compaction field. In this context, this paper presents the families of curves that eloquently illustrate the influence of the discrete changes of the rheological parameters on the dynamic compaction performance.

Keywords: Rheologic model · Technological vibrations · Compactors · Stabilized soil

1 Introduction

The process of compaction by the vibration of the stabilized road structures can be modeled by the Voigt–Kelvin schematization. The mere soil composition determines both the elasticity and the viscosity so that the rigidity and the amortization change discreetly after each passage.

Based on experiments, three situations were highlighted after each passage of the compactor vibrating roller:

- Soils with discrete variable rigidity and amortization. They have 30% clay, 60% sand and gravel, and 10% stabilizer additives (lime, cement, liquid binders).

This article highlights the influence of the modification of rigidity and amortization on the dynamic response. In this context, the response curves of amplitude and transmitted force are represented for an experimental batch in Romania [1–3].

N. Herisanu and V. Marinca (eds.), *Acoustics and Vibration of Mechanical Structures – AVMS-2021*, Springer Proceedings in Physics 274,
https://doi.org/10.1007/978-3-030-96787-1_28

2 Dynamic Parameter Assessment

The vibrating roller–soil interaction is characterized by the k equivalent rigidity and the c equivalent viscous amortization under the action of the perturbator force $\boldsymbol{F} = \boldsymbol{F}(t) = \boldsymbol{F}_0 \boldsymbol{sin\omega t}$, where $\boldsymbol{F}_0 = \boldsymbol{m}_0\boldsymbol{r}\boldsymbol{\omega}^2$, and $\boldsymbol{m}_0\boldsymbol{r}$ is the static moment of the unbalance masses. As the mass of the vibrating compactor roller is m, then the linear dynamic system excited by the harmonic perturbator force $\boldsymbol{F}(t)$ *may be formulated as* <*m, c, k, F*>. Virtually, it is continuously conducted the scan of the excitation pulsation with the hydrostatic actuation of the vibrator inside the compactor roller [4].

In this case, the dynamic response curves can be raised by the continuous variation of the excitation pulsation, namely: the amplitude $\boldsymbol{A}(\boldsymbol{\omega}, \boldsymbol{k}, \boldsymbol{c}) = \boldsymbol{A}(\boldsymbol{\omega})$ and for the maximum force transmitted to the soil $\boldsymbol{Q}_0(\boldsymbol{\omega}, \boldsymbol{k}, \boldsymbol{c}) = \boldsymbol{Q}_0(\boldsymbol{\omega})$ [5–7].

The response curves correspond to the functions in relation to the continuous variable ω and to the discretely variable parameters k and c, as follows:

$$\boldsymbol{A}(\boldsymbol{\omega}, \boldsymbol{k}, \boldsymbol{c}) = \frac{\boldsymbol{m}_0\boldsymbol{r}\boldsymbol{\omega}^2}{\sqrt{\left(k - \boldsymbol{m}\boldsymbol{\omega}^2\right)^2 + \boldsymbol{c}^2\boldsymbol{\omega}^2}} \tag{1}$$

$$\boldsymbol{Q}_0(\boldsymbol{\omega}, \boldsymbol{k}, \boldsymbol{c}) = \boldsymbol{m}_0\boldsymbol{r}\boldsymbol{\omega}^2\sqrt{\frac{\boldsymbol{k}^2 + \boldsymbol{c}^2\boldsymbol{\omega}^2}{\left(k - \boldsymbol{m}\boldsymbol{\omega}^2\right)^2 + \boldsymbol{c}^2\boldsymbol{\omega}^2}} \tag{2}$$

The resonance regime ensures the technological compaction vibrations so that the dynamic parameters $\boldsymbol{A}(\boldsymbol{\omega}, \boldsymbol{k}, \boldsymbol{c})$ and $\boldsymbol{Q}_0(\boldsymbol{\omega}, \boldsymbol{k}, \boldsymbol{c})$ maybe formulated as follows:

(a) Amplitude in post-resonance is given by the relation

$$\boldsymbol{A} = \boldsymbol{A}_\infty = \frac{\boldsymbol{m}_0\boldsymbol{r}}{\boldsymbol{m}} \tag{3}$$

hence, it emerges that in post-resonance the amplitude of the technological vibrations maintains at the same value for any variation of the rigidity or amortization. For this reason, the compaction process is stable and uniform.

(b) Maximum transmitted force $\boldsymbol{Q}_0^p = \boldsymbol{Q}_{0,\infty}$ is

$$\boldsymbol{Q}_0 = \boldsymbol{Q}_{0,\infty} = \frac{\boldsymbol{m}_0\boldsymbol{r}}{\boldsymbol{m}}\boldsymbol{c}\boldsymbol{\omega} \tag{4}$$

or

$$\boldsymbol{Q}_{0,\infty} = \boldsymbol{A}_\infty\boldsymbol{c}\boldsymbol{\omega} \tag{5}$$

hence, it emerges that the transmitted force increases in direct proportion the $\boldsymbol{\omega}$ excitation pulsation [8–10].

For the experimented soils, there were retained the significant values of rigidity and amortization, after each passage on the same layer. In this case, we have: for $i = 1$, 2, 3, 4, 5 successive passages on the same layer, there were measured the following parameters: $\boldsymbol{k}_1 = 2 \cdot 10^7$ N/m, $\boldsymbol{k}_2 = 4 \cdot 10^7$ N/m, $\boldsymbol{k}_3 = 6 \cdot 10^7$ N/m, $\boldsymbol{k}_4 = 8 \cdot 10^7$ N/m, $\boldsymbol{k}_5 = 10 \cdot 10^7$ N/m, $\boldsymbol{m}_0\boldsymbol{r} = 2,5$ kgm, $\boldsymbol{m} = 5 \cdot 10^3$ kg, $\boldsymbol{\omega} = 0 \ldots 500$ rad/s.

3 Dynamic Response Curves Family

It is taken into consideration the situation, where the k_i rigidity as well as amortization $c_i = 2\zeta\sqrt{k_i m}$ change discretely as a result of the change of the fraction of the critical amortization ζ_i after each i passage of the vibrating roller as follows:

$\zeta_1 = 0.10$, $\zeta_2 = 0.15$, $\zeta_3 = 0.20$, $\zeta_4 = 0.25$, $\zeta_5 = 0.30$ [11–13].

The amplitude curves according to the continuous variation of ω and the discrete variation (k, ζ), after each i passage with $i = 1.5$ are presented in Fig. 1.

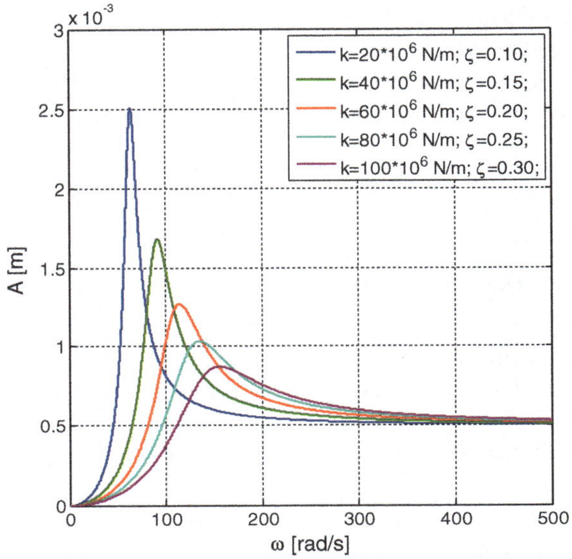

Fig. 1. Variation of vibrations amplitude according to ω and the pair of parameters k, ζ

The maximum transmitted force Q_0 according to pulsation ω and discrete variation (k, ζ), after each i passage with $i = 1.5$ is represented by the curves family in Fig. 2.

If the c ***viscosity coefficient is constant,*** it is found that at resonance the two parameters have different laws of variation, as follows:

(a) For the amplitude of the vibrations at resonance the calculation relation is

$$A^I_{rez} = \frac{m_0 r}{c}\omega_n = \frac{m_0 r}{c}\sqrt{\frac{k}{m}} \tag{6}$$

with the linear variation in relation to $\omega_n = \omega$, as it can be seen in Fig. 1. In this case, all points of resonance of amplitude A_{rez} are on a step in the axes system $A - \omega$

(b) For the amplitude of the force transmitted to the soil at resonance Q_0^{rez}, the calculation relation, from formula (2), can be written as follows

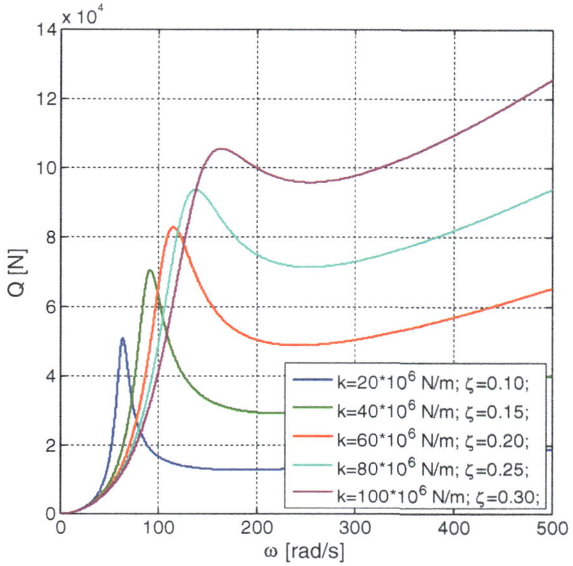

Fig. 2. Variation of maximum force transmitted to the soil according to ω and the pair of parameters k, ζ

.

$$Q_{0,I}^{rez} = \frac{m_0 r}{c}\sqrt{k^2\omega_n^2 + c^2\omega_n^4} \tag{7}$$

with the variation after a monotonically increasing curve with pulsation $\omega = \omega_n$. It is found that all the maximum points of force Q_0^{rez} are on curves parameterized by rigidity k in the $Q_0^{rez} - \omega$ system of axes, when the viscosity coefficient c is constant (Fig. 2) [14–16].

4 Conclusions

The research on several categories of soils enables the rheological and numerical modeling through the established computational relationships as well as the experimental modeling directly in the field and tested in the laboratory. At each passage of the vibrating roller, the rigidity of the k compacted layer and the ζ amortization rate were measured.

(a) For the category of compacted soil, the rigidity and the amortization change after each passage. In this case, the rigidity and the amortization rate ζ have 5 experimentally determined values. It emerges that for each passage the compacted layer of soil shall be characterized by the numerical values of parameters (k_i, ζ_i), where $i = 1, 5$. For the given case, the Voigt–Kelvin rheological model; for each i layer, it will have the discretely variable values k_i, ζ_i where $i = 1.5$.

(b) The $A-\omega$ family of response curves, each passage of the compactor vibrating roller is represented in Fig. 1. As a consequence of the fact that both k and ζ are discretely increasing in a simultaneous way, it emerges that the amplitude of the vibrations at resonance decreases along with the increase of the pulsation. Also, the points of the resonance peaks generate an "envelope" winding curve of resonance.
(c) The family of curves for the dynamic force transmitted to the compacted soil layer has peak values at resonance. The characteristic points belong to the curves according to the nature of the soil. In this case, there are highlighted "envelope" winding curves specific to each layer of soil with k_i, ζ_i as shown in Fig. 2.

It emerges that according to the numerically and experimentally raised curve of response, the viscoelastic behavior of the compacted soil can be evaluated or "read". Based on this analysis there can be established and "calibrated" the rheological model of the experimental batch specific to the road construction work area.

References

1. Bratu, P., Stuparu, A., Leopa, A., Popa, S.: The dynamic analyse of a construction with the base insulation consisting in anti-seismic devices modeled as a Hooke–Voigt–Kelvin linear rheological system. Acta Tech. Napoc. Ser. Appl. Math. Eng. **60**, 465–472 (2018)
2. Bratu, P., Stuparu, A., Popa, S., Iacob, N., Voicu, O., Iacob, N., Spanu, G.: The dynamic isolation performances analysis of the vibrating equipment with elastic links to a fixed base. Acta Tech. Napoc. Ser. Appl. Math. Eng. **61**, 23–28 (2018)
3. Dobrescu, C.F.: Highlighting the change of the dynamic response to discrete variation of soil stiffness in the process of dynamic compaction with roller compactors based on linear rheological modeling. Appl. Mech. Mater. **801**, 242–248 (2015)
4. Adam, D., Kopf, F.: Theoretical analysis of dynamically loaded soils, European workshop: Compaction of Soils and Granular Materials. ETC11 of ISSMGE, Paris, France (2000)
5. Bejan, S.: Analiza Performantei Procesului de Compactare Dinamica Prin Vibratii Pentru Structuri Rutiere. Ph.D. Thesis. "Dunarea de Jos" University of Galati, Galati, Romania (2015)
6. Bratu, P.: The behavior of nonlinear viscoelastic systems subjected to harmonic dynamic excitation. In: Proceedings of the 9th International Congress on Sound and Vibration. University of Central Florida, Orlando, FL, USA, 8–11 July 2002
7. Bratu, P., Debeleac, C.: The analysis of vibratory roller motion. In: Proceedings of the VII International Triennial Conference Heavy Machinery–HM 2011, Session Earth-moving and transportation machinery, Vrnjačka Banja, Serbia, pp. 23–26. 29 June–2 July 2011. ISBN 978-86-82631-58-3
8. Bratu, P., Stuparu, A., Popa, S., Iacob, N., Voicu, O.: The assessment of the dynamic response to seismic excitation for constructions equipped with base isolation systems according to the Newton–Voigt–Kelvin model. Acta Tech. Napoc. Ser. Appl. Math. Eng. **60**, 459–464 (2017)
9. Bratu, P.: Dynamic response of nonlinear systems under stationary harmonic excitation, nonlinear acoustics and vibration. In Proceedings of the 11th International Congress on Sound and Vibration, St. Petersburg, Russia, pp. 2767–2770, 5–8 July 2004
10. Dobrescu, C.F., Braguta, E.: Optimization of vibro-compaction technological process considering rheological properties. In: Springer Proceedings in Physics, Proceedings of the 14th AVMS Conference, Timisoara, Romania,

11. Herisanu, N., Marinca, V. (eds.): Springer: Berlin/Heidelberg, Germany. pp. 287–293, 25–26 May 2017
12. Leopa, A., Debeleac, C., Nastac, S.: Simulation of vibration effects on ground produced by technological equipments. In: Proceedings of the 12th International Multidisciplinary Scientific GeoConference and EXPO-Modern Management of Mine Producing, Geology and Environmental Protection, SGEM 2012, Albena, Bulgaria, vol. 5, pp. 743–750, 17–23 June 2012
13. Morariu-Gligor, R.M., Crisan, A.V., Serdean, F.M.: Optimal design of an one-way plate compactor. Acta Tech. Napoc. Ser. Appl. Math. Mech Eng. **60**, 557–564 (2017)
14. Pintoi, R., Bordos, R., Braguta, E.: Vibration effects in the process of dynamic compaction of fresh concrete and stabilized earth. J. Vib. Eng. Technol. **5**, 247–254 (2017)
15. Mooney, M.A., Rinehart, R.V.: Field monitoring of roller vibration during compaction of subgrade soil. J. Geotech. Geoenviron. Eng. ASCE **133**, 257–265 (2007)
16. Mooney, M.A., Rinehart, R.V.: In-Situ soil response to vibratory loading and its relationship to roller-measured soil stiffness. J. Geotech. Geoenviron. Eng. ASCE **135**, 1022–1031 (2009)

Dynamic Aspect for Self-Propelled Carriages

Gina Diana Muscă Anghelache(✉)

Engineering and Agronomy Faculty in Brăila, Research Centre for Machines and Technological Equipment's Mechanics, "Dunărea de Jos" University of Galai, Galai, Romania
diana.anghelache@ugal.ro

Abstract. Self-propelled carriages are used for the movement of lifting machines and loads. During the movement of lifting machines, dynamic phenomena not only occur during starting and braking periods of such movement, but also when irregularities of the rolling way come up, for example, crawler joints even when operating with rated speed. Vertical oscillations occur when passing over the rail's joints. For the dynamic study of the self-propelled carriage, we assume this can travel horizontally with rolling wheels' hoops and stiff rolling way. The carriages equip the gantry cranes, traveller cranes, etc. For the movement of traveller cranes, we also use carriages. These carriages are different from the general transport carriages through the following features: they only travel on the horizontal rolling way both on longitudinal and transversal directions. Both the carriages with load mechanisms and those for machine travel should not operate on declivity or obstacles rolling ways. Both the hoops of the carriage rolling wheels and the rolling way are stiff thus ensuring a stiff contact. We used a simplified physical model during the research of carriages dynamics. On the basis of this physical model we built the mathematical model, then, we realised the virtual instrument for dynamics of self-propelled carriages.

1 Introduction

The load carriage is an assembly designed to travel with hanged load, rolling on the rails from the steel structure of the overhead crane. The car undercarriage is a steel structure obtained by welding.

The elements of the undercarriage steel structure that supports the mechanical or electrical equipment of the drives are provided with shims welded on the outline having suitable thicknesses for being machined on their seating faces.

The undercarriage is provided with stoppers ensuring the certainty that this won't fall from the carriage rails even if the supporting shaft of the rolling wheel would be broken. The range of use for the carriage on the overhead crane is determined by its main parameters [8].

Besides the general use of machine parts (shafts, bearings, springs, assembly elements, or for movement transmission), the carriages for overhead cranes have a range of parts specific to such machines [7].

In Fig. 1, we noted: (1) undercarriage of the carriage; (2) load hoist motor; (3) cable drum; (4) single hook; (5) load hoist gearbox; (6) self-levelling roll; (7) rolling wheels

N. Herisanu and V. Marinca (eds.), *Acoustics and Vibration of Mechanical Structures – AVMS-2021*, Springer Proceedings in Physics 274,
https://doi.org/10.1007/978-3-030-96787-1_29

of the carriage; (8) translation mechanism motor; (9) translation mechanism gearbox; (10) buffer; (11) wheel driving shaft.

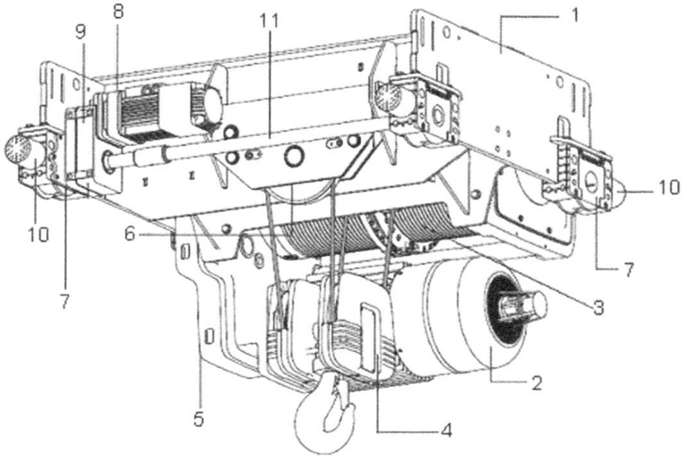

Fig. 1. Load carriage

The travel mechanism of the carriage is an assembly, where the driving unit (with the gearbox in closed housing) is located between the driven wheels, having shafts for carriage travelling.

The shafts are installed to compensate for the sag given by the stresses of the overhead crane and angular deviations caused by the undercarriage.

The suitably chosen buffers reduce the stresses by the impact, which can be transmitted to the steel structure of the overhead crane. The buffers are installed on the carriage. Stoppers are provided to limit the stroke of the carriage.

In Fig. 2, we noted: (1) Electric motor; (2) Coupling; (3) Gearbox; (4) Brake; (5) Axles; (6) Rolling wheels. Driving is done with electric motor 1, coupling with gearbox 3 is done by means of spring coupling 2. Transmission to wheels 7 is provided by intermediate axles 5 and spring coupling with brake wheel whereon the brake with shoes 4 operates.

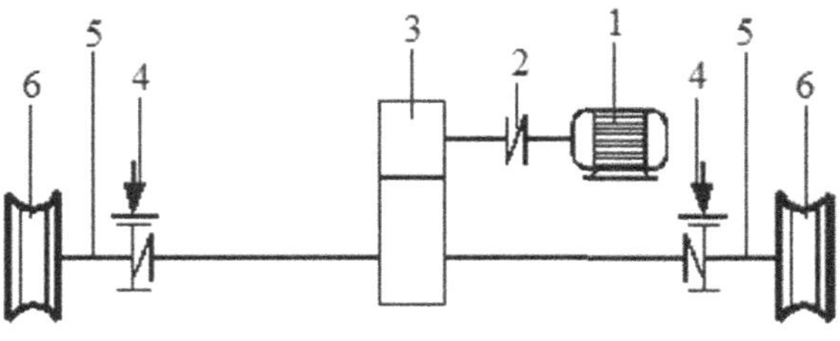

Fig. 2. Travel mechanism of the overhead crane carriage

The study of self-propelled carriages supposes the following stages: determining the physical model, determining the mathematical model and solving differential equations, achievement of a specialized software for experimental determinations.

2 Physical Model

Further to the study of self-propelled carriages, we determined the physical model which can be adopted for research of carriages dynamics and it is shown in Fig. 3 [3, 4].

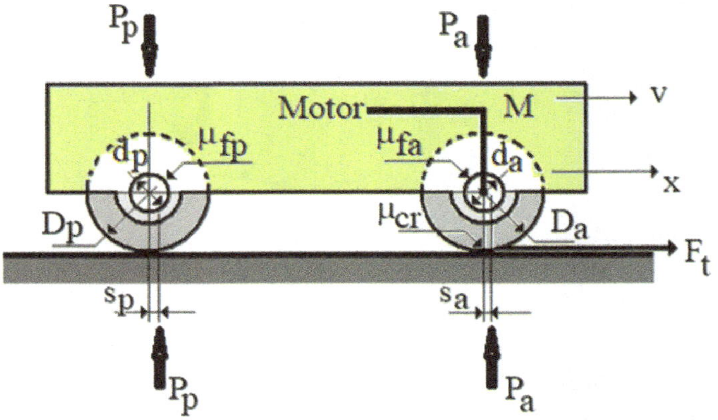

Fig. 3. Self-propelled carriage

In Fig. 3,we made the following notations: M is the total weight of the carriage, P_a is the pressing on the bearings of the active wheels, P_p pressing on the passive bearings D_a and D_p are the rolling diameters of the active and passive wheels, d_a şi, d_p are the diameters of the spindles of the two categories of wheels [5].

Friction coefficients for sliding in the spindles of the two categories of wheels are noted with μ_{fa} and μ_{fp}.

Friction coefficients for roll-over between the wheel and rolling wa are noted with s_a and s_p, different, as they also depend on the wheel diameter.

In this research problem, the friction coefficient for sliding between the wheel and rolling way, μ_{cr} (for drive wheels, it produces the friction force needed for traction, F_t) has a very important role.

As it is known, this force is applied on the wheel and, it opposes the relative movement of the wheel related to the rolling way.

Within the following calculations, the clockwise directions are considered positive for translation movements (travels, speeds, accelerations) and the rotation movements that are in relation with a clockwise rolling are also considered positive.

On the basis of this physical model, the mathematical model is developed, on the basis of which a virtual model was achieved which is the basis for solving the dynamics of self-propelled carriages [6].

3 Mathematical Model

As shown when modelling friction forces, the friction for sliding is of dry type. Depending on the kind of materials in contact, the friction force is strongly dependent on the speed of the friction, either decreasing with the speed as for metal–metal couples, or increasing with the speed as for antifriction material couples. For dry friction between wheel and rolling way, we will use the relation:

$$\mu_{cr} = -\mu_o e^{a\left|\dot{\alpha}\frac{D_a}{2}-\dot{x}\right|} sgn\left(\dot{\alpha}\frac{D_a}{2} - \dot{x}\right) \tag{1}$$

where $\mu_0 = 0,3$ is the friction coefficient for pause (adherence) for steel–steel couple, $a = -0,3$ is an influence coefficient of the relative speed for the same couple, $\left|\dot{\alpha}\frac{D_a}{2} - \dot{x}\right|$ is the module of relative speed for friction (in our case the friction value is influenced by the value of relative speed, not by its direction), α = rotation angle of the wheel, x = carriage travel, $\dot{\alpha}$, $\dot{x}$ the speeds of respective movements. In analytical applications, for the friction to depend, as direction, on the direction of the relative movement, the value so calculated for friction coefficient is multiplied by the sign of relative speed, $sgn\left(\dot{\alpha}\frac{D_a}{2} - \dot{x}\right)$. It is known that the function sgn (the sign of…) has the true values:

$$sgn(x) = \begin{cases} +1, & for\ x > 0 \\ 0, & for\ x = 0 \\ -1, & for\ x < 0 \end{cases}.$$

With these, the movement of the carriage can be described by means of differential equations applying the efforts balance as per d'Alembert principle and grouping within the left member the efforts that oppose to the movement, and, within the right member the efforts that cause the movement [3].

To study the process, we write differential equations that can be solved in two ways, namely: analytical or numerical.

Precise solutions are got by analytical solving of equations, and approximate solutions are got by numerical solving of equations. The analytical method involves complicated calculations being applicable only in particular cases [4].

Analytical methods offer advantages because solutions can be obtained for all types of equations, the calculation is elementary but have a great volume.

Numerical methods have an own algorithm for solutioning. Based on the physical model represented in Fig. 3, we write the system of differential equations that describe the motion of the whole mechanism as follows:

$$\begin{cases} J \cdot \ddot{\alpha} + M_{f1} \cdot sgn(\dot{\alpha}) + F_t \cdot \frac{D_a}{2} = M_M \\ M \cdot \ddot{x} + F_{rez} sgn(\dot{x}) = F_t \end{cases} \tag{2}$$

The dynamic balance of the drive wheel is written in the first equation, where J is their moment of inertia plus all moments of inertia of the drive reduced to their axle, α is the rotation angle of the drive wheels, M_{f1} is the friction moment in bearings, M_M is the motor moment reduced to the axle of drive wheels, F_t is the traction force. The dynamic balance of the car is written in the second equation where F_{rez} is the resisting force to carriage travel. The terms which come up in (2) are calculated with the relations:

$$M_{f1} = P_a \cdot \mu_{fa} \cdot \frac{d_a}{2} \tag{3}$$

$$F_t = P_a \cdot \mu_{cr} \tag{4}$$

$$F_{rez} = \frac{P_p \cdot s_p + P_p \cdot \mu_{fp} \cdot \frac{d_p}{2}}{\frac{D_p}{2}} + \frac{P_a \cdot s_a}{\frac{D_a}{2}} \tag{5}$$

The resisting force of the carriage is produced by roll-over friction of the active and passive wheels as well as by sliding friction in the bearings of passive wheels. Sliding friction in the bearings of active wheels is beaten by the electric motor, in the first equation of (2). The equations system (2) completed with the other expressions cannot be integrated in an analytical way [1]. The motor torque is variable.

The function sgn creates this problem, but without, it a unitary mathematical model could not be written. Non-linearity comes up from the expression of the coefficient of friction with the rolling way. On the basis of numerical methods, we can easily solve this mathematical model [2].

4 Experimental Determinations

For dynamic analysis of a self-propelled carriage, we assume a carriage that is able to travel horizontally with the rigid treads of the rolling wheels and rolling way.

Experience shows that many well-known software products do not solve concrete problems correctly or competitively and that is why a software specialized in numerical simulation has been created.

Experience shows that many commercial software products do not solve concrete problems correctly or competitively and that is why a software specialized in numerical simulation has been created.

We start with the following input data (calculation program asks the user the following data from the keyboard): Diameter of the passive wheels: 400 mm; Diameter of the active wheels: 400 mm; Diameter of the passive spindles: 100 mm; Diameter of the active spindles: 100 mm; Coefficient of friction in the passive spindles: 0.05; Coefficient of friction in the active spindles: 0.05; Roll-over friction of the passive wheels: 1 mm; Roll-over friction of the active wheels: 1 mm; Load on the passive wheels: 2,000 daN; Load on the active wheels: 2,000 daN; Electric motor power: 3 kW; Sync. Motor speed: 1,500 r.p.m.; Nominal motor speed: 1,470 r.p.m.; Carriage nominal speed: 1 m/s; Active wheels moment of inertia: 0.45 kg m^2; Carriage weight: 4,000 kg and movement time: 5 s.

Further to soft running, we obtain the diagram in Fig. 4.

We can notice that the wheels' speed is between 9.288 and 47.746 r.p.m. that leads to a carriage travel up to 4.189 m with 1 m/s speed and wheel spin will be among 0.001 and 0.048 m/s.

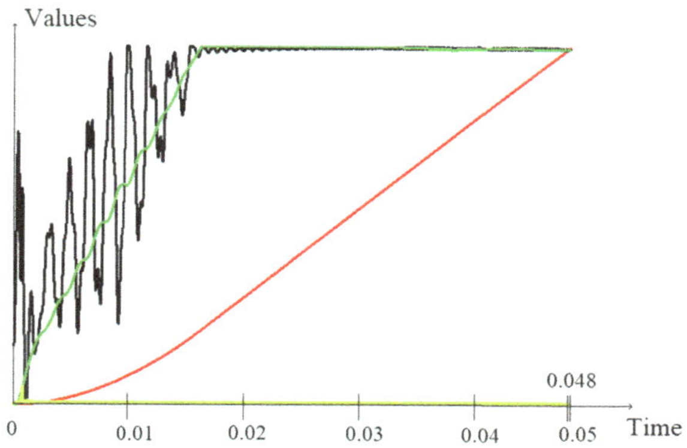

Fig. 4. The results of starting the carriage having 4,000 kg weight

In case we double the carriage weight, becoming 8,000 kg, we will obtain the graph in Fig. 5. Wherefrom we see that despite the motor speed remaining almost unchanged (min. 9.938–max. 47.746), this is stabilized later than in the previous case which will lead to a shorter travel of 3.389 m with a greater spin of the wheel up to 0.057 m/s.

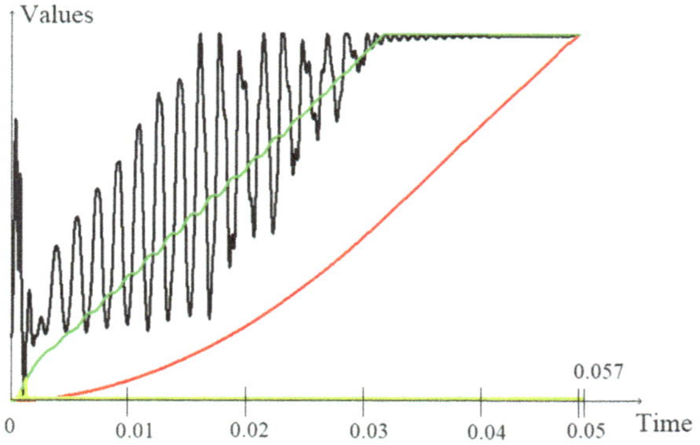

Fig. 5. The results of starting the carriage having 8,000 kg weight

Reverting to the same weight, we will double the wheels' diameter, one by one. In Fig. 6, we will have the active wheels diameter of 800 mm and in Fig. 7 we will modify the passive wheels diameter to 800 mm.

We notice that in the case of active wheel modification, we have a decrease of wheel speed of down to 23.873 r.p.m on a distance a little bit greater than 4.448 m and the wheel spin will have a maximum of 0.149 m/s. In the case of passive wheel modification, the differences are almost neglectable, the travelled distance being 4.240 m.

As a last analysis, we will modify only the motor power to 6 kw, the remaining data coming back to the initial ones and we will obtain the graphs in Fig. 8. For this case we can notice the harshest modification, although the wheel speed keeps the maximum identical to the initial one, this is reached faster and sudden, which has an impact upon

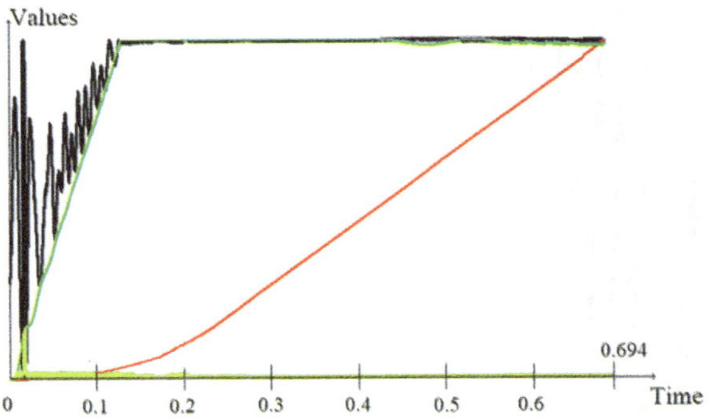

Fig. 6. The results of starting the carriage having the active wheels diameter of 800 mm

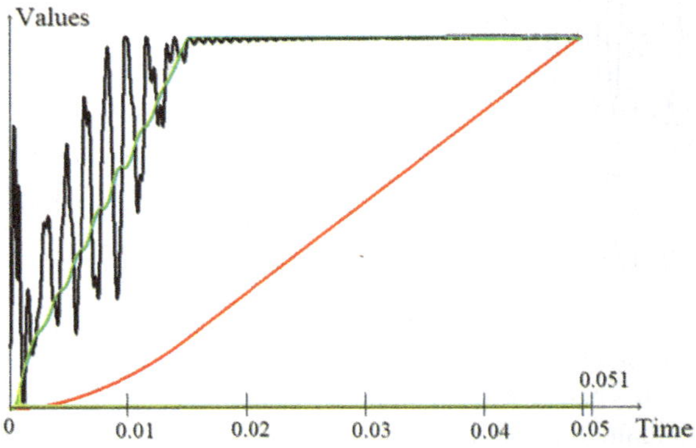

Fig. 7. The results of starting the carriage having the passive wheels diameter of 800 mm

wheel spin, having the maximum of 0.694 m/s and the travel has a very small change in this time, being 4.427 m.

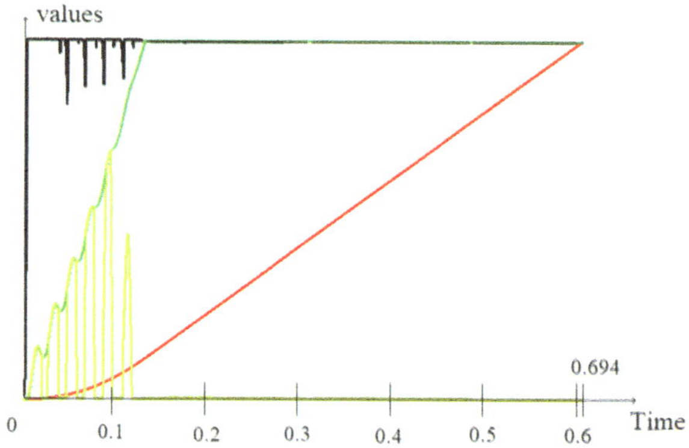

Fig. 8. The results of starting the carriage having motor power of 6 kw

5 Conclusions

Further to experimental research, we notice that by changing the diameter of active wheels, we will obtain smaller speeds but the distance travelled is longer, in exchange, within previous studies, the diameter of passive wheels has a neglectable impact in case their change is not a harsh one.

Its weight has a strong consequence, as when increasing or decreasing we will have an impact on the travelled distance, the heavier it is the travel is smaller and wheel spin increases. And in case of changing the motor power, we will have the opposite impact to the previous one, if we increase the motor power, we will have longer travel distances with greater spins.

References

1. Muscă (Anghelache), G.D., Năstac, S.: Dynamic modelling of overhead crane. In: IOP Conference Series: Materials Science and Engineering, vol. 916, Article number 012071. (2020). https://doi.org/10.1088/1757-899X/916/1/012071 (Published)
2. Nastac, S.: Numerical Analysis With Applications In Mechanical Engineering, p. 100. Impuls Publishing House, Bucuresti (2004). ISBN 973-8132-45-2
3. Oproescu, Ghe.: Modelling of Dynamic Processes on Cable Lifting Machines. Impuls Pulishing House Bucharest (1997)
4. Oproescu, Ghe.: Anghelache G.D.: Dynamics Elements of Lifting and Transport Machines. Galaţi University Press Publishing House, p. 142. ISBN 978-606-8008-71–4
5. Oproescu, Ghe., Năstac, S.: Numerical Modelling Elements, chap. 5, 8.3, 8.4, p. 37. Libertatea Publishing House, Brăila (2000). ISBN 973-99574-5-5

6. Potîrniche, A., Căpățână, G.: Theoretical and experimental analysis for a pendulum-based absorber system used in vibration isolation. In: Proceedings of XIV-th International Symposium "Acoustics and Vibration of Mechanical Structures" AVMS 2017, 25–26 may 2017, Timișoara, published in "Springer Proceedings in Physics", December 2017. ISBN-13: 978-3-319-69823-6
7. Potîrniche, M.A., Căpățână, G.F., Muscă (Anghelache). D.G.: Considerations on geometric modelling and finite element analysis for a trolley travel mechanism of an overhead crane. In: MODTECH International Conference—Modern Technologies in Industrial Engineering, June 23–26 2021
8. Spânu (Ștefan), G.C., Potîrniche, A., Căpățână, G.F.: Dynamic analysis of vibrating conveyors modeled as elastic mechanical systems with three degrees of freedom, Synth. Theor. Appl. Mech. **9**(2) (2018). ISSN 2068–6331

Research on Optimizing the Locations of the Sensors of the Strain Gauge Measurement System for Structures Which Vibrate

Cardei Petru(✉) and Matache Mihai

INMA, 013813 Bucharest, Romania
icsit@inma.ro

Abstract. The article presents the results of theoretical research dedicated to the improvement of measurement systems based on deformation sensors. The increase of the efficiency of the measurement systems based on deformation sensors is also supported by theoretical simulations in order to obtain more precise measurements and to highlight the oscillating characteristics of some mechanical systems. Determining the optimal gluing locations of the deformation sensors on the measured structure (or on an intermediate structure) and highlighting the possibility of obtaining and the necessary conditions for linear relationships between the loads applied to the structure and the stresses or deformations measured by the sensors are the main results presented. Both results lead to the increase of the measurement precision and the possibility of highlighting some mechanical effects of vibratory nature, but also to the increase of the efficiency of the calibration operations of these measurement systems. The authors also suggest the approach of numerical solutions for solving the problems of simulation of measurement systems based on strain gauge sensors, but also emphasize the difficulties associated with these procedures.

Keywords: Optimization · Measurement · Strain · Gauge · System

1 The Statement of the Problem and the Way to Solve It

Numerous experimental researches use measurement systems based on deformation sensors, in order to determine sizes characteristic of metal structures, but not only, [1–11]. The measurements are made for research purposes, validation, design of remedial solutions and improvement or optimization for various industrial structures or devices used in research. Measurement systems based on deformation sensors can be mounted directly on the researched structure or can belong to intermediate metal structures, the measurements being indirect, by measuring the effects of the process studied in the additional structure.

Whatever the measurement system, there are a number of important problems to solve, including identifying optimal locations for measuring the desired effects, establishing the directions of measurement and sensor formations used for this purpose,

N. Herisanu and V. Marinca (eds.), *Acoustics and Vibration of Mechanical Structures – AVMS-2021*, Springer Proceedings in Physics 274,
https://doi.org/10.1007/978-3-030-96787-1_30

optimizing the number of sensors to use efficiently the number of recording channels available. To solve these problems, measuring systems are usually subjected to calibration operations. The calibration operations aim at determining some relations between the quantities measured by the sensors of the measurement system and the parameters to be measured, especially when the measurements are indirect [2, 12].

In general, the structures on which measurements are made are very particular. Bar structures have various components and shapes, other structures have bar-type components but also surfaces, often bodies that require three-dimensional modelling. The application of contact with the environment (border and initial conditions) and loads are also applied very particularly. For these reasons, there is often no algorithmic approach. When the sensors of the measuring system stick directly to the structure that is subjected to the measurements, the experience of the engineers speaks for itself most of the time. This paper gives ways to determine the optimal locations for gluing sensors or at least unquantified indications. Also, methods will be given for obtaining the relations between the measured quantities and the quantities registered by the sensors of the measuring system.

1.1 Working Method

The proposed working method aims to find useful information about the optimal soldering locations of the deformation sensors and to obtain relationships between the quantities to be measured and the quantities recorded by the sensors of the measuring systems. To make the method easier to understand, it will be presented through an elementary application that has an analytical solution, easy to use for both purposes.

It is considered a straight square pipe type bar with an outer side of 0.1 m and a wall thickness of 0.01 m. The material of the bar is steel with the module of linear elasticity $E = 2.1 \cdot 10^{11}$, density $\rho = 7850$ kg/m^3. In these conditions are found the inertial characteristics of the section, the area $A = 0.0036$ m$_2$, the moment of geometric inertia, $I = 0.00000492$ m^4, and the resistance module $W = 0.0000984$ m^3. A sketch of the bar, its request, and support is given in Fig. 1.

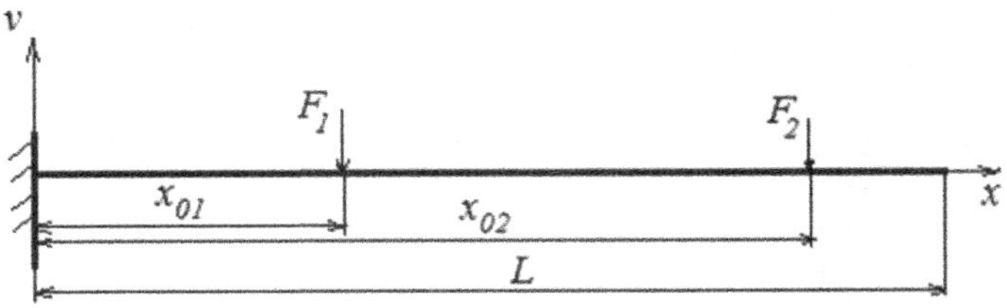

Fig. 1. Sketch of the load and fixing of the bar subjected to a double load of dynamic bending

Two concentrated bending forces stress the bar according to the equations:

$$F_1(x, t) = P_{01}\delta(x - x_{01}) \sin(\omega_{01}t), F_2(x, t) = P_{02}\delta(x - x_{02}) \sin(\omega_{02}t) \quad (1)$$

where P_{01}, P_{02} are the forces magnitudes, x_{01}, x_{02}, the coordinates on the bar, of the points of application of the forces, t is the time, x is the coordinate on the bar, and δ is

Dirac function. The problem of bending vibrations is solved according to [13–15]. The mathematical model is linear (by hypothesis we consider that the bar is required in the linear elastic field) the solution of the problem is the sum of the solutions corresponding to each bending force separately:

$$v(x, t) = v_1(x, t) + v_2(x, t) = \\ = \left(\sum_{k=1}^{N} C_{1k} Y_k(x)\right) \sin(\omega_{01} t) + \left(\sum_{k=1}^{N} C_{2k} Y_k(x)\right) \sin(\omega_{02} t) \tag{2}$$

where v is the vertical variation from the initial position, in time for each point of the vibrating bar, v_1, v_2 are the partial solutions for each of the two forces, $Y_k(x)$ are the eigenfunctions of the bar in Fig. 1, C_{k1}, C_{k2}, $k = 1, \ldots, N$, being the coefficients of these functions in the series that define the solutions, [13, 14]. N is the number of terms considered in the finite solution, because in the calculation one cannot usually consider an infinite number of terms. In Eq. (2), we can consider $N = 2$ or 3, because the coefficients C_{1k} and C_{2k} decrease very quickly with k. The coefficient C_{13} is about 10^5 times smaller than the coefficient C_{11}, and the coefficient C_{23} is about 10^5 times smaller than the coefficient C_{21}. Equation (2) is used, according to [3], for the calculation of the stress in the bar at each point and for each moment of time, according to the formula:

$$\sigma(x, t) = \frac{-EI}{W} \frac{\partial^2 v(x, t)}{\partial x^2} \tag{3}$$

where σ is the axial stress in the bar. According to similar formulas, the stresses generated in the bar by each of the two bending forces are calculated separately. The full calculation is longer, requiring a large space for exposure. We do not reproduce it in the article because it is a routine calculation and can be done by those who want, for example, after [3].

1.2 Results

In this subchapter are given in graphical form some results that help to understand the conclusions of the study.

Figures 2 and 3 show the time dependence of the stress (total, produced by the action of force F_1, respectively, produced by the action of force F_2) on time, in two points on the bar, $x = 0.15$ m (closer to the origin or to the point where the bar is fixed by embedding) and $x = 0.975$, close to the point where the force F_1 acts.

For a more complete understanding of the dependence of the axial stress on the bar coordinate and time, in Fig. 4 is represented (3) as a function of two variables, which represent a surface in three-dimensional space.

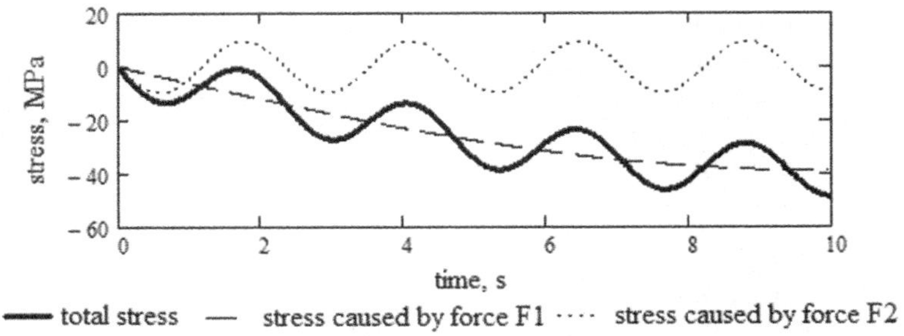

Fig. 2. The variation of the total stress and of the stress generated by each of the bending forces in time at the point on the bar $x = 0.15$ m

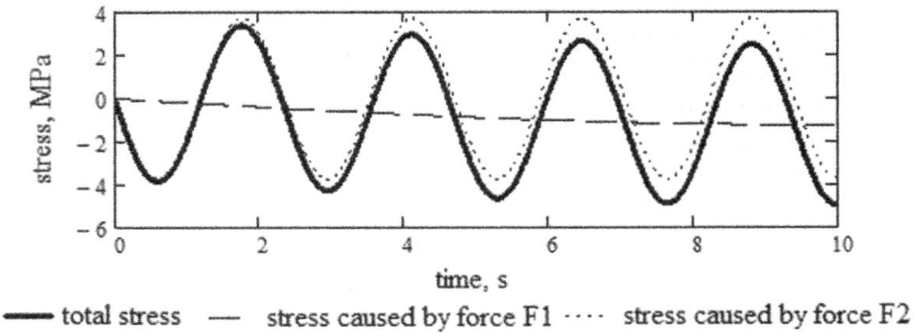

Fig. 3. The variation of the total stress and of the stress generated by each of the bending forces in time at the point on the bar $x = 0.975$ m

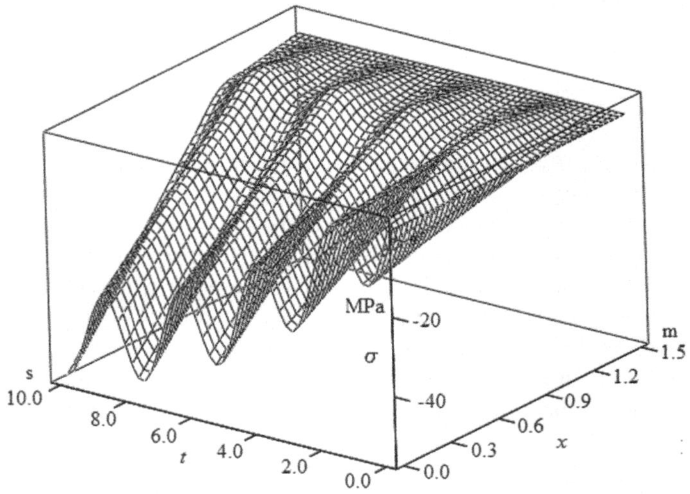

Fig. 4. Time and space dependence of the axial stress in the bar

2 Optimal Points for Gluing Strain Gauge Sensors

Observing the variation of the axial stress in time, in two points on the bar (Figs. 2 and 3), it is natural to ask ourselves if we can highlight points on the bar where the oscillating component can be more easily highlighted. For this aim, it is built the next function:

$$R = \left| \frac{\sigma_2(x,t)}{\sqrt{\sigma_1(x,t)^2 + \sigma_2(x,t)^2}} \right| \tag{4}$$

Function (4) represents the ratio between the axial stress generated by the oscillating force F_2 and the Euclidean norm of the axial stresses generated by the two forces. In order to estimate as accurately as possible, the maximum value of this component, we are looking for a maximum value of the objective function (4), depending on the coordinate on the bar, which has become an optimization parameter. The optimization of function (4) also depends on the time, but considering that the indications of the sensors are recorded for a longer time than the period of oscillation of force F_2, the optimal values will be recorded several times. The coordinates of the optimal points for gluing the sensors on the bar vary in time (Fig. 5).

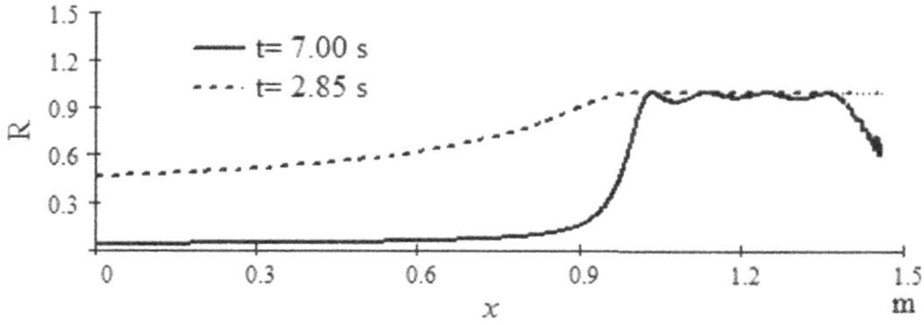

Fig. 5. Optimal points on the bar for gluing the deformation sensors

In Fig. 6, a graphical representation of the objective function (4) shows that the optimal area for highlighting the amplitude of the voltage generated by the force 2 (and the gluing of the deformation sensors), is the area between the points of application of the forces.

3 Linear Relations Between Forces and the Measured Stresses

In order to deduce the calculation relations between the forces applied to the bar and the measured stresses, we start from relation (2), taking into account that P_{01} and P_{02} are constant and can be taken out in front of the sums writing it as follows:

$$v(x,t) = P_{01}\left(\sum_{k=1}^{N} \overline{C}_{1k} Y_k(x)\right)\sin(\omega_{01}t) + P_{02}\left(\sum_{k=1}^{N} \overline{C}_{2k} Y_k(x)\right)\sin(\omega_{02}t) \tag{5}$$

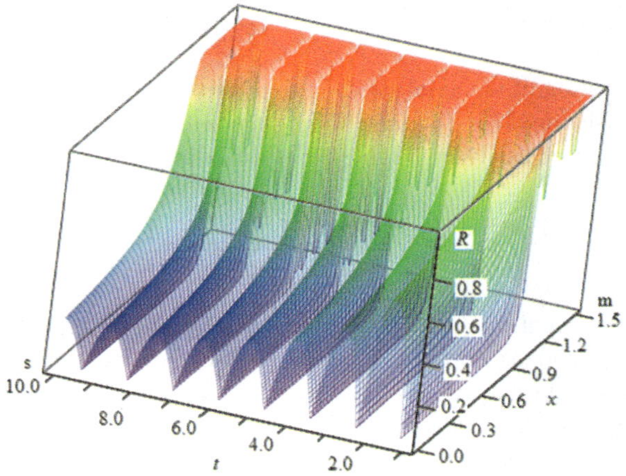

Fig. 6. Dependence of the function R by time and space representation as a surface in three-dimensional space

where

$$C_{1k} = P_{01}\overline{C}_{1k}, C_{2k} = P_{02}\overline{C}_{2k} \tag{6}$$

Noting:

$$\theta_i(x, t) = -\frac{EI}{W}P_{0i}\left(\sum_{k=1}^{N}\overline{C}_{ik}\frac{\partial^2 Y_k(x)}{\partial x^2}\right)\sin(\omega_{0i}t), i = 1, 2 \tag{7}$$

we can write the relationship:

$$\sigma(x, t) = \theta_1(x, t)P_{01} + \theta_2(x, t)P_{02} \tag{8}$$

Assuming that two deformation sensors are glued on the bar, along the bar, in the locations $x = x_1$ and $x = x_2$, at a given time of recording, $t = t^*$, according to (8), there are the relations:

$$\begin{aligned} \sigma_1 = \sigma(x_1, t^*) = \theta_1(x_1, t^*)P_{01} + \theta_2(x_1, t^*)P_{02}, \\ \sigma_2 = \sigma(x_2, t^*) = \theta_1(x_2, t^*)P_{01} + \theta_2(x_2, t^*)P_{02}. \end{aligned} \tag{9}$$

If the condition

$$\theta_1(x_1, t^*)\theta_2(x_2, t^*) - \theta_1(x_2, t^*)\theta_2(x_1, t^*) \neq 0 \tag{10}$$

is met, then the solution of the linear system of Eq. (9) is

$$\begin{aligned} P_{01} = \frac{\theta_2(x_2, t^*)\sigma_1 - \theta_2(x_1, t^*)\sigma_2}{\theta_1(x_1, t^*)\theta_2(x_2, t^*) - \theta_1(x_2, t^*)\theta_2(x_1, t^*)}, \\ P_{02} = \frac{\theta_1(x_1, t^*)\sigma_2 - \theta_1(x_2, t^*)\sigma_1}{\theta_1(x_1, t^*)\theta_2(x_2, t^*) - \theta_1(x_2, t^*)\theta_2(x_1, t^*)} \end{aligned} \tag{11}$$

Noting:

$$\begin{aligned}
c_{11} &= \frac{\theta_2(x_2, t^*)}{\theta_1(x_1, t^*)\theta_2(x_2, t^*) - \theta_1(x_2, t^*)\theta_2(x_1, t^*)}, \\
c_{12} &= \frac{-\theta_2(x_1, t^*)}{\theta_1(x_1, t^*)\theta_2(x_2, t^*) - \theta_1(x_2, t^*)\theta_2(x_1, t^*)}, \\
c_{21} &= \frac{\theta_1(x_1, t^*)}{\theta_1(x_1, t^*)\theta_2(x_2, t^*) - \theta_1(x_2, t^*)\theta_2(x_1, t^*)}, \\
c_{22} &= \frac{-\theta_1(x_2, t^*)}{\theta_1(x_1, t^*)\theta_2(x_2, t^*) - \theta_1(x_2, t^*)\theta_2(x_1, t^*)}
\end{aligned} \tag{12}$$

the linear relations between the intensities of the forces applied to the bar and the stresses measured in two points of it can be written as follows:

$$P_{01} = c_{11}\sigma_1 + c_{12}\sigma_2, P_{02} = c_{21}\sigma_1 + c_{22}\sigma_2 \tag{13}$$

4 Conclusions

Theoretical research conducted leads to some useful conclusions for users who measure mechanical forces using measurement systems based on deformation sensors.

(1) Mechanical force measurement systems based on deformation sensors can be perfected by numerical simulation. This procedure prepares the calibration operation and the effective measurement operation. By theoretical calculation, optimal bonding locations for strain gauge sensors can be chosen based on various objective functions. We have given the example in this article of an objective function that aims to highlight in measurement the effect of one of the forces applied to the structure. Since that shape is oscillating, determining the optimal solder location of the sensor also facilitates spectral analysis of the signal, which allows a more accurate knowledge of the vibration phenomenon.

(2) The theoretical analysis allows the determination of some linear relations between the intensities of the applied forces, at least in the case of the harmonic excitations used in the example given in this article. For forces with non-harmonic intensity, one can try to develop them in Fourier evenings. The theoretically determined linear relations facilitate the control in the calibration operation that will confirm or correct the theoretical relations.

(3) In the case of complex structures, for example, a load-bearing structure of agricultural vehicles or machines, the theoretical solution is very difficult to obtain. Numerical solutions are recommended, for example, those based on finite difference schemes or the finite element method. Numerical solutions can be used for both purposes in which we used the theoretical solution. The main problems of obtaining numerical solutions are related to: geometric modelling, physical modelling (boundary and initial conditions), and exact knowledge, in detail of the possibilities of numerical simulation of the programmes used, correct interpretation of results.

(4) The linear relations between the intensities of the forces and the measured stresses can result from a well-defined system of linear equations (which meet not only the condition of the main determinant nonzero, but also additional conditions imposed on the coefficients, for a precise solution), but also from a large set of experimental data using the least-squares method, for example. If the theoretical solution is not possible and the desired linear relations are deduced by calibration, the last calculation method is recommended.

Acknowledgements. This work was supported by the Romanian Research and Innovation Ministry, through the Project entitled "Researches on achieving integrated systems for the bio economy field according to the concept of intelligent agriculture" – PN 19 10 01 01 – Ctr. 5N/07.02.2019, and by a grant of the Ministry of Research, Innovation and Digitization (MRID), CCCDI – UEFISCDI, project number PN-III-P1-1.2-PCCDI-2017-0560 "Eco-nanotechnologies and intelligent equipment for soil properties mapping and evaluating the dynamics of the plant in order to improve agricultural production and environmental protection," within PNCDI III and was done by "NUCLEU" Programme, developed with the support of MRID project PN 19 10 01 02.

References

1. Tripa, P.: Metode experimentale pentru determinarea deformatiilor si tensiunilor mecanice, Editura Mirton, Timisoara (2010)
2. Vanwalleghem, J., de Baere, I., Loccufier, M., Van Paepegem, W.: Dynamic calibration of a strain gauge based handlebar force sensor for cycling purposes, 7th Asia-Paccific Congress on Sports Technology, APCST 2015. Proceedia Engineering **112**, 219–224 (2015)
3. Hoffman, K..: An Introduction to Stress Analysis and Transducer Design using Strain Gauges. HBM, Darmstadt (2012)
4. Hoffman, K.,: An Introduction to Measurement using Strain Gauges. HBM, Darmstadt (1989)
5. Zhao, Y., Liu, Y., Li, Y., Hao, Q.: Development and application of resistance strain force sensors. Sensors **20**, 5826 (2020)
6. dos Santos, F., L., Peeters, B.: On the use of strain sensor technologies for strain modal analysis: Case studies in aeronautical applications. Rev. Sci. Instrum. **87**, 102506 (2016)
7. Ene, T., A., Mocanu, V.: Dynamics of oversowing machine MSPD 2.5 related to working and constructive parameters. INMATEH-Agric. Eng., **44**(3), (2014)
8. Roeber, J., B., W.: Tractor measurement and data acquisition system for hydraulic power, draft force, and power take-off torque, Thesis, University of Nebraska (2016)
9. Drobota, V.: Rezistenta materialelor, Editura Didactica si Pedagogica, Bucuresti (1982)
10. Kostić, M.,M., Rakić, D., Z., Ličen, H., H, Malinović, N., C.: Design and construction of three point hitch device for measuring draft tillage implement - Data aquisition and post processing analysis. J. Food, Agric. & Environ., vol. **12**(2), pp. 1300–1307, (2014)
11. Kim, Y., S.,Kim, T., J., Kim, Y., J., Lee, S., D., Park, S.,U., Kim, W.,S.: Development ao a Real-Time Tillage Depth Measurement System for Agricultural Tractors:Application to the Effect Analysis of Tillage Depth on Draft Force during Plow Tillage, Sensors, **20**(3), (2020)
12. Swiech, L.: Calibration of sa Load Mesurement System for an Unmanned Aircraft Composite Wing Based on Fibre Bragg Gretings and Electrical Strain Gauges. Aerospace, vol. **72** (2020)
13. Buzdugan, Gh., Fetcu, L., Rades, M.: Vibratii mecanice, Editura Didactica si Pedagogica, Bucuresti (1982)

14. Han, S., M., Benaroya, H., Wei, T.: Dynamics of transversely vibrating beams using four engineering theories. J. Sound Vib., **225**(5), 935–988(1999)
15. Timoshenko, S., P.: On the transverse vibrations of bars of uniform cross-section. Philos. Mag., pp. **125**(1922)

Structural Vibration

Nonlinear Vibrations Control of Rotating Thin-Walled Piezo-Composite Beams

Jerzy Warminski(✉), Jaroslaw Latalski, and Zofia Szmit

Department of Applied Mechanics Faculty of Mechanical Engineering, Lublin University of Technology, Nadbystrzycka 36, 20618 Lublin, Poland
j.warminski@pollub.pl, j.latalski@pollub.pl, z.szmit@pollub.pl

Abstract. A nonlinear model of a rotating structure comprising a cylinder hub and three clamped thin-walled composite box beams is studied in this paper. The piezoelectric material layers are embedded into the specimens laminate to suppress vibrations of the rotor near resonance states. The reduced mathematical model of the rotor takes into account nonlinear electro-mechanical properties of the active blades as well as hub dynamics considered as a rigid body. Based on former authors research the governing equations of motion are written and resonance curves are obtained considering two system configurations, namely a fully symmetric design with identical beams and a non-symmetric one. The accounted for detuning results from the variation of the reinforcing fibres orientation and change in blade stiffness. To suppress vibrations a control unit is introduced to the system. This is represented in the mathematical formulation as an additional governing equation with quadratic coupling to the rotor generalized variables. This approach corresponds to the nonlinear saturation control strategy. Effectiveness of the proposed control method for selected operational states and excitation conditions is presented.

1 Introduction

Rotating structures play a very important role in many branches of mechanical and aerospace engineering. Representative examples of actual designs might be steam and gas turbines, wind power generators, compressors, aircraft propellers, helicopter rotor blades, etc.

Recent trends in modern designs are aimed at enhancing their functionality beyond the regular tasks. This may refer to structural health monitoring or actively responding to the changes in operational conditions. This goal may be achieved by incorporating active materials, like piezoceramics, shape memory alloys or magnetostrictive ferro fluids to tailored hosting structures usually made of lightweight composites [1]. Further enhancement is structural control technology—an approach combining the master structure with control unit and suitable control strategy.

N. Herisanu and V. Marinca (eds.), *Acoustics and Vibration of Mechanical Structures – AVMS-2021*, Springer Proceedings in Physics 274,
https://doi.org/10.1007/978-3-030-96787-1_31

One of the numerous control algorithms is a nonlinear saturation control method. This vibration suppression technique is based on the internal resonance phenomenon. The controller is added to the structure to represent a damped second order system which is coupled to the plant via quadratic nonlinearities. At internal resonance, the nonlinear terms act as a platform transferring the energy between the plant and the controller mode [2].

This approach was postulated by Nayfeh et al. [3] to control roll and pitch modes in ships motions. Later was adopted by Oueini et al. [4] to vibration absorber design for cantilever beams. The feedback signal was generated by a strain gage, and the actuation was achieved through piezoceramic patches. The strategy was tested experimentally for the first and second modes of two different specimens.

Ashour and Nayfeh [5] studied the feasibility of saturation technique to control a cantilever rectangular plate. Finite-element methods were used to extract the eigenmodes and frequencies of the system. This contributed to the frequency tuning of the control algorithm. A numerical study was conducted to optimize the location of the actuators to maximize design controllability.

The performance of the saturation control method in time delayed systems was studied by Saeed et al. [6]. With the varying time delay but keeping fixed controller parameters, it was seen that the vibration could be suppressed at some values of time delay only which was called "vibration suppression region". Time delay effect was studied also by Sayed et al. [7] with reference to nonlinear vibrating system via 1:4 internal resonance. The cases of external and parametric forcing were analysed.

Also Warminski et al. [8] studied the performance of saturation control in particular with reference to self-excited strongly nonlinear beam structure driven by an external force.

The provided literature review on the saturation control technique reveals its applications of the rotating structures to be rather sparse. Therefore, this research is a continuation of former authors' studies regarding rotors with passive/active beams and different control strategies to suppress structural vibrations. In particular, in [9] a rotating structure consisting of a rigid hub and three laminated passive beams was considered. Forced responses of the system under regular and chaotic excitations were investigated. Simulations were performed accounting for the different manufacturing tolerances in the laminate breaking its intended macroscopic symmetric structural behaviour. In research [10] a saturation adaptive control strategy was tested to suppress vibrations of a rotor consisting of a hub and a single composite active material blade. In [11] dynamics of a three-bladed rotor was discussed. A linear analytical model of the plant was formulated and effectiveness of a saturation control strategy was tested. The examined cases considered different magnitudes of flexural–torsional deformations couplings. Next, Warminski and Latalski [12] studied the nonlinear dynamic characteristics of the three-bladed active rotor. In the analytical model of the system the nonlinearities in constitutive relations of the piezoceramic material were accounted for. Both free and forced vibrations cases were examined. Finally, in the recent research [13] Warminski et al. considered an active beam in a geometrically nonlinear regime. The proposed extended Euler–Bernoulli model incorporated a nonlinear curvature and coupled transversal-longitudinal deformations. The control inputs were represented in boundary conditions as non-homogenous terms. Classical linear proportional (P) control and nonlinear cubic (C) strategy as well as mixed (P–C) one with time delay were analysed for vibration reduction.

In the current paper the dynamics of the active three-bladed rotor is examined. The reduced mathematical model of the rotor takes into account nonlinear electro-mechanical properties of the active blades as well as hub dynamics considered as a rigid body. To suppress vibrations a nonlinear saturation control strategy is adopted. Effectiveness of the proposed control method for selected operational states and excitation conditions is presented.

2 Model of a 3-Bladed Rotor with Active Elements

The system under investigation consists of three elastic blades clamped to the rigid cylinder-like hub—see Fig. 1a. The beams have thin-walled box-shape profile and are made of a composite material. The webs and flanges of the cross section are laminated following the antisymmetric orientation of reinforcing fibres in opposite walls (Fig. 1b). This configuration results in the so called circumferentially asymmetric stiffness (CAS) lamination scheme. It exhibits the mutual coupling of beam bending occurring about minor axis of the cross section and torsional deformations. By fixing the beams in a way the minor axis of the

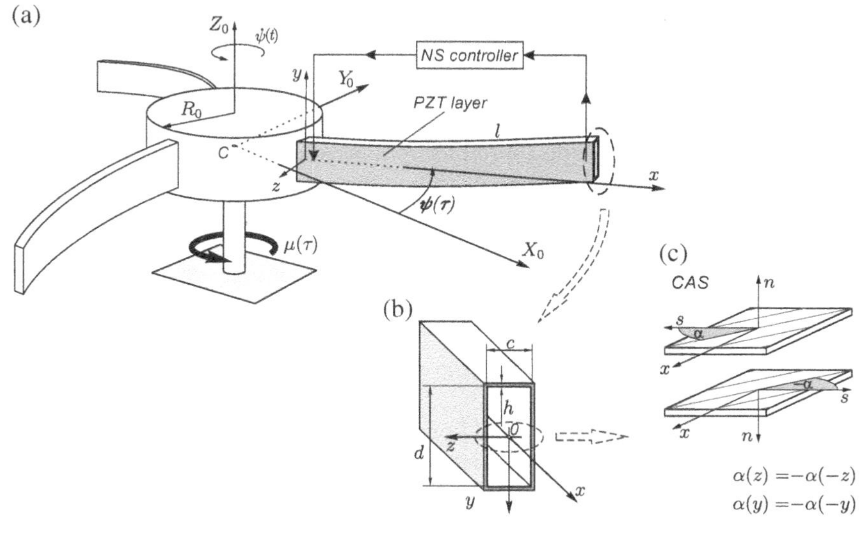

Fig. 1. Rotor composed of a rigid hub and three flexible blades with active piezo-layers

cross-section is parallel to the axis of rotation (presetting angle equals $\pm\pi/2$—Fig. 1a) the lead-lag bending combined to torsion occurs. The magnitude of this modal coupling depends on the reinforcing fibres orientation [14,15]. Moreover, the outer surfaces of the each single beam flanges are covered by active PZT layers exhibiting nonlinear constitutive properties. These active material superficial transducers can be induced to control structural oscillations.

According to the previous authors' papers [16,17] and presented therein detailed derivation procedure by means of the Hamilton's principle and the Galerkin's discretization method the state of the controlled system is governed by a set of five ordinary differential equations

$$
\begin{aligned}
&\left(J_{\mathrm{h}}+\sum_{j=1}^{3}J_{\mathrm{b}j}+\sum_{j=1}^{3}\alpha_{\mathrm{h}j2}q_j^2\right)\ddot{\psi}+\zeta_{\mathrm{h}}\dot{\psi}+\sum_{j=1}^{3}\left(\alpha_{\mathrm{h}j1}\ddot{q}_j+\alpha_{\mathrm{h}j3}q_j\dot{q}_j\dot{\psi}\right)=\mu,\\
&\qquad\ddot{q}_j+\zeta_j\dot{q}_j+\alpha_{j2}\ddot{\psi}+\left(\alpha_{j1}+\alpha_{j3}\dot{\psi}^2\right)q_j+\alpha_{j4}q_j\dot{q}_j\dot{\psi}\\
&\qquad\qquad+\alpha_{j5}\mathrm{sgn}(q_j)q_j^2+\alpha_{j6}q_j^3=q_c^2\sum_{j=1}^{3}g_{1j},\\
&\qquad\qquad\ddot{q}_c+\zeta_c\dot{q}_c+\omega_{0c}^2q_c=q_c\sum_{j=1}^{3}g_{2j}q_j
\end{aligned}
\tag{1}
$$

where subscript $j = 1, 2$ or 3 refers to the subsequent beam of the rotor. The coordinate q_j is a generalized beam coordinate accounting for coupled bending-torsional deformation and a combined electric component attributed to active layer; J_{h}, $J_{\mathrm{b}j}$ denote a dimensionless mass moment of inertia of the hub and each individual beam, respectively. The variable $\psi(\tau)$ is an angular position of the rotor, so $\dot{\psi}$ and $\ddot{\psi}$ correspond to its angular velocity and acceleration, respectively. Coefficients ζ_{h}, ζ_j are viscous damping ratios of the hub and each individual beam. The terms $\alpha_{\mathrm{h}j1}$, $\alpha_{\mathrm{h}j2}$, $\alpha_{\mathrm{h}j3}$, α_{j1}, α_{j2}, α_{j3}, α_{j4}, α_{j5}, α_{j6} are the ones obtained from the modal reduction procedure for each referenced beam j. Finally, the variable μ present on the right-hand side of the first formula represents the dimensionless driving torque applied to the hub that can be—in general—any time dependent function $\mu = \mu(\tau)$. Specifically, in this research, the torque is supposed to be represented as a sum of a constant component μ_0 and a second variable component—$\rho\sin\omega\tau$—to provide external excitation to the blades, thus $\mu(\tau) = \mu_0 + \rho\sin\omega\tau$. In the later expression ρ, ω and τ denote amplitude and frequency of the periodic component and dimensionless time, respectively.

Studying the governing system (1) one observes it is coupled by inertia terms appearing in the hub and beams equations. Moreover, the nonlinearities present in blades equations originate from the nonlinear constitutive relations of the transducers piezoceramics.

The control method adopted in this research is based on nonlinear coupling of the master structure (plant) and an additional oscillator governed by the generalized coordinate q_c in the analytical model. The signal from the controller is squared and forwarded to actuating devices affecting the plant—compare right hand side of the second equation in the (1) set. This kind of coupling contributes to strong nonlinear interaction in the system providing the controller natural frequency ω_{0c} is tuned in 1 : 2 ratio to the natural frequency of the master structure. Performance of the controller can be tweaked by adjusting the magnitudes of gains g_{1j} and g_{2j}. Obviously, if both gains g_{1j} and g_{2j} are equal to zero the controller is switched off and the system performs non-controlled forced vibrations. The inherent damping of the control device characteristic for any real system is modelled by an independent linear viscous damping represented by the coefficient ζ_c.

Numerical analysis of the studied active rotor dynamics and control strategy is performed for the graphite-epoxy laminate beams with embedded layers of PZT-3203HD piezoceramic. Physical properties of these materials as well as geometric dimensions of the structure are specified in [16].

3 Dynamics and Control of a Symmetric Rotor

In the first studied case all the blades are presumed to be identical and the rotor structure is fully symmetric. Setting the reinforcing fibres orientation to 75° (measured from the profile perimeter direction) and considering the first deformation mode the values of dimensionless coefficients α_{ij} in Eqs. (1) have been computed—Table 1. It should be noted the accepted reinforcing fibres orientation of 75° corresponds to strong coupling between bending and twisting deformations—see Fig. 2c.

Table 1. Dimensionless coefficients of the symmetric rotating structure—75°

Hub equation Eq. $(1)_1$			
	$J_{\mathrm{h}} = 5$	$J_{\mathrm{b}1} = J_{\mathrm{b}2} = J_{\mathrm{b}3} = 1$	$\zeta_{\mathrm{h}} = 0.1$
	$\alpha_{\mathrm{h}j1} = 0.5323$	$\alpha_{\mathrm{h}j2} = -0.4042$	$\alpha_{\mathrm{h}j3} = -0.8085$
Beams equations Eq. $(1)_2$			
	$\alpha_{j1} = 10.8636$	$\alpha_{j2} = -1.7723$	$\alpha_{j3} = 0.3484$
	$\alpha_{j4} = 1.5498$	$\alpha_{j5} = -2.3268$	$\zeta_j = 0.02 \times \sqrt{\alpha_{j1}}$

To start the analysis let us study the linear free vibrations case accounting for the rotor as a complete system—i.e. the hub and three piezo-blades. Regarding the angular speed of the structure as an independent but constant parameter the governing equations Eqs. (1) get simplified ($\ddot{\psi} = 0$ and $\dot{\psi} = \Omega_h$). Then, the Campbell diagram can be plotted representing the fundamental frequency of the structure as a function of its angular speed Ω_h—Fig. 3.

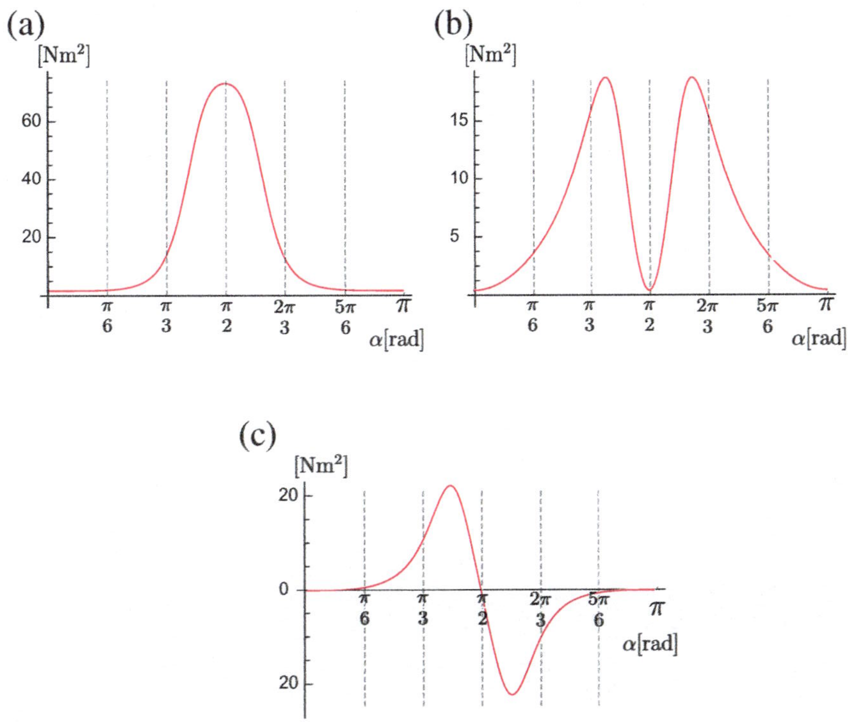

Fig. 2. CAS beam stiffness with respect to reinforcing fibres orientations: bending (**a**), twisting (**b**), and bending-twisting (**c**)

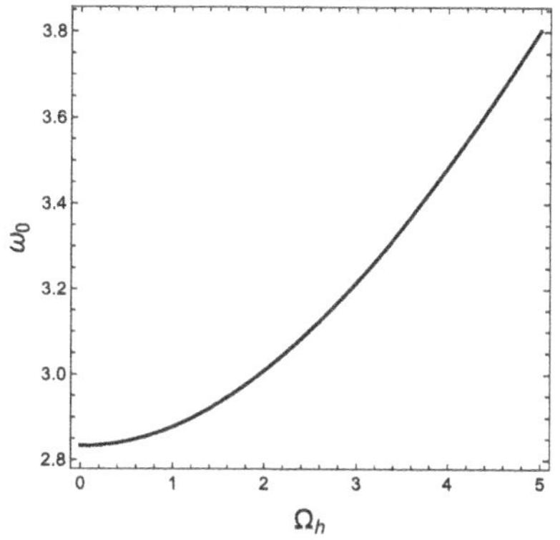

Fig. 3. Natural frequency ω_0 of a symmetric rotor with piezo-blades against hub angular velocity Ω_h.

It's worth noting when all the blades are identical and the hub does not rotate ($\Omega_h = 0$) the fundamental frequency of the structure is $\omega_0 = 2.83276$. This is different from the first natural frequency of a separated non-rotating beam $\omega_{01(\mathrm{beam})} = \sqrt{\alpha_{11}} = 3.296$ due to the influence of the hub inertia J_h. This observation is important for the proper design of the vibration absorber since precise tuning of NS control system to plant frequency in 1:2 ratio ($\omega_{0c} = \frac{1}{2}\omega_0$) is critical for the effectiveness of vibrations mitigation [10,11].

Studying the forced vibrations case the resonance curves of the system are plotted assuming solely the harmonic part of the torque excitation, i.e. the constant torque component μ_0 is zero. This corresponds to the oscillations of the structure about its rest position. The resonance curves for selected values of excitation amplitude ρ for the blades as well as for the hub are presented in Fig. 4a and b, respectively. Since the rotor is symmetric, responses of all the blades are the same. As we see the nonlinearity of the softening type is apparent for large oscillations with unstable solutions occurring if ρ reaches 0.2 value—note green and pink curves representing $\rho = 0.2$ and $\rho = 0.3$, respectively.

To check effectiveness of the NS control strategy for the rotating system we consider the case with large response amplitudes corresponding to the above mentioned green resonance curve $\rho = 0.2$ in Fig. 4. Since motion of all the blades is fully synchronized (symmetric rotor) we set the identical gains $g_{11} = g_{12} = g_{13}$ for control signals supplied to individual beams. Similarly, the same gains $g_{21} = g_{22} = g_{23}$ of the input signals from beams to the controller are assumed—see Eqs. (1). The gains attain values $g_{1j} = 0.01$, $g_{2j} = 1.0$, $j = 1, 2, 3$ and the damping coefficient of the controller is $\zeta_c = 0.001$. This data has been adopted from the former authors research [10] where a single rotating beam was studied.

The outcomes of NS control are presented in Fig. 5. The black colour resonance curves represent the uncontrolled system responses. In blue the domains of active NS control are indicated. It's readily seen the controller works effectively for the narrow resonance zone. As a result of NS controller activation the

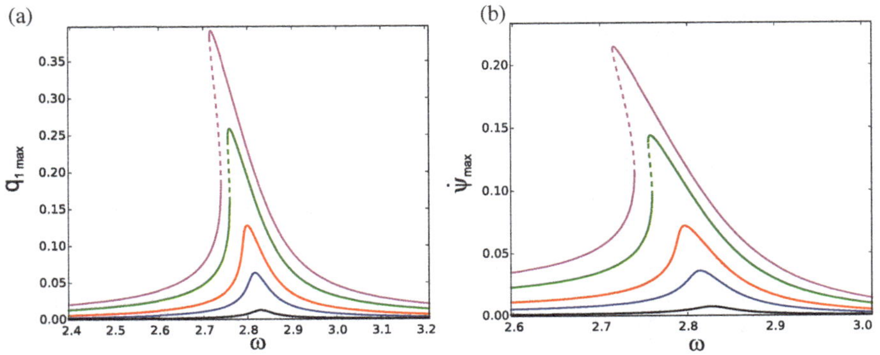

Fig. 4. Resonance curves for a symmetric rotor with piezo-blades for selected amplitudes of the excitation $\rho = 0.01$—black, $\rho = 0.05$—blue, $\rho = 0.1$—red, $\rho = 0.2$—green, $\rho = 0.3$—pink: **a** amplitudes of the blades q_1, q_2, q_3 and **b** amplitude of angular hub velocity; oscillations of the structure about its rest position ($\mu_0 = 0$)

former solutions on the resonance curve become unstable and oscillations amplitudes are suppressed close to zero. The zoomed bifurcation plots around this zone are shown in Fig. 5b. It should be underlined the amplitudes suppression effect occurs both for the beams and for the hub oscillations as well—Fig. 5d. The corresponding signal of the control unit is presented in Fig. 5c.

Further analysis of these results reveals some potential drawbacks of the presented solution. In the frequency ranges a bit distant from the resonance zone the amplitudes are getting high (blue branches out of resonance) and the control strategy is counter-effective. This undesired effect can be restrained by optimisation of the controller gains with respect to the excitation magnitude and by triggering the controller only in the relevant frequency zone.

For lower excitation magnitudes (ρ), oscillations of the rotor are smaller and the additional branches are less pronounced. The responses of the beams, the hub as well as the controller input and an activation scenario are presented in Figs. 6, and 7. These figures correspond to the cases $\rho = 0.1$ and for $\rho = 0.01$, respectively.

A further benefit of the adopted nonlinear control strategy is the so-called saturation phenomenon [5,18,19]. This effect can be explained by the bifurcation diagram given in Fig. 8. In this plot responses of the system are recorded for a fixed frequency of the excitation $\omega = 2.8$ with respect to varied excitation amplitude ρ which is regarded thereby as a bifurcation parameter.

When increasing the free parameter ρ, responses of the blades and the hub Fig. 8a–c increase proportionally (black line) till the bifurcation point occurring about at $\rho = 0.035$. At this point the controller is activated (Fig. 8d) and the resonance solutions loose their stability (Fig. 8b). From now amplitudes of the beams and the hub do not increase (blue lines) but rather stay on a certain level (amplitude saturation) or might even decrease. The saturation level can be tuned according to the operational conditions constraints or other requirements.

In the studied above cases all the embedded transducers were activated and collaboratively used in the control algorithm. The interesting question arises how effective is the control strategy if only certain active devices are operated and the remaining ones stay idle. The results of such a scenario are presented in Fig. 9 where only the elements embedded in blade No. 1 have been triggered.

The obtained results indicate the considered NS controller is still capable of suppressing structural vibrations. However, the intended operational mode introduces a detuning to the system and responses of the active blade No. 1 and passive ones Nos. 2 and 3 turn different—Fig. 9a–c). In this case the NS controller interacts directly with the blade No. 1 (coordinate q_1 in Fig. 9a) reducing its vibrations but due to inertial coupling in the system via the hub it influences the other beams as well. Finally, it should be noted, the zone of vibration suppression is narrower while comparing to the fully symmetric case discussed earlier.

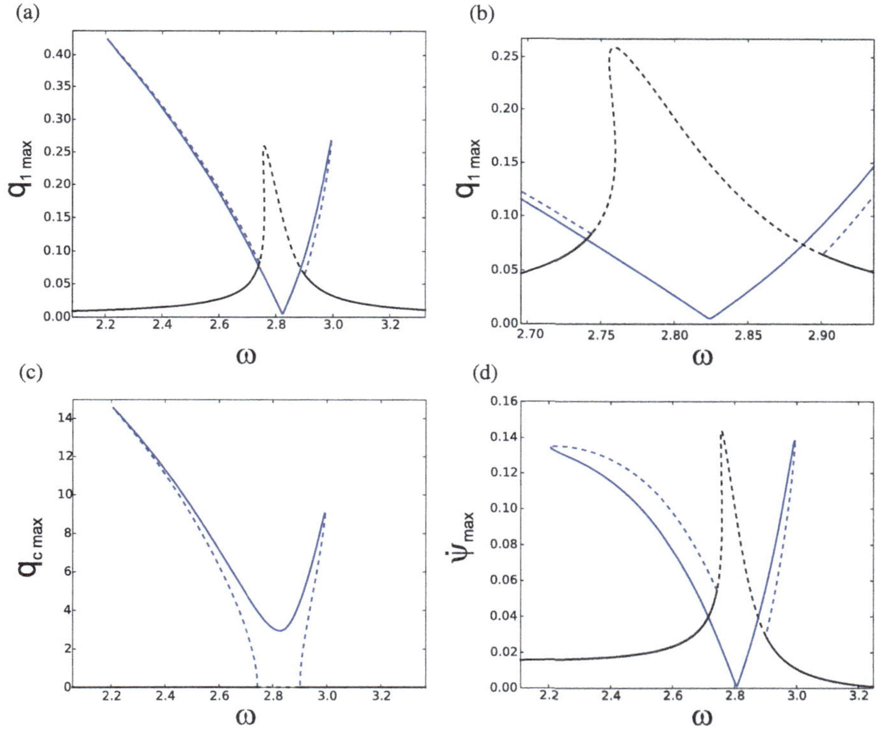

Fig. 5. Resonance curves for a symmetric rotor with piezo-blades and nonlinear saturation controller: **a** amplitudes of blades coordinates q_1, q_2, q_3, **b** zoom next to bifurcation points, **c** response of the controller q_c, **d** amplitude of angular velocity of hub oscillations $\dot{\psi}$ of the structure about its rest position; $\mu_0 = 0$, $\rho = 0.2$.

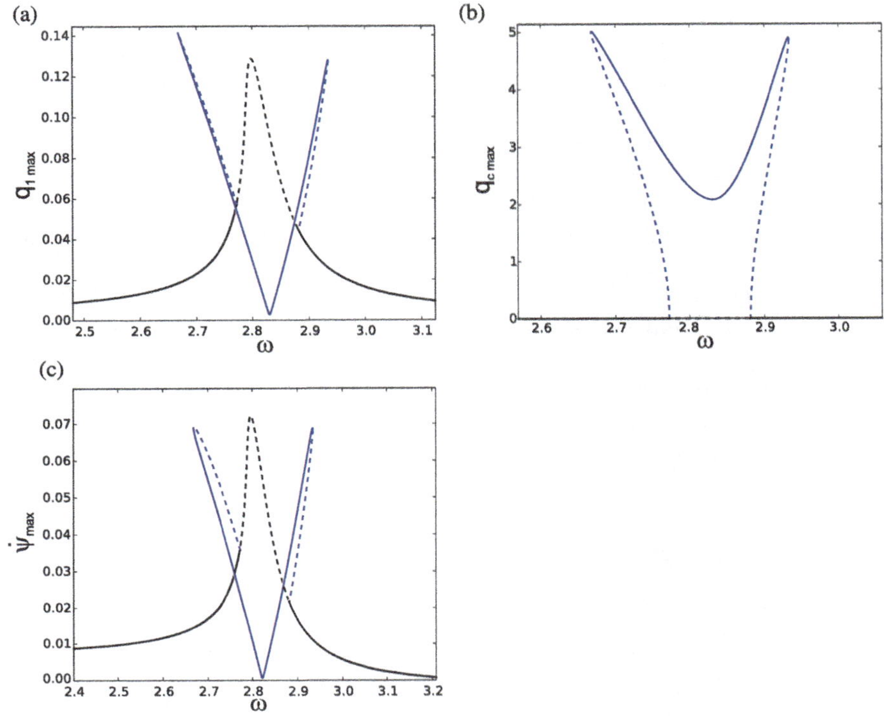

Fig. 6. Resonance curves for a symmetric rotor with piezo-blades and nonlinear saturation controller: **a** amplitudes of blades coordinates q_1, q_2, q_3, **b** zoom next to bifurcation points, **c** response of the controller q_c, **d** amplitude of angular velocity of hub oscillations $\dot{\psi}$ of the structure about its rest position; $\mu_0 = 0$, $\rho = 0.1$.

4 Mistuned Rotor Control

Next we consider the case of a mistuned rotor. To this aim the blade No. 3 is replaced by a new one. This alternative design keeps the same geometry and active layers placement but exhibits slightly different mechanical properties due to a small change in the orientation of the reinforcing fibres. Namely, the fibres are set at 70° in contrast to the former orientation 75°. The blades Nos. 1 and 2 stay unchanged. The updated dimensionless coefficients α for the mistuned rotor are presented in Table 2. Analogous to the first studied case of the symmetric rotor all the beams are controlled by the control unit.

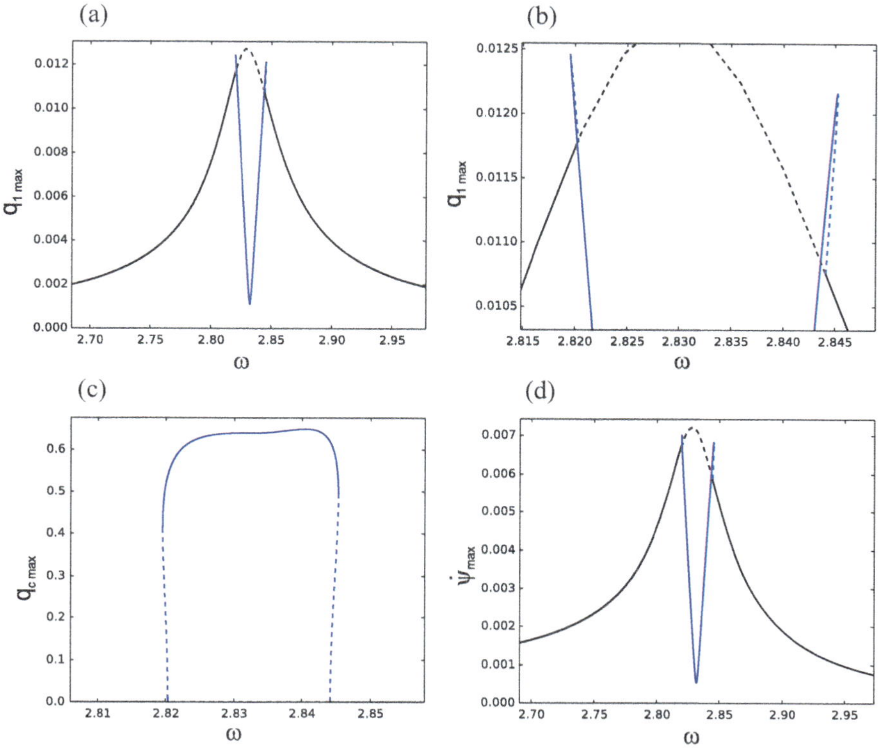

Fig. 7. Resonance curves for a symmetric rotor with piezo-blades and nonlinear saturation controller: **a** amplitudes of blades coordinates q_1, q_2, q_3, **b** zoom next to bifurcation points, **c** response of the controller q_c, **d** amplitude of angular velocity of hub oscillations $\dot{\psi}$ of the structure about its rest position; $\mu_0 = 0$, $\rho = 0.01$

The resonance curves for high excitation amplitudes $\rho = 0.2$ are presented in Fig. 10 (uncontrolled system). Due to the rotor mistuning the second resonance zone emerges relevant to changed stiffness of the blade No. 3. Studying the first resonance zone it can be observed the amplitudes of the beam No. 3 are much higher (Fig. 10c). This stems from the fact this specimen is significantly less stiff than the remaining beams Nos. 1 and 2. Moreover, one should keep in mind the CAS beam stiffness is highly sensitive to variations in fibres orientations near the 75° range—compare Fig. 2. This explains the sharp change in beam stiffness (and relatively distant second resonance frequency) caused by a relatively small change in composite fibres setting.

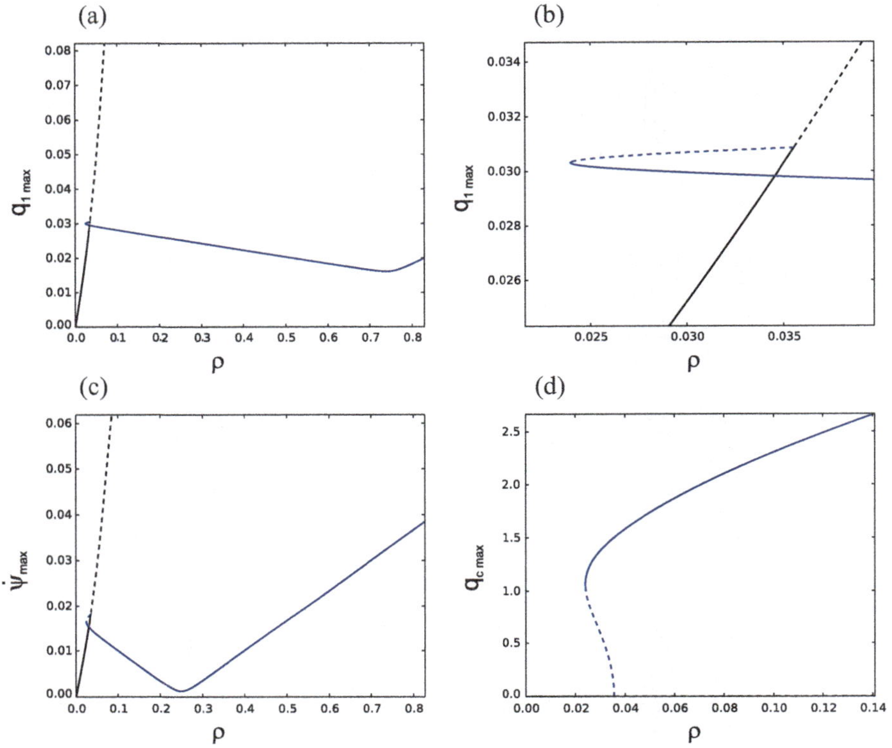

Fig. 8. Bifurcation diagrams of the blades (**a**) with zoom next to bifurcation point (**b**), hub (**c**) and controller response (**d**); $\mu_0 = 0$, $\omega = 2.8$

As reported in the literature review the NS control requires precise tuning to the natural frequency of the whole structure. Thus we have to consider both resonance states occurring at $\omega_{01} = 2.70861$ and $\omega_{02} = 3.14363$, respectively. The controller natural frequency ω_{0c} is tuned to the half of ω_{01} or half of ω_{02} depending on the excitation frequency value and is automatically activated if amplitudes of the blades and hub increase.

The resonance curves with triggered controller are presented in Fig. 11. The amplitudes in both resonance zones are reduced for all rotor blades. The results of the controller activation are presented by a blue line for the first resonance zone and a green line for the second resonance zone (Fig. 11e). Because amplitudes in the second resonance are much smaller the vibrations suppression is less spectacular comparing to the first resonance area.

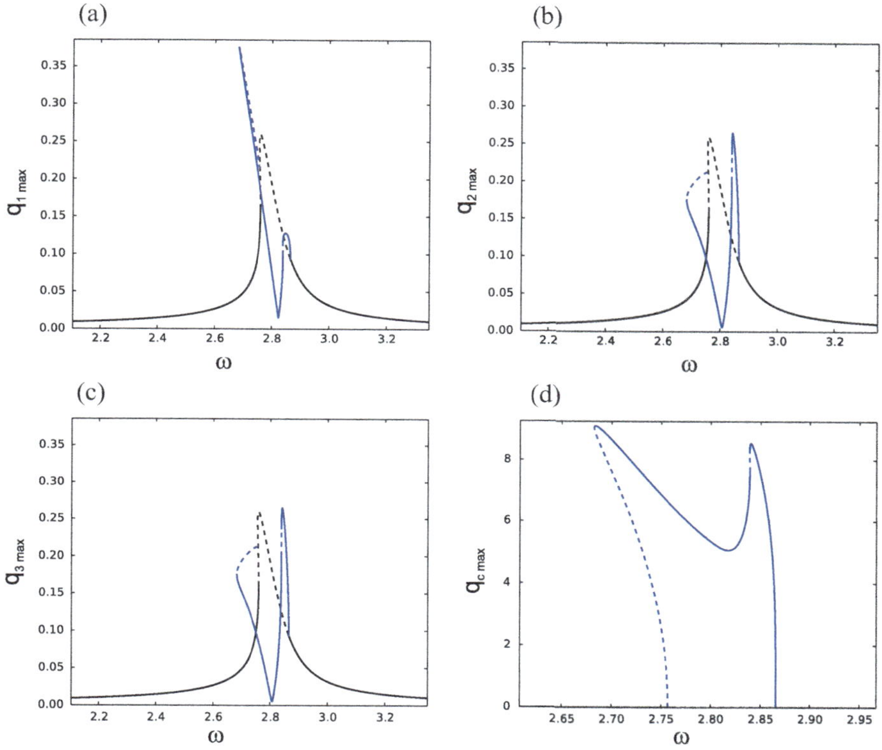

Fig. 9. Resonance curves for a symmetric rotor with controlled piezo-blade No. 1, nonlinear saturation controller, blade No. 2 and No. 3 are not controlled: **a** amplitudes of blade No. 1 coordinate q_1, **b** bade No. 2 - q_2, **c** bade No. 3 - q_3, **d** response of the controller q_c; the structure oscillates about its rest position; $\mu_0 = 0$, $\rho = 0.2$

The final test of the control algorithm is run for the system performing full rotation. To this aim the constant component of the torque μ_0 is set to $\mu_0 = 0.02$ which corresponds to approx. 700 rpm in physical units. This value is large enough to reveal any potential differences in the proposed algorithm efficiency comparing to the former non-rotating design analysis.

Table 2. Dimensionless coefficients of the mistuned rotating structure: beam No. 1–75° deg, beam No. 2–75°, beam No. 3–70°

Hub equation Eq. $(1)_1$			
	$J_h = 5$	$J_{b1} = J_{b2} = J_{b3} = 1$	$\zeta_h = 0.1$
	$\alpha_{hj1} = -0.5323$	$\alpha_{hj2} = -0.4042$	$\alpha_{hj3} = -0.8085$, $j = 1, 2$
	$\alpha_{h31} = -0.5318$	$\alpha_{h32} = -0.4037$	$\alpha_{h33} = -0.8074$
Beams equations Eq. $(1)_2$; Beam No. 1 and No. 2, $j = 1, 2$			
	$\alpha_{j1} = 10.8636$	$\alpha_{j2} = -1.7723$	$\alpha_{j3} = 0.3484$
	$\alpha_{j4} = 1.5498$	$\alpha_{j5} = -2.3268$	$\zeta_j = 0.02 \times \sqrt{\alpha_{j1}}$
Beams equations Eq. $(1)_2$; Beam No. 3; $j = 3$			
	$\alpha_{31} = 9.0351$	$\alpha_{32} = 1.7739$	$\alpha_{33} = 0.3489$
	$\alpha_{34} = -1.5493$	$\alpha_{35} = -1.5696$	$\zeta_3 = 0.02 \times \sqrt{\alpha_{31}}$

The controller should be tuned to the first and the second natural frequency of the whole structure considering the influence of the angular velocity and centrifugal stiffening effect. The free vibrations characteristics are obtained analytically from (1) by neglecting controller equations and nonlinear terms and constant angular speed as independent parameter. The course of two first natural frequencies curves against angular speed are shown in Campbell diagram in Fig. 12 for ω_{01} in black and ω_{02} in red. It's clear—for the rotating structure—the controller has to be tuned to the shifted resonance zones, respectively.

Response of the rotor is presented in Fig. 13. If the controller is properly tuned its effectiveness for $\mu = 0.02$ is similar to the case for $\mu = 0$ (Fig. 11). The difference is observed in the diagram of angular velocity (Fig. 13d) where, due to the torque constant component μ_0, the characteristic is shifted to about 0.2 value, which corresponds to average angular speed of the rotor.

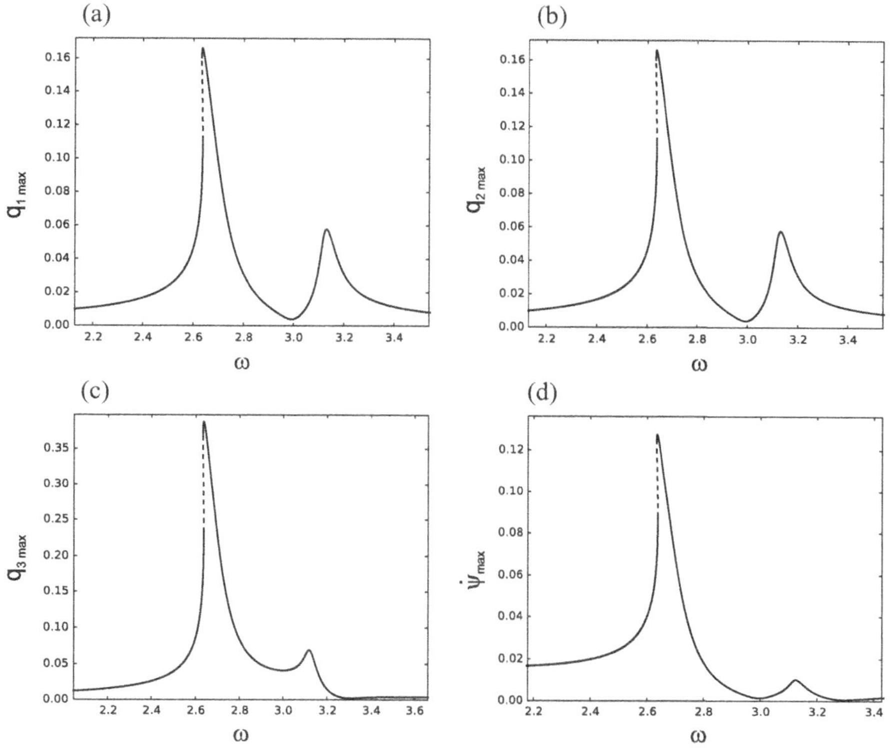

Fig. 10. Resonance curves for a mistuned rotor with piezo-blades. **a** blade No. 1–75°, **b** blade No. 2–75°, **c** blade No. 3–70° and **d** amplitude of angular hub velocity; $\rho = 0.2$, $\mu_0 = 0$.

The changes in the natural frequencies characteristics shown in Fig. 12 are very small for low angular speed. Therefore, it can be concluded for the threshold value of $\Omega_h \approx 1$, the controller tuned to non-rotating system natural frequencies works sufficiently well and does not need angular speed corrections. Nevertheless, for larger Ω_h values the controller settings should be tuned properly to account for the centrifugal stiffening effect and shifted resonance zones.

5 Conclusions

A nonlinear model of the rotating structure comprising the cylinder hub and three clamped thin-walled CAS composite box beams is presented in this paper. The piezoelectric material layers are embedded into the specimens laminate to control vibrations of the rotor in near resonance states. The presented research comprises the analysis of the fully symmetric rotor with three identical blades with 75° orientation of reinforcing fibres and for the mistuned rotor with one

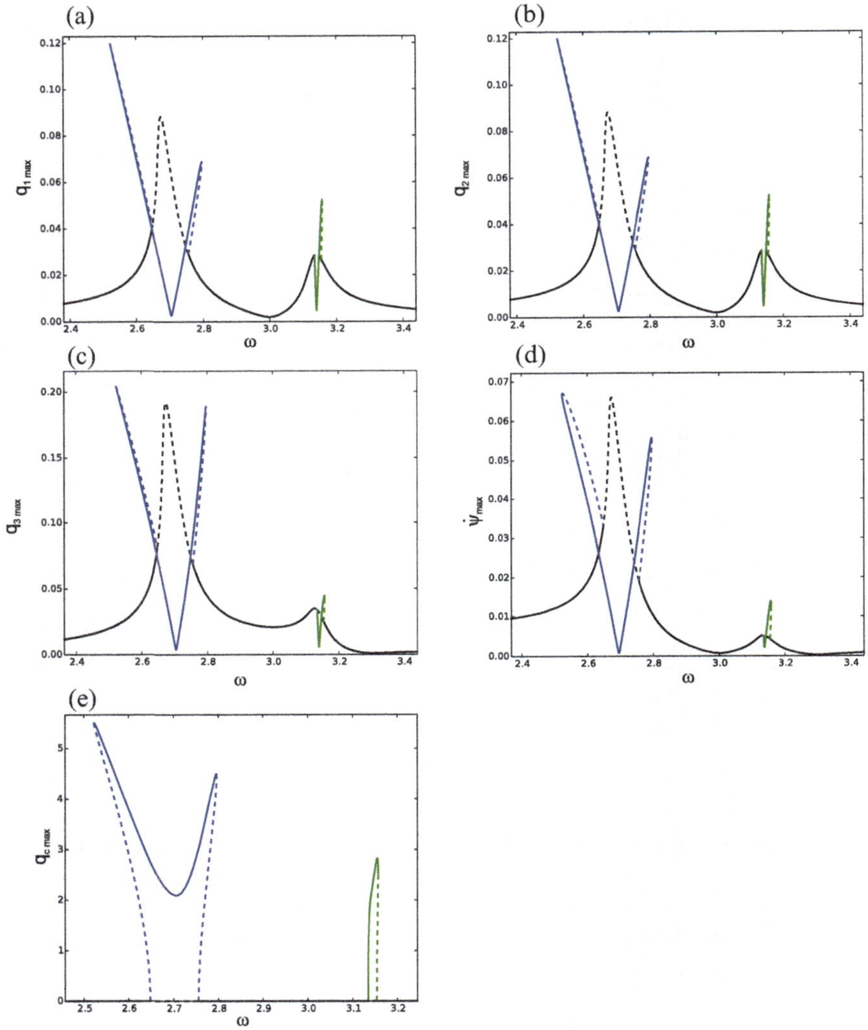

Fig. 11. Resonance curves for a mistuned rotor with controlled piezo-blades, **a** amplitudes of blade No. 1 coordinate q_1, **b** bade No. 2 - q_2, **c** blade No. 3 - q_3, **d** amplitude of hub angular velocity $\dot{\psi}$, **e** response of the controller q_c; the structure oscillates about its rest position; $\mu_0 = 0$, $\rho = 0.1$

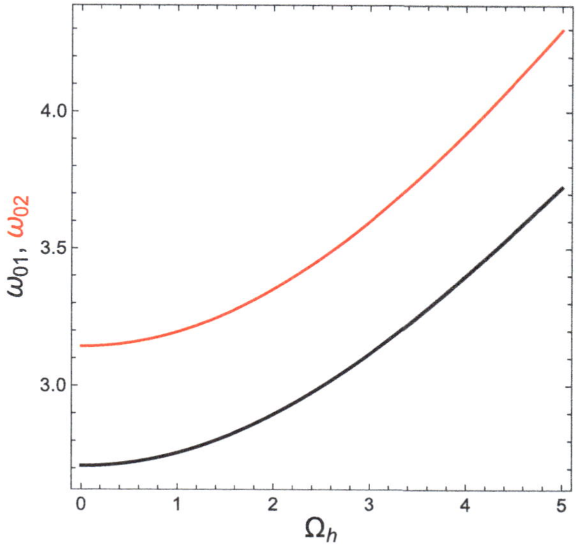

Fig. 12. Natural frequencies ω_{01} - black and ω_{02} - red of a mistuned rotor with piezo-blades against hub angular velocity Ω_h

blade substituted by the new one made of the composite material with 70° fibres. Vibrations generated by torque supplied to the hub are reduced following the nonlinear saturation control strategy and actuated active elements on the beams. It is shown that the controller needs to be precisely tuned to the combined hub-beams structure characteristics and actual rotating angular velocity. Tuning the controller to the individual blades frequencies does not give expected results.

For the fully symmetric rotor—due to the nonlinear coupling of the structure with the control unit—the resonance curve for large oscillations becomes unstable and beams and hub vibrations are nearly completely suppressed. To avoid additional intrusive branches on the resonance curve and eventual undesired increased system response the controller gains g_1 and g_2 must be optimized to the vibrations level. Moreover, it has been observed if the controller is optimized for the given excitation frequency then an increase of the excitation amplitude leads to the saturation of the blades and hub amplitudes.

The performed simulations indicate the saturation control works properly also when the control signal is provided to a single blade only while keeping the other blades passive. But, similarly to the fully controlled system, the controller must be tuned to the complete structure and not for a single blade.

Acknowledgments. The work is financially supported by grant 2016/23/B/ST8/01865 from the National Science Centre, Poland.

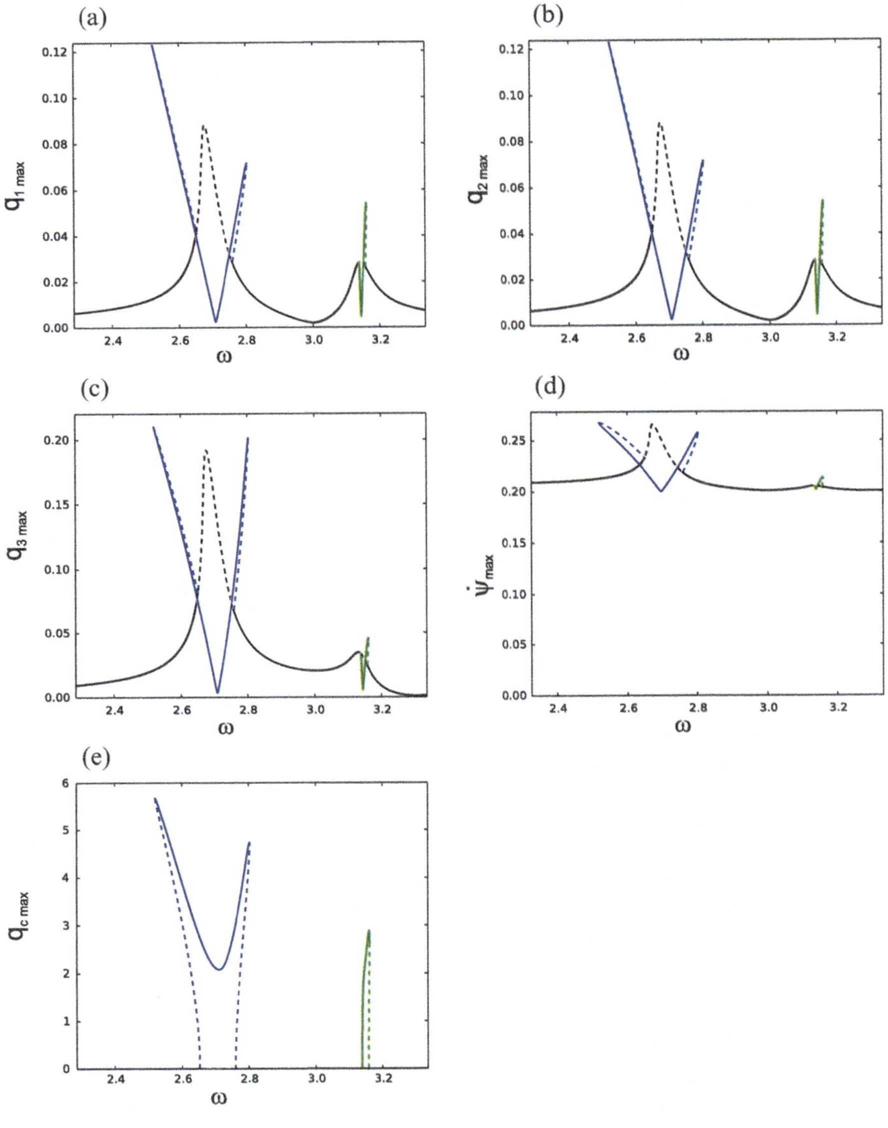

Fig. 13. Resonance curves for a mistuned rotor with controlled piezo-blades, **a** amplitudes of blade No. 1 - coordinate q_1, **b** bade No. 2 - q_2, **c** blade No. 3 - q_3, **d** amplitude of hub angular velocity $\dot{\psi}$, **e** response of the controller q_c; full rotation of the structure; $\mu_0 = 0.02$, $\rho = 0.1$

References

1. Chopra, I., Sirohi, J.: Smart Structures Theory. Cambridge Aerospace Series, vol. 35. Cambridge University Press, New York, NY (2014)

2. Oueini, S.S., Golnaraghi, M.F.: Experimental implementation of the internal resonance control strategy. J. Sound Vib. **191**(3), 377–396 (1996)
3. Nayfeh, A.H., Mook, D.T., Marshall, L.R.: Nonlinear coupling of pitch and roll modes in ship motions. J. Hydronautics **7**(4), 145–152 (1973)
4. Oueini, S.S., Nayfeh, A.H., Pratt, J.R.: A nonlinear vibration absorber for flexible structures. Nonlinear Dyn. **15**(3), 259–282 (1998)
5. Ashour, O.N., Nayfeh, A.H.: Adaptive control of flexible structures using a nonlinear vibration absorber. Nonlinear Dyn. **28**(3/4), 309–322 (2002)
6. Saeed, N.A., El-Ganini, W.A., Eissa, M.H.: Nonlinear time delay saturation-based controller for suppression of nonlinear beam vibrations. Appl. Math. Model. **37**(20–21), 8846–8864 (2013)
7. Sayed, M., Mousa, A.A., Alzaharani, D.Y.: Non-linear time delay saturation controller for reduction of a non-linear vibrating system via 1:4 internal resonance. J. Vibroengineering **18**(4), 2515–2536 (2016)
8. Warminski, J., Cartmell, M.P., Mitura, A., Bochenski, M.: Active vibration control of a nonlinear beam with self- and external excitations. Shock Vib. **20**, 1033–1047 (2013)
9. Warminski, J., Latalski, J., Szmit, Z.: Vibration of a mistuned three-bladed rotor under regular and chaotic excitations. J. Theor. Appl. Mech. **56**(2), 549–566 (2018)
10. Latalski, J., Warminski, J.: Dynamics and saturation control of rotating composite beam with embedded nonlinear piezoelectric actuator. In: Belhaq, M. (Ed.) Recent Trends in Applied Nonlinear Mechanics and Physics, vol. 199. Springer Proceedings in Physics. Springer International Publishing, Cham, pp. 129–152 (2018)
11. Warminski, J., Latalski, J.: Nonlinear control of flexural–torsional vibrations of a rotating thin-walled composite beam. Int. J. Struct. Stab. Dyn. **17**(5), 1740003–1–1740003–17 (2017)
12. Warminski, J., Latalski, J.: Vibrations of rotating thin-walled composite beams with nonlinear piezoelectric layers. In: Kovacic, I., Lenci, S. (Eds.), IUTAM Symposium on Exploiting Nonlinear Dynamics for Engineering Systems, vol. 37. IUTAM bookseries. Springer, Cham, Switzerland, pp. 291–302 (2020)
13. Warminski, J., Kloda, L., Latalski, J., Mitura, A., Kowalczuk, M.: Nonlinear vibrations and time delay control of an extensible slowly rotating beam. Nonlinear Dyn. **103**(4), 3255–3281 (2021)
14. Latalski, J., Warminski, J., Rega, G.: Bending-twisting vibrations of a rotating hub-thin-walled composite beam system. Math. Mech. Solids **22**(6), 1303–1325 (2017)
15. Song, O., Librescu, L.: Free vibration of anisotropic composite thin-walled beams of closed cross-section contour. J. Sound Vib. **167**(1), 129–147 (1993)
16. Latalski, J.: A coupled-field model of a rotating composite beam with an integrated nonlinear piezoelectric active element. Nonlinear Dyn. **90**(3), 2145–2162 (2017)
17. Warminski, J., Latalski, J.: Saturation control for a rotating thin-walled composite beam structure. Procedia Eng. **144**, 713–720 (2016)
18. Ashour, O.N., Nayfeh, A.H.: Nonlinear control of plate vibrations. In: AIAA/ASME/ASCE/AHS/ASC Structures, Structural Dynamics, and Materials Conference (Reston, 2001), vol. AIAA-2001-1233, American Institute of Aeronautics and Astronautics, pp. 400–410
19. Oueini, S.S., Nayfeh, A.H., Golnaraghi, M.F.: A theoretical & experimental implementation of a control method based on saturation. Nonlinear Dyn. **13**(2), 189–202 (1997)

19. Oueini, S.S., Nayfeh, A.H., Golnaraghi, M.F.: A theoretical & experimental implementation of a control method based on saturation. Nonlinear Dyn. **13**(2), 189–202 (1997)

An Analysis of the Inertial Platform Behavior of the ELI-NP Project in Case of an Earthquake

Polidor Bratu[1], Nicusor Dragan[2], Calin Itu[3], Paul Nicolae Borza[3], Ovidiu Vasile[4], Sorin Vlase[3](✉), and Adrian Eracle Nicolescu[5]

[1] ICECON, Bucharest, Romania
[2] University Dunarea de Jos, Galati, Romania
[3] University Transilvania of Brasov, Brasov, Romania
svlase@unitbv.ro
[4] University Politehnica Bucharest, Bucharest, Romania
[5] University Ovidius, Constanta, Romania

Abstract. This paper studies the vibration of the inertial platform developed in Magurele-Bucharest, in the frame of the ELI-NP project. Two main components of this project are positioned on a concrete inertial insulating platform: a very high intensity laser system (10–30 PW) and a very brilliant, intense gamma beam (19 meV). This paper analyzes the behavior of the inertial platform of the ELI-NP laser project in the event of an earthquake. In order to obtain some results that would interest the researchers, the model proposed for the vibration calculation is subjected to a load similar to the one that appeared during the earthquake of 1977. The obtained results can be used to improve the project design.

Keywords: Gamma beam · Inertial platform · Earthquake · Vibration

1 Introduction

Civil structures, such as buildings and bridges that are considered an integral part of modern society, were originally designed to withstand static loads, but in everyday reality, these structures are still subject to a variety of dynamic loads from external natural sources. or artificial ones such as earthquakes, winds, or traffic actions. These dynamic loads can cause sustained and high-intensity vibrational movements at a time that can be detrimental to the structures as well as the occupants of the building. In the past century, natural phenomena such as hurricanes, tsunamis, earthquakes, and even floods have caused major disasters, leaving behind collapsed structures, large abandoned regions, and more importantly, great loss of life. For the reasons presented, in recent decades, engineers have been put in a position to develop numerical or analytical calculation algorithms for large structures as well as engineering technological solutions to reduce intensities from external disturbing sources leading to fast and efficient solutions to problems.

The requirement to increase the degree of protection for these structures is imposed by the criteria related to the safe operation of the building, human comfort, and durability

N. Herisanu and V. Marinca (eds.), *Acoustics and Vibration of Mechanical Structures – AVMS-2021*, Springer Proceedings in Physics 274,
https://doi.org/10.1007/978-3-030-96787-1_32

of the structure. In this sense, the protection of civil structures is undoubtedly a global priority and the challenge for engineers in the same direction is to develop safe civil structures for the best possible resistance to such natural phenomena.

The gamma source under construction at ELI-NP is especially well suited for these studies because of the small beam size at the target position which allows matching the full-beam flux with a very small target. Nevertheless, this advantage can quickly turn into an experimental quagmire if the overlap between the beam spot at the target position and the target is not very good or at least measurable. To achieve this, it is necessary to realize good insulation and to reduce significantly the vibration of the ground transmitted to the platform of the laser and gamma beam system. The researcher must realize a very precise target alignment with the gamma beam and this is not possible if the transmissibility of the vibration is not satisfactorily controlled. The design of the inertial platform imposes the implementation of a high precision of positioning the devices, the pieces, the alignment system for the experiments due, in the more general view, to the huge impact on the experiments. This engineering achievement is unique in its size and the results obtained in the design and study of this structure should be known for future projects (Fig. 11).

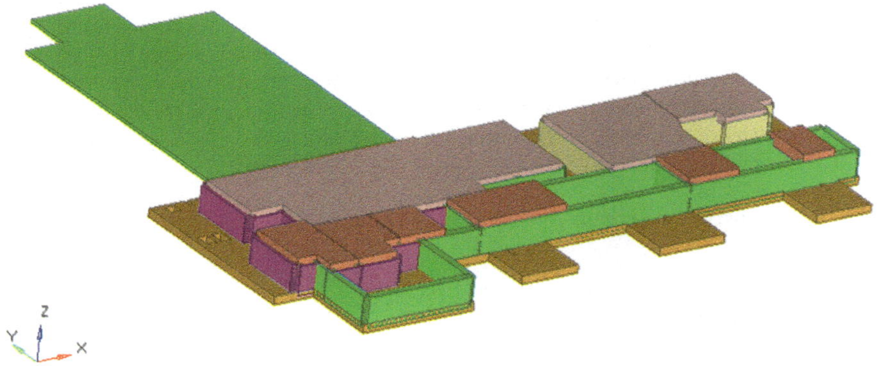

Fig. 1. Inertial platform with different constructive elements

The complexity of the study comes from the fact that the ELI-NP will deliver a beam of high-energy photons of unprecedented intensity within a very small opening angle, and for this, precision becomes an important thing. Insulation of the base platform is a means of reducing the transmission of vibrations in buildings, and the first attempt to address the problem in this way was in the 60s (past century) [1–3]. Since then, many types of springs have been fitted to many constructions in order to reduce the effects of anthropogenic source vibrations. In most cases, the building rests on steel springs or rubber elements. The aim is to reduce the transmission of vibrations by at least 10 dB for frequencies above 10 Hz, but this goal is rarely achieved. An important element in insulation is the type of spring used but also important are the cost of operation and the difficulty of implementing the system. More accurate modeling allows the choice of parameters for the elastic and dissipative elements that correspond, as well as possible, to the final purpose of vibration isolation.

2 Description of the Platform

The design of the inertial platform used for ELI-NP laser project represents a unique achievement realizing the insulation of the laser and gamma beam installations from the excitations that may come from the environment. Such projects and their design are very rare, and the results obtained within them are even more important, through the information they bring, to the scientific community.

If the platform is considered rigid with six-degrees-of-freedom (Stewart platform), the isolation methods have been intensively studied. There are several parameters that influence the ability of such a system to be isolated. A study method using a genetic sorting algorithm is developed in [4]. In these cases, genetic algorithms have been shown to have higher optimization efficiency, better computational accuracy, and reduced optimization time.

A dynamic model for a six-axis hybrid vibration isolation system based on a cubic configuration of the Stewart platform is proposed. The displacement excitations and the loads are simultaneously considered using a Newton–Euler formulation. A force feedback control principle is proposed and studied. Numerical simulation strengthens confidence in the proposed method [5].

An analysis of a six-degrees-of-freedom (DOF) active vibration isolation system is presented in [6]. The system is applied to isolate a Stewart platform, for better and more precise control and a wide range of earth-based systems. The paper presents the design and analysis of the platform as a performing and robust vibration isolation system. A prototype has been realized and tested (Fig. 2).

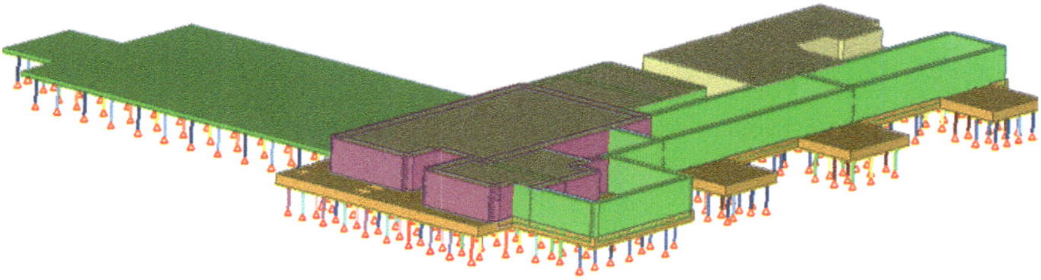

Fig. 2. Inertial platform supported by springs

The inertial platform is suspended on a number of, almost 1000, spring and shock absorber batteries. All installations in the ELI-NP structure are located on a platform that should be isolated from the influence of any environmental vibrations. The mass of the entire concrete platform, without partitions, roofs, and equipment, is approximately 25,000 tons. By adding the partitions, a mass of approximately 35,000 tons is reached, and by adding all the equipment and roofs, a mass of 54,000 tons is obtained. The platform is placed on a series of springs and dampers that have the role of isolating the equipment on the platform from any excitations that could come from the external environment. Due to the size of the platform and particularly, large mass, the concrete can deform under the action of its own weight or the weight of the walls and equipment placed over the platform. The numerous springs located at a short distance from each other have the role of uniformizing the loads that appear in the elastic supports of the platform. In this way, the loads that press on the platform, as well as their own weight will not give large deformations to the platform plate. In the configuration of the batteries, it is possible to act on the springs and shock absorbers so that some are active or inactive, and others do not allow an adjustment and control of the uniformity of the platform support. In essence, the platform consists of two concrete blocks, rigidly connected to each other, placed on batteries with springs that distribute the weight evenly on the floor (Fig. 3).

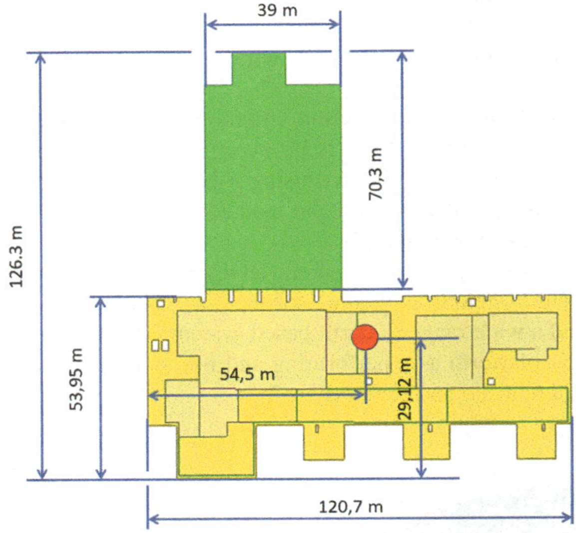

Fig. 3. Mass center position of the platform

3 Response Spectra Method

The elastic model of the platform, obtained using the finite element method, offers us the system of differential equations describing the vibration:

$$[m]\{\ddot{x}(t)\} + [c]\{\dot{x}(t)\} + [k]\{x(t)\} = -[m]\{r\}\ddot{x}_g(t) \tag{1}$$

here, $[m]$ represents the mass matrix (a square matrix with dimension n); $[k]$ is the stiffness matrix; $[c]$ is the damping matrix; $\{r\}$ represents the influence coefficient vector; $\{x(t)\}$ is the relative displacement vector (with its derivatives $\{\dot{x}(t)\}$ and $\{\ddot{x}(t)\}$) and $\ddot{x}_g(t)$ is the earthquake ground acceleration.

The classic modal analysis, used for an undamped system, studies the system:

$$\{\ddot{x}(t)\} + [d]\{x(t)\} = 0 \tag{2}$$

where

$$[d] = [m]^{-1}[k] \tag{3}$$

Shape solutions are sought:

$$\{ x(t)\} = \{X\}cos(\omega t + \psi) \tag{4}$$

Providing that the solution verifies the system, we obtain

$$([\mathrm{K}]-\omega^2[M])\{X\}\sin(\omega t + \psi) = 0 \tag{5}$$

at any time t. This requires

$$\Big([\mathrm{K}]-\omega^2[M]\Big)\{X\} = 0 \tag{6}$$

a relationship that represents a linear, homogeneous system. It will admit other solutions besides the banal one, zero, if and only if

$$det\Big([K] - \omega^2[M]\Big) = 0 \tag{7}$$

If n is the size of the system (number of differential equations), developing the determinant gives a polynomial of degree n în ω^2

$$P(\omega^2) = det\Big([K] - \omega^2[M]\Big) \tag{8}$$

which represents the characteristic polynomial of the system. The values $\omega_i = \sqrt{\lambda_i}$ are the system's eigenvalues.

The spectrum of eigenvalues defines the natural characteristics of the system and depends only on the mass distribution and the elastic properties of the elements that make up the studied ensemble. It is not influenced in any way by the initial values of positions and speeds, nor by the forces that require the system.

We denote the solution of the homogeneous linear system (6) in which we introduced the eigenvalue ω_i previously determined, respectively, with $\{X_i\}$

$$\Big([K] - \omega_i^2[M]\Big)\{X_i\} = 0 \quad ; \quad i = \overline{1, n} \tag{9}$$

This solution is called the i-rank eigenvector of the system. Whereas

$$det\Big([K] - \omega_i^2[M]\Big) = 0 \tag{10}$$

the n components of the vector $\{X_i\}$ will not be independent, so we will not be able to obtain the size of the vector but only its direction. The eigenmode of rank i can be obtained if a mode of scaling of its components is established. Note, after scaling, the eigenmodes with $\{\Phi_i\} = \mu_i\{X_i\}$ and their whole form the modal matrix $[\Phi]$:

$$[\Phi] = [\Phi_1\ \Phi_2\ \ldots\ \Phi_n] \tag{11}$$

4 Results

Table 1 compares the first 20 frequencies obtained for different platform configurations (only the 2 blocks—25,000 tons, the blocks with the addition of vertical walls—35,000 tons, and the blocks with vertical walls and roofs—54,000 tons) (Figs. 4, 5).

Table 1. Eigenfrequencies [Hz]

No	25,000 tons	35,000 tons	54,000 tons
1	2,27	1,90	1,53
2	2.43	2,03	1,64
3	2,49	2,08	1,68
4	3,38	2,82	2,28
5	3,40	2,84	2,29
6	3,45	2,89	2,33
7	3,60	3,01	2,42
8	3,66	3,06	2,47
9	3,89	3,26	2,62
10	3,92	3,28	2,64
11	4,00	3,34	2,69
12	4,04	3,38	2,73
13	4,14	3,46	2,78
14	4,38	3,66	2,95
15	4,42	3,69	2,96
16	4,43	3,71	2,98
17	4,52	3,78	3,04
18	4,57	3,82	3,07
19	4,60	3,85	3,10
20	4,64	3,88	3,12

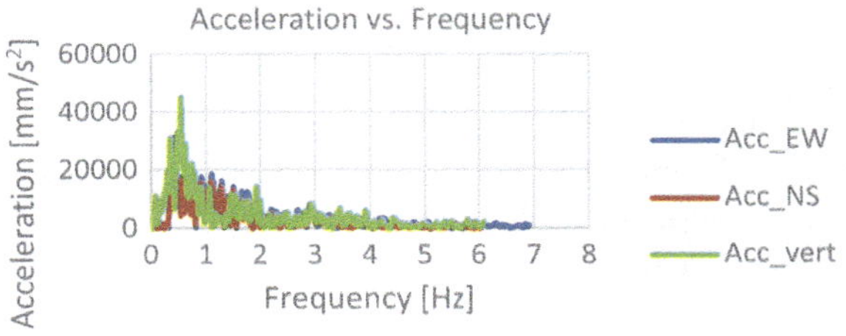

Fig. 4. Acceleration spectrum

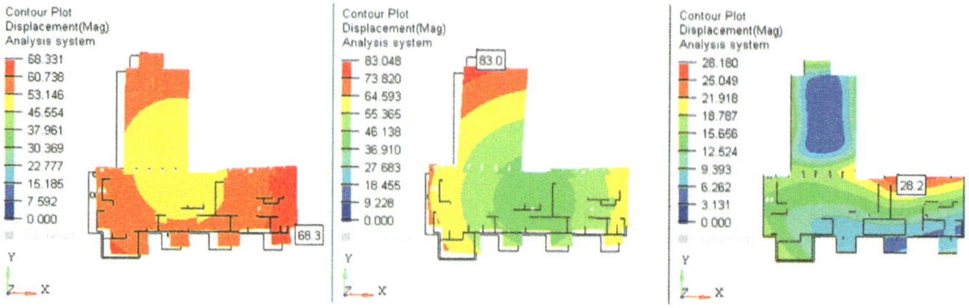

Fig. 5. The first three eigenmodes of vibration

5 Discussion

This paper studies the response of the inertial platform supported on sets of spring batteries on the ground base to identify if excessive vibrations exist and how are the different ground vibrations transmitted to the platform. On the platform, a lot of devices and instruments are distributed implying in the scientific research and the necessary annexes. Usually, vibrations coming from the vehicles and other anthropic activities and transmitted through the ground are nonstationary vibrations. To reduce the influence of these vibrations on the inertial platform, this is suspended on a set of spring batteries and dampers that can be activated or not. This suspension constitutes an excellent tool to reduce the transfer of the unwanted vibrations to the studied platform.

The platform consists of two plates, one main, on which are found most of the instruments and devices and where the massive partition walls are raised, which is about 1600 mm, and the second platform, which is called secondary, smaller and thinner, is only 600 mm, and weighs about half the weight of the unloaded mainboard. If the mainboard is loaded with all the equipment and partitions, the platform weighs about 1/6 of the total weight of the platform. This platform is found to have significant amplitudes in its eigenmodes, so in the case of external excitations with frequencies close to the natural frequencies of the entire platform, we can expect an amplification of these excitations at the secondary platform. The problem can be solved by activating additional springs from the spring batteries that support the secondary platform. Additionally, another problem that can be mentioned is the fact that the eigenfrequencies of the platform are found in the range, where the earthquakes that take place in the area have excitation frequencies.

6 Conclusions and Future Works

The dynamic analysis of the platform indicates a large number of natural frequencies below 5 Hz, which is a negative aspect from the point of view of the platform's behavior in earthquakes. Anthropic activities will not influence the behavior of the system, taking into account the frequencies generated by these activities. Instead, the effects of a catastrophic event (earthquake) could amplify the possibilities of vibration in certain directions of the platform, which could lead to destruction and damage to the installation. For this reason, a careful calculation is required to the structure's response in case of an earthquake and its permanent monitoring for the analysis of the transmissibility of external excitations. The ELI-NP project has a decoupling system for earthquakes, so the need for an additional isolation system of the entire mass should be studied. Such a study, the necessity of which has been demanded by the research made, is to be carried out.

References

1. Arnold, C.: Seismic Design & Devices for Detaching Building from the Ground. Archit. AIA J. **76**, 64–67 (1987)
2. Tarics, A.G.: Base-Isolation—A New Strategy for Earthquake Protection of Buildings. J. Archit. Plan. Res. **4**, 64–76 (1987)
3. Teramoto, T.: Recent Current of Nase Isolation and Seismic Control-System for Building Structure. Tetsu Hagane J. Iron Steel Inst. Jpn. **73**, S348 (1987)
4. Jiang, M.; Rui, X.; Zhu,W.; Yang, F.; Zhang, Y.: Optimal design of 6-DOF vibration isolation platform based on transfer matrix method for multibody systems. Acta Mech. Sin. 1–11 (2020)
5. Han, P.;Wang, T.;Wang, D.H.: Modeling and control of a Stewart platform based six-axis hybrid vibration isolation system. In Proceedings of the 2008 7thWorld Congress on Intelligent Control. and Automation, Chongqing, China, 25–27 June 2008; Volume **1–23**, pp. 1613–1618
6. Geng, Z., Haynes, L.S.: Six-degree-of-freedom active vibration isolation using a stewart platform mechanism. J. Robot. Syst. **10**, 725–744 (1993)

Spectral Approach of Reinforced Concrete Column Dynamics

Maciej Dutkiewicz[1(✉)] and Marcela R. Machado[2]

[1] Faculty of Civil, Architecture and Environmental Engineering and Architecture, Bydgoszcz University of Science and Technology, Al. Prof. S. Kaliskiego 7, 85-796 Bydgoszcz, Poland
macdut@pbs.edu.pl

[2] Department of Mechanical Engineering, University of Brasilia, Brasilia 70910-900, Brazil

Abstract. In this paper, the spectral element method (SEM) was used for determining the dynamic characteristics of the reinforced concrete column. For selected concrete strengths and modulus of longitudinal elasticity, the structure responsible for column cross-section reduction, change of its height, and the value of the compressive force were analyzed. The frequency response functions for each case were performed. General conclusions on damage detections technique were presented.

Keywords: Reinforced concrete column · Spectral element method · Frequency response function

1 Introduction

Due to the structural safety and durability, the monitoring of the object plays an important role. In the first stage of the facility service, it is crucial to relate the actual structure response to the response resulting from the design model parameters. In the subsequent stages, the results of monitoring allow for the current assessment of the technical condition of the structure.

The aim of this paper is to show that the analysis of frequency response functions (FRF) based on the spectral element method (SEM) is a useful tool in the damage detection technique.

It is obvious that damage can cause changes in local and global properties of a structure, affecting local stiffness, density, mass, thermal, electrical, and magnetic conductivity, electromechanical resistance, and strain energy. These changes can be identified during the dynamic signal analysis of the excited structure.

Structural dynamic analysis plays an important role in damage detection. The introduction of this paper highlights the following issues: (i) application of the SEM as the complementary to the laboratory or in situ detection technique performance, (ii) damage detection technique in the context of the comprehensive study, (iii) structural health monitoring as a general procedure that was recommended in the service of civil engineering structures.

N. Herisanu and V. Marinca (eds.), *Acoustics and Vibration of Mechanical Structures – AVMS-2021*, Springer Proceedings in Physics 274,
https://doi.org/10.1007/978-3-030-96787-1_33

The motivations for the research devoted to the dynamic analysis of the reinforced concrete column were: (i) the use of wave solutions in dynamics, (ii) the use of SEM in structural analysis as an efficient method based on a strict solution of the dynamic equation, (iii) analysis of various parameters of the reinforced concrete structure, which allow for inference in the field of damage detection and health monitoring, (iv) the authors' ongoing research and laboratory testing on safety, durability, and interaction between concrete and high-grade rebars, grade B600B in columns with the described, in the next paragraph of the paper, parameters such as geometrical dimensions, elasticity modulus, grade of the concrete, length, section area, and compression force acting on the column.

The current state of the art, referring to the need to ensure safety, durability, and integrity of the structure, required the development of nondestructive evaluation techniques (NDE). The basic premise of NDE techniques is to measure the signal on an undamaged structure, which is the reference to the failed structure.

The beginnings of SEM were included in the paper [1], where the dynamic stiffness influence coefficients were used in the analysis of the beam. These coefficients in the table form for a uniform free beam and the dynamic response with matrix inversion in the Fourier transform plane and a numerical inversion, based on the Cooley–Tukey algorithm were presented. The extension of this work was presented in [2], where the response of the analyzed structure was received with the application of the fast Fourier transform algorithm. The formulation was based on the exact solution of the transformed governing equation of motion of a beam element, and it led to the exact solution of the problem. The fundamentals of SEM were also shown in [3]. The author presented the spectral analysis of the wave motion, waves in rods, beams, and plates, as well as structure–fluid interaction. Special attention was paid to membrane spectral elements, folded plate structures, and segmented cylindrical shells. Lee in [4] introduced the fundamentals of discrete Fourier transform, and various methods of spectral element formulation such as the force–displacement, relation method, the variational method, and the state-vector equation method. Part of Doyle's research presented in [4] addressed the SEM-based linear and nonlinear structural damage identification methods to identify multiple local damages generated in a structure from experimentally measured FRF data. Cho and Lee [5] proposed a fast Fourier transforms based spectral analysis method for the dynamic analysis of the linear discrete dynamic systems. The forced vibration of a three-degrees-of-freedom system was considered an illustrative problem. Kim et al. [6] presented the modal spectral element formulated for thin plates moving with constant speed under a uniform in-plane axial tension. The concept of the Kantorovich method was used to formulate the modal spectral element matrix in the frequency domain. The effects of the moving speed and the in-plane tension on the dynamic characteristics of a moving plate were investigated. The paper [7] developed a spectral element model for the axially moving beam-plates subjected to sudden external thermal loads. In paper [8], the spectral elements were used to transform a large periodic lattice beam into a structurally equivalent continuum beam model. Lee [9] applied the frequency-domain approach to identify the dynamic characteristics of the mechanical joints. The bolt-joint was represented by the equivalent joint model with four parameters.

In [10], the FEM-SEM approach for wave propagation analysis in cable structures was performed. The methodology allowed for the study of the defect or damage in the cable. The results showed how damage affected the high-frequency response. In [11], the authors developed the model of the transmission line via the spectral element method. Numerical analysis investigated the natural frequency of the model due to different parameters such as a tension force and the area of the cross-section of the conductor. The purpose of the investigations presented in [12] was to analyze the natural vibrations of the transmission line and compare them with the results obtained in the numerical simulations. Sensitivity of the wavenumber was performed for changing the tension force, as well as for the different damping parameters. Two numerical models based on the spectral element method for transmission lines considering hysteretic and aerodynamics damping with analyses that consider dispersion diagrams and frequency response functions were shown in [13]. As SEM lead to a transcendental eigenvalue problem, to obtain the natural frequencies of the conductor, the Wittrock–Williams algorithm was used. The numerical models have been compared with the analytical solution of the cable and experimental tests with the overhead conductor Grosbeak cable.

In paper [14], the principles, equations, and applications of damage modeling and elastic waves propagation were presented. Krawczuk [15] presented the use of the wave propagation approach combined with a genetic algorithm and the gradient technique for damage detection in beam structures. Paper [16] considered the finite spectral element of a cracked Timoshenko beam for modal and elastic wave propagation analysis. The changing location of the crack, on the wave propagation, was examined. Research [17] provided the knowledge of damage identification technique using plate waves, also called Lamb waves, and conclusions were applied to other forms of guided waves. In paper [18], the approach for identifying structural damage in an aluminum plate with piezoelectric materials to sense the dynamic response of the structure was presented. The lamb wave propagations and impedance methods were shown. In the literature, structural health monitoring was defined as nondestructive and continuous monitoring characteristics using the sensors to diagnose anomalous structural behavior. It involves measuring and evaluating the state properties of structure [19].

The presented research shows how the response of a reinforced concrete column changes when its cross-section decreases in the event of the column's failure, when the net cross-section is reduced. The analysis was performed for selected concrete strengths and modulus of longitudinal elasticity. The dynamic response of the column was researched with the change of its height and the value of the compressive force.

2 Mathematical Model of the Column

Considering a simplified model, the governing differential equation for the undamped free vibration can be written as [20–22]

$$EI\frac{\partial^4 v(x,t)}{\partial x^4} - T\frac{\partial^2 v(x,t)}{\partial x^2} + \rho A\frac{\partial^2 v(x,t)}{\partial t^2} = 0 \tag{1}$$

where ρA is mass per unit length, EI the uniform bending rigidity, T is compression force, and $v(x,t)$ is the column displacement as a function of the position x and time t.

The undamped Euler–Bernoulli equation of motion subjected to axial force and under bending vibration is governed by Eq. (1). Figure 1 shows an elastic two-node element with a uniform rectangular cross-section subjected to an axial force, where the properties are assumed to be deterministic variables.

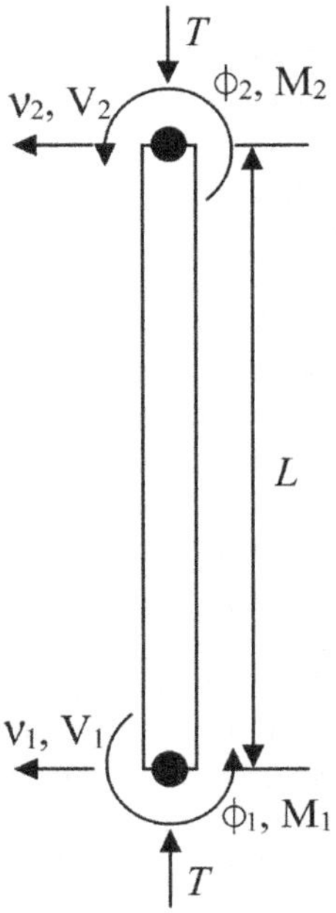

Fig. 1. Two-node spectral element

By considering a constant coefficient, a displacement solution can be assumed in the form:

$$v(x, t) = v_0 e^{-i(kx-\omega t)} \tag{2}$$

where v_0 is the amplitude, ω is the frequency, and k is the wavenumber. Substituting it in Eq. (1), the dispersion equation is given by

$$k^4 EI + k^2 T - \omega^2 \rho A = 0 \tag{3}$$

In further analysis, k_1 wave number is taken into consideration, as k_2 is the imaginary number. The general solution for the spectral element subjected to an axial load of length, L, can be expressed in the form:

$$\begin{aligned} v(x, \omega) &= a_1 e^{-ikx} + a_2 e^{-kx} + a_3 e^{-ik(L-x)} + \\ &+ a_4 e^{-k(L-x)} = \boldsymbol{s}(x, \omega)\boldsymbol{a} \end{aligned} \tag{4}$$

where

$$s(x,\omega) = \left\{ e^{-ikx}, e^{-kx}, e^{-ik(L-x)}, e^{-k(L-x)} \right\} \tag{5}$$

$$a(x,\omega) = \{a_1, a_2, a_3, a_4\}^T \tag{6}$$

The spectral nodal displacements and slopes of the element are related to the displacement field at node 1 ($x = 0$) and node 2 ($x = L$) by

$$d = \begin{bmatrix} v_1 \\ \phi_2 \\ v_2 \\ \phi_2 \end{bmatrix} = \begin{bmatrix} v(0) \\ v'(0) \\ v(L) \\ v'(L) \end{bmatrix} \tag{7}$$

By substituting Eq. (4) into the right-hand side of Eq. (7) and written in a matrix form gives

$$d = \begin{bmatrix} s(0,\omega) \\ s'(0,\omega) \\ s(L,\omega) \\ s'(L,\omega) \end{bmatrix} a = G(\omega)a \tag{8}$$

where

$$G(\omega) = \begin{bmatrix} 1 & 1 & e^{-ikL} & e^{-kL} \\ -ik & -k & ike^{-ikL} & ke^{-kL} \\ e^{-ikL} & e^{-kL} & 1 & 1 \\ -ike^{-ikL} & -ke^{-kL} & ik & k \end{bmatrix} \tag{9}$$

The frequency-dependent displacement within an element is interpolated from the nodal displacement vector $\boldsymbol{d}$, by eliminating the constant vector $\boldsymbol{a}$ from Eq. (4). The dynamic stiffness matrix for the spectral beam element under axial tension can be determined as

$$S(\omega) = K(\omega) - \omega^2 M(\omega) \tag{10}$$

where

$$K(\omega) = \int_0^L \left(EIg''(x)^T g''(x) + Tg'(x)^T g'(x) \right) dx \tag{11}$$

$$M(\omega) = \rho A \int_0^L g(x)^T g(x) dx \tag{12}$$

where expresses the spatial partial derivative. By solving the integral, the dynamic stiffness matrix is

$$S(\omega) = \frac{EI}{\Delta}\begin{bmatrix} s_{11} & s_{12} & s_{13} & s_{14} \\ & s_{22} & s_{23} & s_{24} \\ & & s_{33} & s_{34} \\ sym & & & s_{44} \end{bmatrix} \tag{13}$$

where $\Delta = \cos(kL)\cosh(kL) - 1$ and the components of element matrix (Eq. 13) are given as

$$\begin{aligned} s_{11} &= -k^3(\cos(kL)\sinh(kL) + \sin(kL)\cosh(kL)) \\ s_{1_2} &= -k^2\sin(kL)\sinh(kL) \\ s_{13} &= k^3(\sin(kL) + \sinh(kL)) \\ s_{14} &= k^2(\cos(kL) - \cosh(kL)) \\ s_{22} &= k(\cos(kL)\sinh(kL) - \sin(kL)\cosh(kL)) \\ s_{23} &= k^2(\cosh(kL) - \cos(kL)) \\ s_{24} &= k(\sin(kL) - \sinh(kL)) \\ s_{33} &= -k^3(\cos(kL)\sinh(kL) + \sin(kL)\cosh(kL))s_{34} = k^2\sin(kL)\sinh(kL) \\ s_{44} &= k(\cos(kL)\sinh(kL) - \sin(kL)\cosh(kL)) \end{aligned} \tag{14}$$

As far as the structure of the column is uniform without any sources of discontinuity, it can be represented by a single spectral element with very accurate solutions [3]. However, if there exist sources of discontinuity such as the point loads, the column should be spatially discretized into spectral elements. Analogous to the finite element method (FEM) [23], the spectral elements can be assembled to form a global structure matrix system [4].

3 Numerical Analysis and Results

In the simulations, the reinforced concrete column with basic parameters of length L = 3 m, area of A = 4 × 10^{-2} m^2 is considered. The material properties are modulus of elasticity E = 29.7 GPa, the material density of ρ = 2,500 kg/m^3, and the compression load T = 2,500 kN. These are the average values of the parameters of reinforced concrete columns commonly used in the construction industry, in particular, the compression load means the maximum force that a reinforced concrete column with the previously indicated L, A, and E parameters can bear. For comparison, the following changes of parameters were analyzed: elasticity modulus E, compression load T, section area A, and length L. Changes of parameters, in particular modulus E and section area A, change during element damage and cracking at the corresponding compression force. The analyzed parameters are presented in Table 1.

Figure 2 shows the relationship between compression strength and modulus of elasticity [24].

Table 1. Analyzed parameters

No	E (GPa)	A (m^2)	T (kN)	L (m)
1	14.9	0.0081	0.001	0.5
2	29.7	0.0121	0.1	1.0
3	33.2	0.0225	10	3.0
4	36.4	0.04	100	5.0
5	39.3	0.0625	1,000	
6	42.0	0.09	2,500	
7	44.6	0.123	25,000*	
8	47.0	0.16	50,000*	

*For comparison

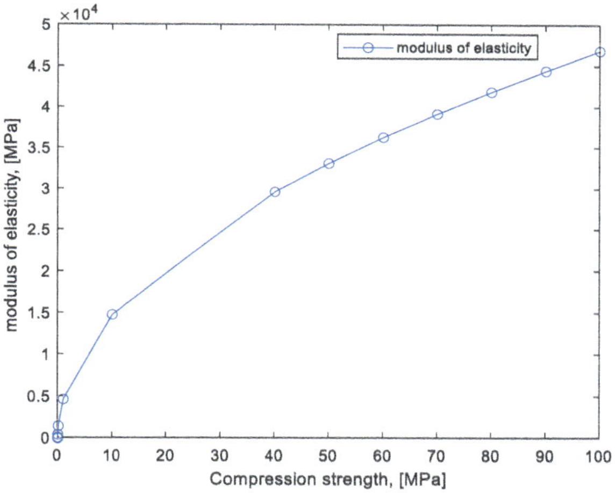

Fig. 2. The relationship between compression strength and modulus of elasticity [24]

Figure 3 presents the inertance response obtained for elasticity modulus from E1 to E3, L = 3.0 m, T = 2,500 kN, A = 0.04 m^2. The frequency band of analysis varies from 0 to 15 kHz approaching the resonance frequencies and mode shapes. From Figs. 3 and 4, it can be concluded that with the increase in elasticity modulus, the eigenfrequency for all modes increases. The average increase in elasticity modulus by 8% causes an increase in eigenfrequency by 4% for all frequencies.

Figure 5a presents the inertance response obtained for compression force from T_1 to T_6, L = 3.0 m, E = 29.7 GPa, A = 0.04 m^2. The frequency band of analysis varies from 0 to 15 kHz approaching the resonance frequencies and mode shapes. Figure 5b presents the first three resonance frequencies for compression force from T_1 to T_6 change. From Fig. 5, it can be concluded that with the increase in compression force, the values of eigenfrequency for all modes decreases.

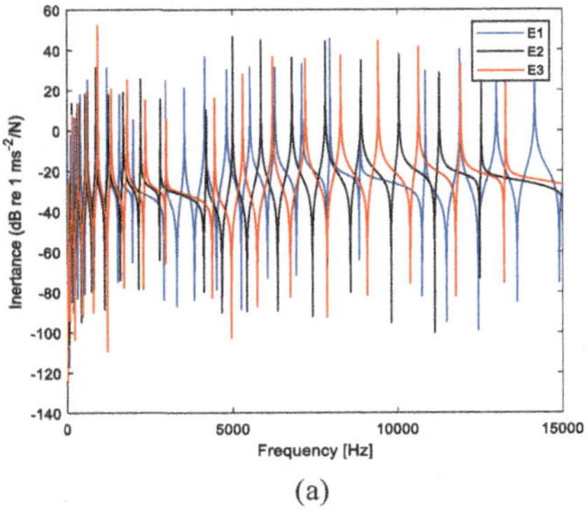

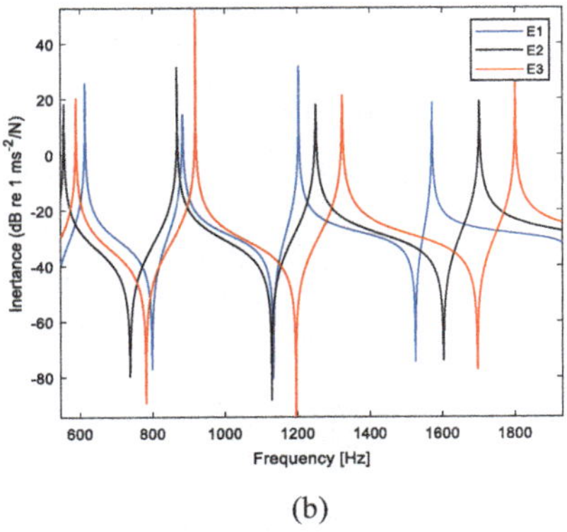

Fig. 3. Inertance response for elasticity modulus $E_1..E_3$ (parameters accord. to the Table 1), L = 3.0 m, T = 2,500 kN, A = 0.04 m^2, **a** full range of analyzed frequency, **b** zoom for frequency range <600,1800> Hz

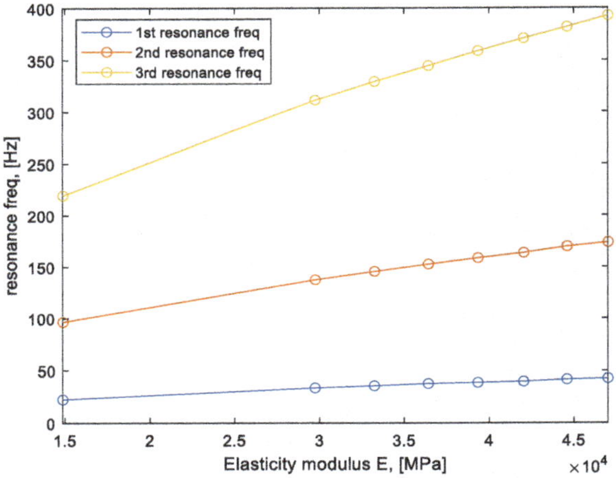

Fig. 4. First three resonance frequencies for elasticity modulus $E_1..E_8$ change. Values of modulus are presented in Table 1

Figure 6a presents the inertance response obtained for section areas A_1, A_4, A_8, L = 3.0 m, E = 29.7 GPa, T = 2,500 kN. The frequency band of analysis varies from 0 to 15 kHz approaching the resonance frequencies and mode shapes. Figure 6b presents the first three resonance frequencies for section area from A_1 to A_6 change. From Fig. 6, it can be concluded that with the increase in column's section area, the eigenfrequency for all modes increases. The average increase in section area by 54% causes an increase in eigenfrequency by 33% for all frequencies.

Figure 7a presents the inertance response obtained for column's length from L_1 to L_4, E = 29.7 GPa, T = 2,500 kN, A = 0.04 m^2. The frequency band of analysis varies from 0 to 15 kHz approaching the resonance frequencies and mode shapes. Figure 7b

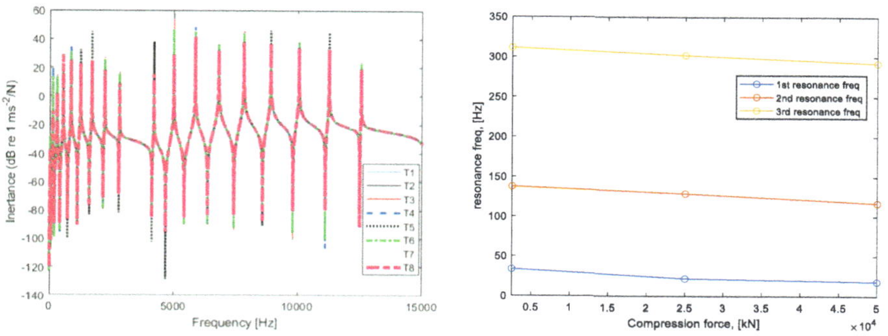

(a) (b)

Fig. 5. Inertance response for compression force $T_1..T_6$ (parameters according to Table 1), L = 3.0 m, E = 29.7 GPa, A = 0.04 m^2, **a** full range of analyzed frequency, **b** first three resonance frequency for compression force T1..T6 change

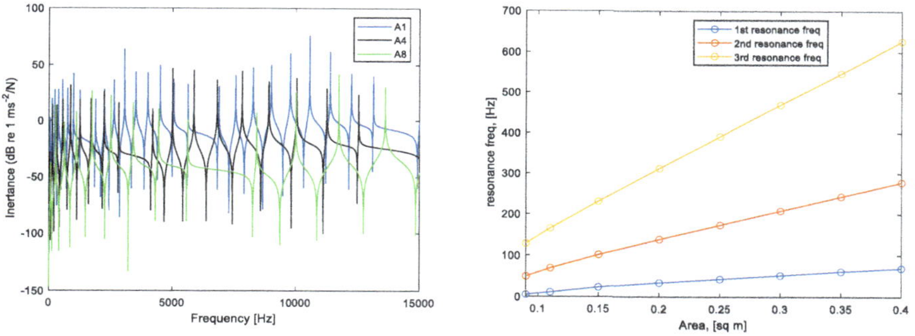

(a) (b)

Fig. 6. Inertance response for section areas A_1, A_4, A_8 (parameters according to Table 1), L = 3.0 m, E = 29.7 GPa, T = 2,500 kN, **a** full range of analyzed frequency, **b** first three resonance frequency for section areas $A_1..A_6$ change

presents the first three resonance frequencies for length from L_1 to L_4 change. From Fig. 7, it can be concluded that with the increase in column's length, the eigenfrequency for all modes decreases. The average increase in column length by 22% causes a decrease in eigenfrequency by 24% for all frequencies.

4 Conclusions

In this paper, the application of the spectral element method and frequency response functions analysis showed that this methodology could be a tool for determining the dynamic characteristics of a column, as a part of a damage detection technique. The numerical analysis with the use of the SEM could be complementary to the laboratory or in situ detection technique performance.

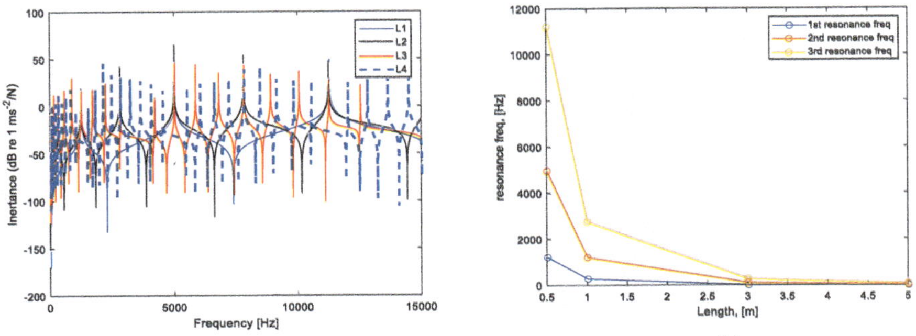

(a) (b)

Fig. 7. Inertance response for column's lengths $L_1..L_4$ (parameters accord. to the Table 1), E = 29.7 GPa, T = 2,500 kN, A = 0.04 m^2, **a** full range of analyzed frequency, **b** first three resonance frequencies for lengths $L_1..L_4$ change

In the research, the structure responsible for the selected concrete strengths and elasticity modulus, column cross-section reduction, change of its height, and the value of the compressive force were analyzed. The frequency response functions for each case were performed. General conclusions on damage detections technique were presented.

The next aspect constituting the purpose of this paper was to present the wide possibilities of modeling and simulation of concrete structures using the spectral method. Based on the analysis performed in this paper and the literature review on SEM, FEM, and other methods of analysis, it should be emphasized that compared with other methods, SEM has the advantages:

- Reduction of the number of nodes and elements;
- Fast convergence;
- High accuracy of results.

References

1. Beskos, D., Narayanan, G.: Use of dynamic influence coefficients in forced vibration problems with the aid of fast Fourier transform. Comput. Struct. **9**(2), 145–150 (1978)
2. Spyrakso, C., Beskos, D.: Dynamic response of frameworks by fast Fourier transform. Comput. Struct. **15**(5), 495–505 (1982)
3. Doyle, J.F.: Wave Propagation in Structures-Spectral Analysis Using Fast Discrete Fourier Transforms (1997). https://doi.org/10.1007/978-1-4684-0344-2_6
4. Lee, U.: Spectral Element Method in Structural Dynamics. Wiley (2009)
5. Cho, J., Lee, U.: An FFT-based spectral analysis method for linear discrete dynamic systems with nonproportional damping. Shock. Vib. **13**(6), 595–606 (2006)
6. Kim, J., Cho, J., Lee, U., Park, S.: Modal spectral element formulation for axially moving plates subjected to in-plane axial tension. Comput. Struct. **81**(20), 2011–2020 (2003)
7. Kwon, K., Lee, U.: Spectral element modeling and analysis of an axially moving thermoelastic beam plate. J. Mech. Mater. Struct. **1**(4), 605–632 (2006)

8. Lee, U.: Equivalent continuum representation of lattice beams: spectral element approach. Eng. Struct. **20**(7), 587–592 (1998)
9. Lee, U.: Dynamic characterization of the joints in a beam structure by using spectral element method. Shock. Vib. **8**(6), 357–366 (2001)
10. Dutkiewicz, M., Machado, M.: Spectral element method in the analysis of vibrations of overhead transmission line in damping environment. Struct. Eng. Mech. **71**(3), 291–303 (2009). https://doi.org/10.12989/SEM.2019.71.3.291
11. Dutkiewicz, M., Machado, M.R.: Spectral approach in vibrations of overhead transmission lines. IOP Conf. Ser.: Mater. Sci. Eng. **471**(5), (2019). https://doi.org/10.1088/1757-899X/471/5/052029
12. Dutkiewicz M., Machado M.: Spectral element method in the analysis of vibrations of overhead transmission line in damping environment. Struct. Eng. Mech. **71**(3), 291–303 (2009). https://doi.org/10.12989/SEM.2019.71.3.291
13. Machado, M.R., Dutkiewicz, M., Matt, C.F.T., Castello, D.A.: Spectral model and experimental validation of hysteretic and aerodynamic damping in dynamic analysis of overhead transmission conductor. Mech. Syst. Signal Process. **136**(106483) (2019). https://doi.org/10.1016/j.ymssp
14. Ostachowicz, W.M.: Damage detection of structures using spectral finite element method. Comput. Struct. **86**, 454–462 (2008). https://doi.org/10.1016/j.compstruc.2007.02.004
15. Krawczuk, M.: Application of spectral beam finite element with a crack and iterative search technique for damage detection. Finite Elem. Anal. Des. **80**, 1809–1816 (2002). https://doi.org/10.1016/S0168-874X(01)00084-1
16. Krawczuk, M., Palacz, M., Ostachowicz, W.: The dynamic analysis of a cracked Timoshenko beam by the spectral element method, J. Sound Vib. **264**(5), 1139–1153 (2003). https://doi.org/10.1016/S0022-460X(02)01387-1. ISSN 0022-460X
17. Su, Z., Ye, L.: Identification of damage using lamb waves. Springer, New York (2009)
18. Wait, J.R., Park, G., Sohn, H., Farrar, C.R.: Plate damage identification using wave propagation and impedance methods. Proc. SPIE **5394**, 53–65 (2004)
19. Moyo, P., Brown, J.M.W.: Detection of anomalous structural behavior using wavelet analysis. Mech. Syst. Signal Process. **16**, 429–445 (2002)
20. Clough, R., Penzien J.: Dynamics of Structures. McGraw Hill, New York
21. Yu, B.Y.J., Soliman, M.: Estimation of cable tension force independent of complex boundary conditions. ASCE J. Eng. Mech. **60**, 1–8 (2014)
22. Rao, S.S.: Mechanical Vibration. Person Prentice-Hal (2008)
23. Zienkiewicz, O.C., Taylor, R.L.: The Finite Element Method, 4th edn. McGraw-Hill, London (1991)
24. ACI-318-99, Building Code Requirements For Structural Concrete

Gough-Stewart Platform Stiffness and Eigenmodes Evaluation

Ciprian Dragne(✉) and Veturia Chiroiu

Solid Mechanics Institute of the Romanian Academy, Bucharest, Romania
ciprian_dragne@yahoo.com

Abstract. This paper presents an evaluation of the stiffness and eigenmodes of a Stewart platform during changing positions in the workspace. Stiffness and eigenmodes are important in the design phase and even during service they can be useful for the detection of damages that could appear in the structure. Computer simulation is used to solve the mathematical model and could be a base for any structural optimization.

Keywords: Gough-Stewart · Stiffness · Eigenmodes · Damage detection

1 Introduction

Hard-to-resect tumor structures in some internal human organs can be treated by brachytherapy medical methods using ionizing radiation emitted by seeds that have a radioactive material inside. The placement of these seeds requires very high precision. This is why the idea of using robotic systems to carry out these operations has become increasingly popular [1]. The placement of these seeds inside the affected organs with minimal damage to other adjacent structures or those in the path of medical intervention still requires research over robotic system kinematics and into control strategy of the robotic arm to ensure maximum efficiency of the methodology and medical procedure. One of the robotic systems capable to perform such a task could be a Gough-Stewart platform. The concept of a robotic system related to all design constraints involves many performance aspects for a complete tool ready for practice: cinematics, biomechanical, medical, robot control strategy, system performances for scene visualization and 3D reconstruction techniques, etc.

The objective of this research is to provide an evaluation of changes in the stiffness of the mechanical structure and the eigenmodes and eigenvalues during changes in robot positioning in the workspace. This evaluation is important in the concept phase of the designing process of a new robot based on the Gough-Stewart platform. But such evaluation is useful also during service. Using proper sensors and special IT programming tools can be used for performing tasks like structural integrity evaluation. The weakening of the assembly points and damage of parts imply changes in stiffness at eigenmodes and eigenvalues. These damages can be even localized precisely if they are evaluated before the total collapse of the structure [2].

N. Herisanu and V. Marinca (eds.), *Acoustics and Vibration of Mechanical Structures – AVMS-2021*, Springer Proceedings in Physics 274,
https://doi.org/10.1007/978-3-030-96787-1_34

2 Development of the Mathematical Model

2.1 Stewart Platform

The Stewart platform is also known as the "Gough-Stewart" platform because it was first introduced in 1956 by Eric V. Gough as an application in the automotive domain and then in 1965 by D. Stewart as a mechanism to simulate the movement of the aircraft in flight simulators [3, 4].

The Gough-Stewart platform consists of two platforms, one fixed and the other a moveable platform, connected to each other by a number of actuators (axial or not). In the most common case, there are 6 identical links as shown in the rough concept from Fig. 1. Because here are used 6 links, the system has 6 DOFs. Changes in the actuator's length led to changes in the mobile-platform position in the workspace that can be fully imposed with required coordinates (translational and rotational values) (Fig. 2).

Inverse kinematics of the Stewart platform are governed by Eq. (1):

$$[X] = [T][X_0] \tag{1}$$

where $[X_0]$ is initial coordinates of specific points and [X] is the complete vectors of the coordinates of the points from a mobile platform and in generalized coordination has the next form:

$$[X] = \begin{bmatrix} p_x\ p_y\ p_z\ \alpha\ \beta\ \gamma \end{bmatrix} \tag{2}$$

where $\begin{bmatrix} P_x, P_y, P_z \end{bmatrix}$ are the Cartesian coordinates of the point {P}—center of the mobile platform coordinates relative to a global coordinate system located in point {O} and $\begin{bmatrix} \alpha, \beta, \gamma \end{bmatrix}$ are the Euler rotational angles of the mobile platform, coordinates relative to a local coordinate system located in point {P} are the center of the mobile platform.

Global matrix [T] of the transformation is presented in the next equation:

$$[T] = \begin{bmatrix} [R^P] & [T_0^P] \\ 0 & 1 \end{bmatrix} \tag{3}$$

which includes rotational and translational matrix.

$$\begin{bmatrix} R^P \end{bmatrix} = \begin{bmatrix} c\beta c\gamma & c\gamma s\alpha s\beta - c\alpha s\gamma & s\alpha s\gamma + c\alpha c\gamma s\beta \\ c\beta s\gamma & c\alpha c\gamma + s\alpha s\beta & s\gamma c\alpha s\beta s\gamma - c\gamma s\alpha \\ -s\beta & c\beta s\alpha & c\alpha c\beta \end{bmatrix} \tag{4}$$

where s() and c() are sine and cosines functions.

Translation vector is the position of the center of the mobile platform relative to the global coordinate system:

$$\begin{bmatrix} T_0^P \end{bmatrix} = \begin{bmatrix} dX\ dY\ dZ \end{bmatrix}^T \tag{5}$$

The vector for defining the link direction is

$$\left\{ \overrightarrow{L_i} \right\} = \left\{ \overrightarrow{B_i P_i} \right\} = \left[\overrightarrow{T} \right] + \left[R^P \right] \cdot \left\{ \overrightarrow{PP_i} \right\} - \left\{ \overrightarrow{OB_i} \right\}, i = 1 \div 6 \tag{6}$$

The length of each link is described by Eq. (7):

$$Leg_i = \left\| \left\{ \overrightarrow{L_i} \right\} \right\| = \left\| \left\{ \overrightarrow{B_i P_i} \right\} \right\|, i = 1 \div 6 \tag{7}$$

2.2 Model Workspace

This study was considered a model defined by: fixed nodes from a base platform located on a circle with 135 mm radius, nodes of links from mobile platform distributed on a circle with 100 mm radius, and initial length of the links equal to 200 mm. Displacement constraints of the center of the mobile platform, between $[-200 \div 200]$ mm in x- and y-directions and between $[100 \div 280]$ mm in the z-direction, obtained by the workspace are presented in Fig. 3. This research considered only Cartesian changes at coordinate values. Evaluation of the workspace is based also on the Euler-angular changes at the mobile platform involving a more complex space evaluation that needs to use an additional procedure for elimination of the duplicate points from the results.

Current workspace evaluation takes into consideration the limits of the actuators and also an evaluation of the collision detection between links.

2.3 Static and Eigen Model

Mathematical model of the structure can be represented only by mass and stiffness matrices into the simplest Eq. (8):

$$[M]\{\ddot{X}\} + [K]\{X\} = \{F\} \tag{8}$$

where [M] and [K] are the mass and stiffness global matrices of the assembled system, respectively, {F} is the vector of the external or reaction forces and $\{X\}$, $\{\ddot{X}\}$ are displacement and acceleration vectors of nodes that describe the structure, respectively. The mass and stiffness global matrices are assembled according to FEA rules. The stiffness and mass matrices are sparse matrices and have dimensions depend on number of nodes of the model.

Real structures are not rigid. Actuators have an internal stiffness. But this stiffness does not represent the entire stiffness of the links. A complete stiffness of the links should also include terms related to the spherical or universal joint stiffness or other values of the assembly parts into the joints area until the platform. Also, the stiffness matrix of the complete assembly includes the stiffness of the platform parts. This model for platform parts considered an aluminum platform with 10 [mm] thickness. For the fixed platform, the stiffness should not be used if the model is considered ground-fixed at the location of the base points of the links ($\{B_i\}$, $i = 1 \div 6$). This study used also the stiffness of the fixed platform, because fixation points of the entire system are different than fixation points on the links into the fixed platform.

The stiffness of the links is constant, and does not depend on the length of the actuator. This stiffness has a value equal to 1000 [N/mm]. Further research will include an equation of the stiffness of the links depending on the length of the actuators.

Evaluation of the stiffness of the Stewart platform is based on Eq. (9):

$$[K]\{X\} = \{F\} \tag{9}$$

where $F = (\{F_x\}, \{F_y\}, \{F_z\})$ depends on the direction that should be evaluated.

For evaluation of the normal eigenmodes on the structure, Eq. (10) is used:

$$[M]\{\ddot{X}\} + [K]\{X\} = 0 \tag{10}$$

Finally, it is necessary to explain that it was not established yet an analytical relationship between stiffness, dimensions, and the free shape of the Stewart mechanism [11]. A numerical model based on the discrete definition is more appropriate for an evaluation of the concept during design. Using a more complex definition, a model that could take into consideration also the contact between parts, friction, damping, and stress and strength evaluation of the robot structure will be necessary for further structural checking and optimization.

3 Simulation Model and Results

3.1 Parametric FEA Model

For evaluation based on mobile platform, location used a parametric FEA model where the location of the new nodes is defined based on the inverse kinematics equation of the Stewart platform (Fig. 4).

3.2 Numerical Results

Model is ground-fixed at three points (BC-1, BC-2, and BC-3) with constraints ($x = 0$, $y = 0$, and $z = 0$). The distance between fixed and mobile platforms is 187.75 mm in the Z-direction and is considered fixed at each initial model.

For stiffness evaluation, each model was loaded at the center point of the mobile platform with 1000 N. Dynamic model is overloaded by an additional mass at the center of the mobile platform with 5 kg (Figs. 5, 6, 7, 8, 9, 10, 11 and 12).

Eigenmodes identification:

- Mode-1 is the X-translation eigenmode;
- Mode-2 is the Y-translation eigenmode;
- Mode-3 is the Z-translation eigenmode;
- Mode-4 is the torsional eigenmode;
- Mode-5 is the pitch rotation of the mobile platform around the Y-axis;
- Mode-6 is the roll rotation of the mobile platform around the X-axis.

3.3 Damage Detection

Based on a numerical model developed initially, a structure can be monitored during service. Changes in eigenmodes that overflow a specific factor should be considered warnings and involved after a structural inspection and evaluation for damage. The location of the damaged part can be even estimated based on numerical model results.

In case of a more detailed model, for dynamic and damage detection can be used a model reduction technique with master nodes that take into consideration also the locations of the used sensors [6].

4 Conclusions

1. The maximum changes in stiffness appear in the Z-direction from 4255[N/mm] at the home-position until 840[N/mm] at a point position of platform center with (x, y, z) = (100,0,0) [mm] far away, that reveal changes during plane Y = 0 until 80%;
2. Movement in Z-plane reveals a loss in stiffness even higher. At maximum (dx = −100, dy = −100, dz = 0), it shows a 347 [N/mm] that represents a 91.84% decrease in the stiffness relative to the home-position of the platform;
3. The maximum changes in resonance frequency are between 33.6[Hz] and 27[Hz] for fundamental mode, which could have changed until 25%;
4. A dynamic model that could consider also the damping matrix of the system will show more accurate results for a more complex assembly. But damping matrix is not easy to evaluate. A dynamic model identification based on experimental measurements is required. This model will be evaluated in another study.

Further Research

1. A complete evaluation should also take into consideration the determination of the all-load cases and necessary robots structure strength during service.
2. Evaluation of the Stewart platform stiffness at torsional and bending moment loadings will be useful.
3. An evaluation of the Stewart platform strength depending on the end-effector location will be considered for further research.
4. Transient response of complete assembly at ground motion could reveal a more complex behavior during service.

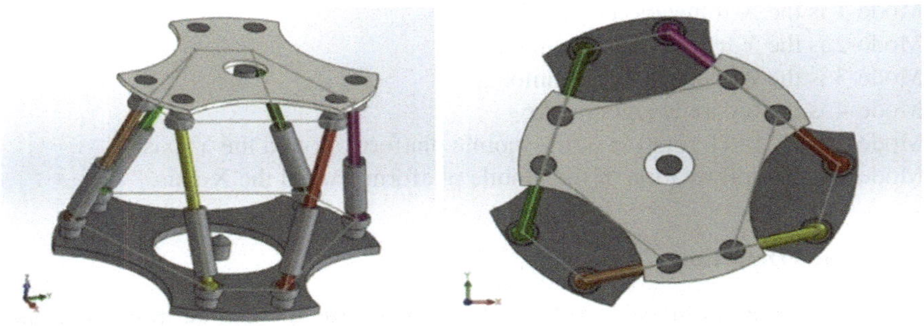

Fig. 1. Stewart Platform—initial concept. (left—isometric view, right—top view)

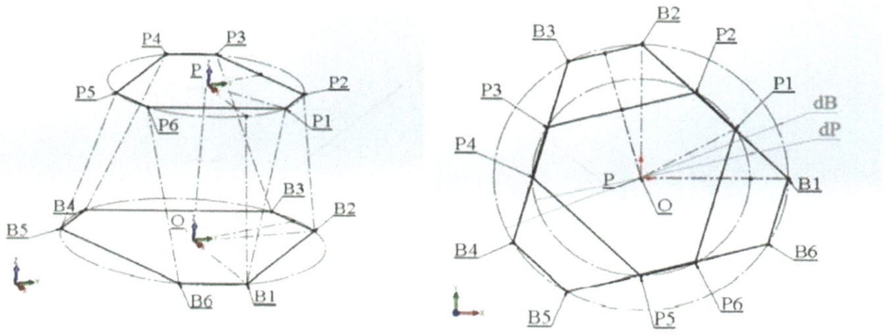

Fig. 2. Stewart Platform—symbolic representation. (left—isometric view, right—top view)

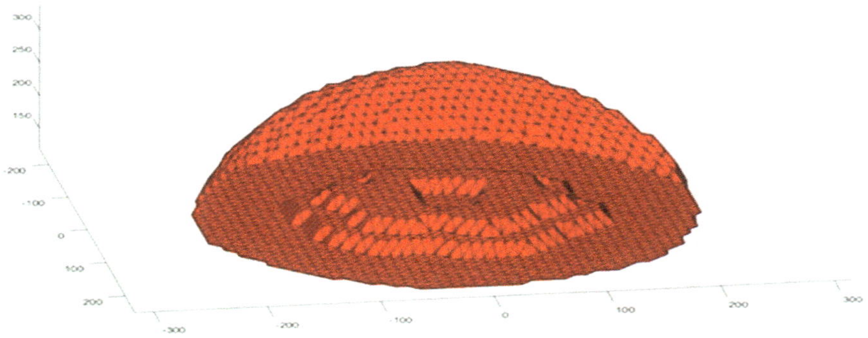

Fig. 3. Stewart Platform workspace

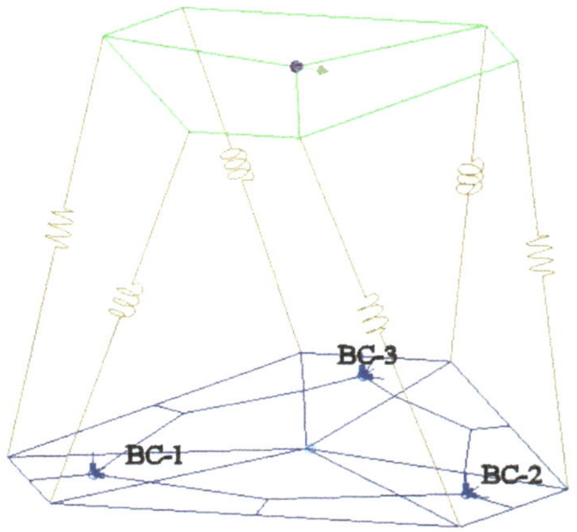

Fig. 4. Stewart Platform numerical model at home-position

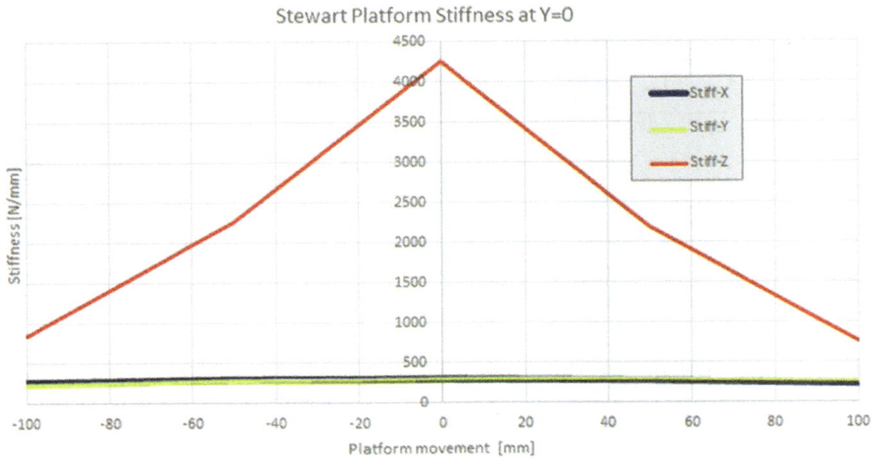

Fig. 5. Stewart platform stiffness in plane y = 0

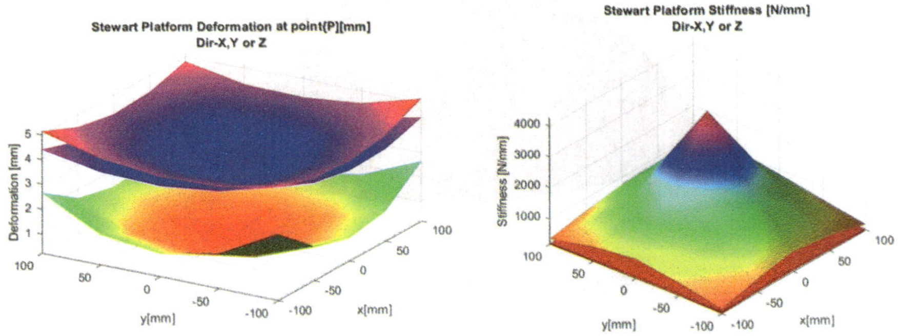

Fig. 6. System stiffness evaluation for entire workspace (left-deformation, right-stiffness)

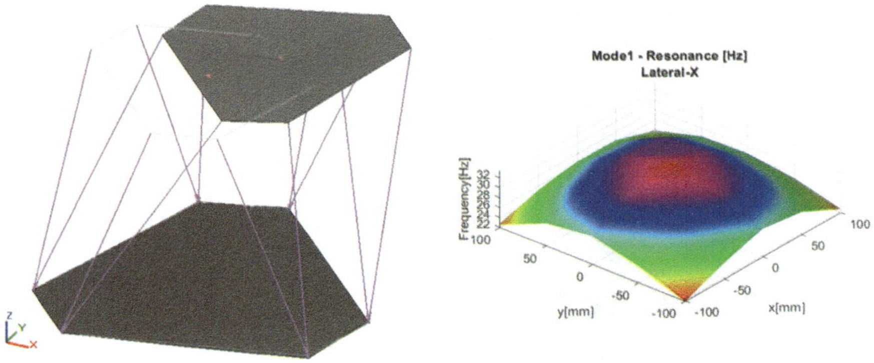

Fig. 7. Normal mode-1. Resonance frequency function of (x, y)

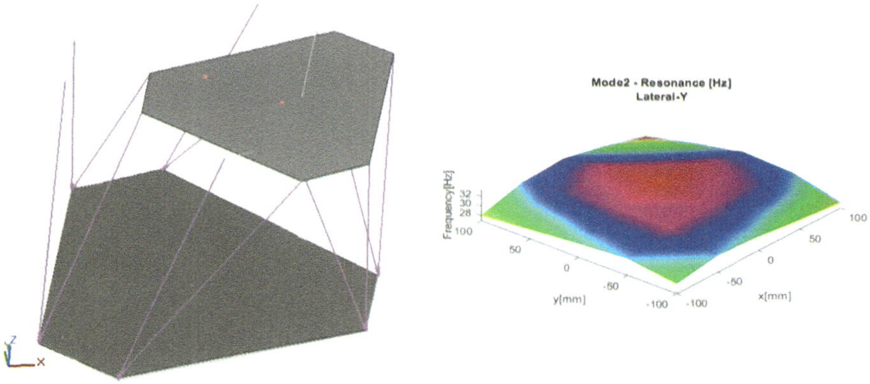

Fig. 8. Normal mode-2. Resonance frequency function of (x, y)

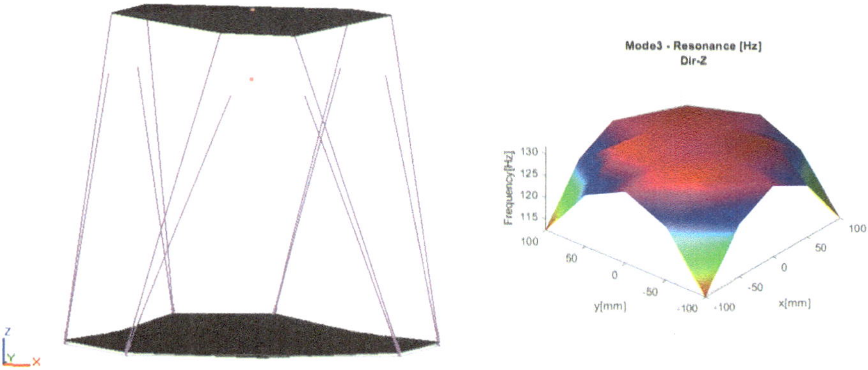

Fig. 9. Normal mode-3. Resonance frequency function of (x, y)

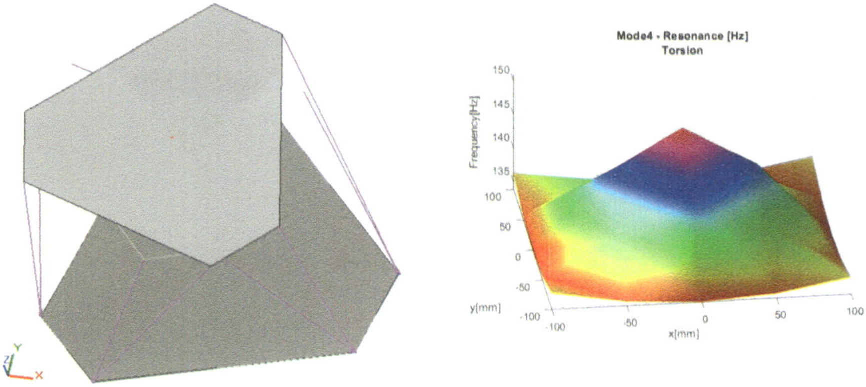

Fig. 10. Normal mode-4. Resonance frequency function of (x, y)

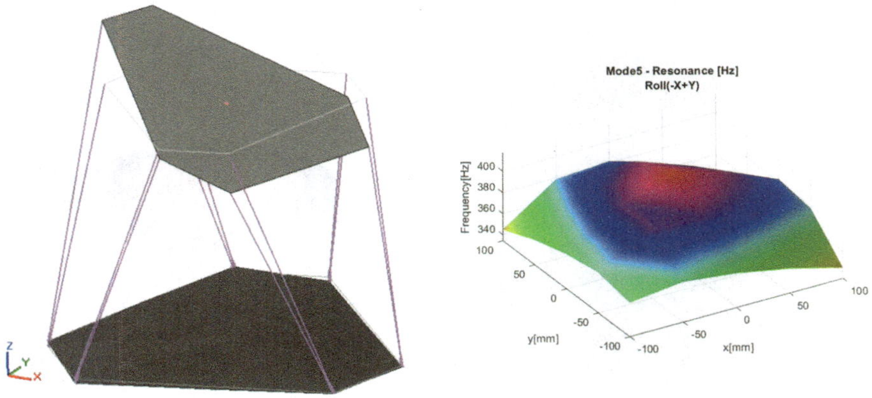

Fig. 11. Normal mode-5. Resonance frequency function of (x, y)

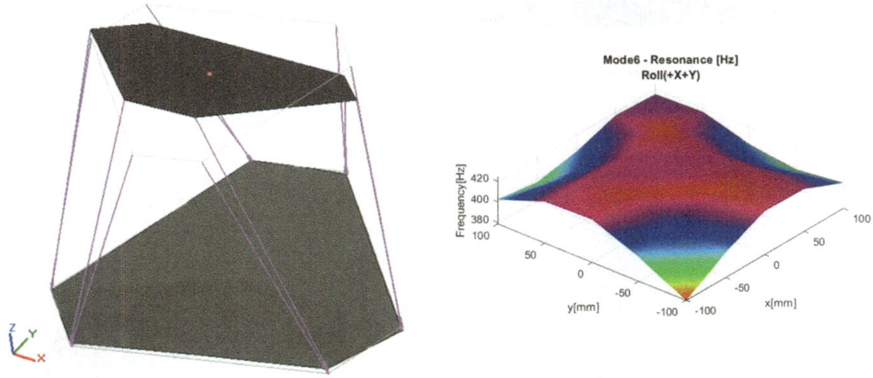

Fig. 12. Normal mode-6. Resonance frequency function of (x, y)

Acknowledgements. This work was supported by a grant of the Romanian Ministry of Research and Innovation, CCCDI–UEFISCDI, PROIECT PN-III-P2-2.1-PED-2019-0085, Contract no. 447PED/2020, "Sistem robotic modular si reconfigurabil pentru simularea cutremurelor (Acronim: POSEIDON)".

References

1. Fiorini, P.: History of robots and robotic surgery. In: The SAGES Atlas of Robotic Surgery (2018)
2. Ewins, D.J.: Modal testing: theory and practice. Wiley, New York (1985)
3. Gough, V.E.: Contribution to discussion of papers on research in Automobile Stability, Control and Tire performance. Proc. Auto Div. Inst. Mech. Eng. 392–394 (1956)
4. Stewart, D.: A platform with six degrees of freedom. Proc. Inst. Mech. Eng. **180**, 371–386 (1965)
5. Wavering, A., Guangming, Z., Clinton, C.M.: Stiffness modeling of a Stewart-platform-based milling machine. NIST Publ. (1998)
6. Boyin, D. et al.: Stiffness analysis and control of a Stewart Platform-based manipulator with decoupled sensor-actuator locations for ultrahigh accuracy positioning under large external loads. J. Dyn. Syst. Meas. Contr. https://doi.org/10.1115/1.4027945 (2014)
7. Verhoeven, R., Hiller, M., Tadokoro, S.: Workspace, stiffness, singularities and classification of tendon-driven stewart platforms. In: Advances in robot kinematics: analysis and control, pp. 105–114, Springer Netherlands (1998), ISBN=978–94–015–9064–8.
8. Lee, Y.-T., Choi, H.-R., Chung, W.-K., Youm, Y.: Stiffness control of a coupled tendon-driven robot hand. IEEE Control Syst. Mag. **145**, 10–19 (1994)
9. Raghavan, M.: The Stewart platform of general geometry has 40 configurations. ASME J. Mech. Des. **115**, 277–282 (1993)
10. Sefrioui, J., Gosselin, C.M.: On the quadratic nature of the singularity curves of planar three-degree-of-freedom parallel manipulators. Mech. Mach. Theory **304**, 533–551 (1995)
11. Ruggiu, M.: Cartesian stiffness matrix mapping of a translational parallel mechanism with elastic joints. Int. J. Adv. Rob. Syst. (2017)

Vibration Characterisation of a Damaged Cable by the Spectral Element Method

Maciej Dutkiewicz[1(✉)] and Marcela R. Machado[2]

[1] Faculty of Civil, Environmental Engineering and Architecture, Bydgoszcz University of Science and Technology, 85-796 Bydgoszcz, Poland
macdut@pbs.edu.pl

[2] Department of Mechanical Engineering, University of Brasilia, Brasilia 70910-900, Brazil

Abstract. The vibration characterisation is directly associated with the physical properties of the system, such as mass, damping and stiffness. The presence of structural damage like a crack reduces structural stiffness, shifting its eigenfrequencies and changing its response. This paper brings applications of damage modelling in cable structural elements. The proposed approach deals with the spectral element method (SEM), a wave-based numerical technique used to model structures. The analysis of the vibration responses for healthy and damaged cables allows one to indicate a change in response and indicate the existence of anomalies such as the crack in the system. Numerical simulation demonstrates the efficacy of the method in the model of the crack effect on the dynamic response of the cables.

Keywords: Damaged cable · Spectral element method · Crack effect

1 Introduction

Cable is a structural element widely utilised in industrial structures and equipment due to its numerous advantages. Local and global failures are known as the cascade effect; it happens when a failure spreads its impact to the whole system. Excessive vibration is one of the common causes of damages and fatigues in cables, therefore characterising the dynamic response of such a structure is important. Cable, e.g. transmission line conductor, is a complex structure composed of multi-wire that must be considered in the numerical analysis.

The vibration in cables was performed in [1], Barbieri et al. [2, 3], McClure and Lapointe [4] and Cardou [5] and Spak et al. [6–8]. Therefore, effective and practical tools are required to model conductor cables since an analytical solution does not exist. The efficiency of SEM to calculate the dynamic cable response has been explored in [9–12]. The results of the SEM showed very good precision as compared to the experimental measurements. Machado et al. [13] developed a spectral element for the transmission conductor cable, including the hysteretic and aerodynamic damping effect using the spectral element method. Cracked rod spectral element was presented in [14], beams in [15], and stochastic rod and beams in [16].

N. Herisanu and V. Marinca (eds.), *Acoustics and Vibration of Mechanical Structures – AVMS-2021*, Springer Proceedings in Physics 274,
https://doi.org/10.1007/978-3-030-96787-1_35

The dynamic response is directly associated with the mechanical components and systems' physical properties such as mass, damping, and stiffness. Vibration-based methods to detect damage are based on the fact that the presence of structural damage reduces structural stiffness, shifts eigenfrequenciesand changes frequency response function and mode shapes. This method's advantages and applications are simple, low cost and particularly effective for detecting larger damages in larger infrastructure or rotating machinery. The limitations include insensitivities to smaller damage, damage growth, excitation in high frequencies, large number of measurement points, hypersensitivity to the boundary and environmental changes. An extensive review of vibration-based damage detection methods has been published in [17–19].

The main objective of this paper is to characterise the vibration of the cable structure with an open-crack modelled by SEM. The main issue is the formulation of the spectral cable with an open crack. Simulated examples are performed. The results demonstrated the numerical model's efficiency in simulating the dynamic response of the healthy and damaged cable. Aside from checking the accuracy of the technique including damage crack into the model.

2 Cracked Cable Spectral Element

The cable spectral element with a transversal and non-propagating crack is treated in this paper and illustrated in Fig. 1. The cable element is based on an undamped Euler–Bernoulli beam equation of motion subjected to axial force and under bending vibration. The element contains two nodes with two degrees of freedom per node, where L is the beam length, L_1 is the crack position from node-one and $'a'$ is the crack depth related to the cross-section high. The crack is modelled by dimensionless local flexibility represented by θ, which is calculated by Castigliano's theorem and the laws of fracture mechanics.

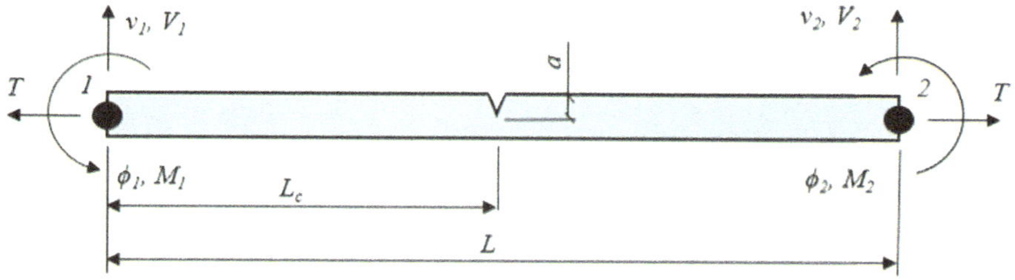

Fig. 1. Two-node cracked cable spectral element.

Considering a simplified model, the governing differential equation for the undamped free vibration can be written as [21]

$$EI\frac{\partial^4 v(x,t)}{\partial x^4} - T\frac{\partial^2 v(x,t)}{\partial x^2} + \rho A\frac{\partial^2 v(x,t)}{\partial t^2} = 0 \tag{1}$$

where ρA is mass per unit length, EI the uniform bending rigidity, T is compression force and $v(x,t)$ is the column displacement as a function of the position x and time t.

By considering a constant coefficient, the vertical displacement solution can be assumed of the form

$$v(x,t) = v_0 e^{-i(kx-\omega t)} \tag{2}$$

where v_0 is the amplitude, ω is the circular frequency and k is the wavenumber. The dispersion equation is obtained as $k^4 + \alpha k^2 - k_F^4 = 0$, where $\alpha = T/EI$ and $k_F = \sqrt{\omega}(\rho A/EI)^{1/4}$ is the wavenumber for the flexural wavemode. By solving the dispersion equation, four wavenumbers are described by

$$k_{1,3} = \pm\sqrt{\left[\left(\frac{\alpha}{2}\right)^2 + k_F^4\right]^{1/2} - \frac{\alpha}{2}}, \text{ and} k_{2,4} = \pm i\sqrt{\left[\left(\frac{\alpha}{2}\right)^2 + k_F^4\right]^{1/2} + \frac{\alpha}{2}} \tag{3}$$

The solution applied to this element is divided into two parts as for the beam element [15, 16], the left $\hat{v}^l(x)$ and the right $\hat{v}^r(x)$ side crack. Therefore, the nodal displacements are

$$\hat{v}^l(x) = a_1 e^{-i(kx)} + a_2 e^{-(kx)} + a_3 e^{-ik(L_1-x)} + a_4 e^{-k(L_1-x)} [0 \le x \le L_1], \tag{4}$$

$$\begin{aligned} \hat{v}^r(x) &= \mathfrak{a}_5 e^{-ik(L_1+x)} + \mathfrak{a}_6 e^{-k(L_1+x)} + \mathfrak{a}_7 e^{-ik(L-(L_1+x))} \\ &\quad + \mathfrak{a}_8 e^{-k(L-(L_1+x))} [0 \le x \le L - L_1] \end{aligned} \tag{5}$$

The coefficients $\mathfrak{a}_i$, $(i = 1:8)$ can be calculated as a function of the nodal spectral displacements considering the boundary conditions for the element. Hence, for the cracked cable element, the boundary conditions at the ends of the cable and the crack position must be considered on the left side of the element for $x = 0$, where $\hat{v}^l(x) = \hat{v}_1$, $\partial\hat{v}^l(x)/\partial x = \hat{\phi}_1$; In the crack position, displacement and rotation are considered for $\hat{v}^l(x)$. Thus, $x = L_1$ where $\hat{v}^r(x)$ and $x = 0$, it has

$$\hat{v}^l(x) = \hat{v}^r(x) \frac{\partial\hat{v}^r(x)}{\partial x} - \frac{\partial\hat{v}^l(x)}{\partial x} = \theta \frac{\partial^2\hat{v}^r(x)}{\partial x^2} \frac{\partial^2\hat{v}^l(x)}{\partial x^2} = \frac{\partial^2\hat{v}^r(x)}{\partial x^2} \frac{\partial^3\hat{v}^l(x)}{\partial x^3} = \frac{\partial^3\hat{v}^r(x)}{\partial x^3} \tag{6}$$

On the right side of the element for $x = L - L_1$, it has $\hat{v}^r(x) = \hat{v}_2$, $\partial\hat{v}^r(x)/\partial x = \hat{\phi}_2$. By considering Eqs. (5) and (6) for describing the nodal spectral displacements to the left and right of the crack and applying the boundary conditions in matrix form, one has

$$\boldsymbol{H}_C(\omega) = \frac{EI}{L^3}\begin{bmatrix} 1 & 1 & \mathfrak{a} & \mathfrak{b} & 0 & 0 & 0 & 0 \\ -ik & -k & ik\mathfrak{a} & k\mathfrak{b} & 0 & 0 & 0 & 0 \\ \mathfrak{a} & \mathfrak{b} & 1 & 1 & -\mathfrak{a} & -\mathfrak{b} & -\mathfrak{c} & -\mathfrak{d} \\ \left(ik + \theta k^2\right)\mathfrak{a} & \left(k - \theta k^2\right)\mathfrak{b} & \theta k^2 - ik & -k - \theta k^2 & -ik\mathfrak{a} & -k\mathfrak{b} & ik\mathfrak{c} & k\mathfrak{d} \\ -ik^2\mathfrak{a} & k^2\mathfrak{b} & -k^2 & k^2\mathfrak{a} & k^2 & -k^2\mathfrak{b} & k^2\mathfrak{c} & -k^2\mathfrak{d} \\ -ik^3\mathfrak{a} & -k^3\mathfrak{b} & -ik^3 & k^3 & -ik^3\mathfrak{a} & k^3\mathfrak{b} & ik^3\mathfrak{c} & -k^3\mathfrak{d} \\ 0 & 0 & 0 & 0 & \mathfrak{f} & \mathfrak{g} & 1 & 1 \\ 0 & 0 & 0 & 0 & -ik\mathfrak{f} & -k\mathfrak{g} & ik & k \end{bmatrix} \tag{7}$$

where $\mathfrak{a} = e^{ikL_1}$, $\mathfrak{b} = e^{-kL_1}$, $\mathfrak{c} = e^{-ik(L-L_1)}$, $\mathfrak{d} = e^{-k(L-L_1)}$, $\mathfrak{f} = e^{-ikL}$, $\mathfrak{g} = e^{-kL}$

Equation (8) falls into the general formulation of the element, where the spectral nodal displacement is $\boldsymbol{d} = \boldsymbol{H}_C(\omega)\mathbf{a}$. The coefficients can be related to nodal spectral displacements by $a_i = H_{ci1}^{-1}\hat{v}_1 + H_{ci2}^{-1}\hat{\phi}_1 + H_{ci3}^{-1}\hat{v}_2 + H_{ci4}^{-1}\hat{\phi}_2$. At this point, the $H_C(\omega)$ matrix, originally of 8×8 dimension, is reduced to an 8×4 matrix. This process occurs according to Eq. (8), where $\mathrm{H}_{\mathrm{cij}}^{-1}$ represents the elements of the inverse matrix of Eq. (7). The nodal spectral forces can be represented by differentiating the spectral displacements $\hat{v}^l$ and $\hat{v}^r$ related to x, and, in this case, generally represented by $\boldsymbol{f} = \boldsymbol{G}(\omega)\boldsymbol{a}$. In the matrix form, we have

$$\boldsymbol{G}\begin{Bmatrix} \hat{f}_1 \\ \hat{f}_2 \\ \hat{f}_3 \\ \hat{f}_4 \end{Bmatrix} = \begin{bmatrix} ik^3 & -k^3 & -ik^3\mathfrak{a} & k^3 & 0 & 0 & 0 & 0 \\ -k^2 & k^2 & -k^2\mathfrak{a} & k^2 & 0 & 0 & 0 & 0 \\ 0 & 0 & 0 & 0 & ik^3\mathfrak{f} & -k^3\mathfrak{g} & -ik^3 & k^3 \\ 0 & 0 & 0 & 0 & -k^2\mathfrak{f} & k^2\mathfrak{g} & -k^2 & k^2 \end{bmatrix} \begin{Bmatrix} a_1 \\ a_2 \\ a_3 \\ a_4 \\ a_5 \\ a_6 \\ a_7 \\ a_8 \end{Bmatrix} \tag{8}$$

From the nodal forces and displacements, Eqs. (9) and (7), the frequency-dependent dynamic stiffness matrix of the Euler–Bernoulli beam element with an open, non-propagating crack related to the nodal forces and displacements is

$$\boldsymbol{f}_{[4x1]} = \boldsymbol{G}_c(\omega)_{[4x8]}\boldsymbol{H}_c^{-1}(\omega)_{[8x4]}\boldsymbol{d} = \boldsymbol{S_c}(\omega)_{[4x4]}\boldsymbol{d} \tag{9}$$

where $\boldsymbol{S}_c(\omega)$ is a 4×4 cracked beam dynamic stiffness matrix.

3 Local Crack Flexibility

The crack flexibility coefficient θ is calculated using Castigliano's theorem, where the flexibility in the crack position for a one-dimensional cable spectral element can be obtained by $c = \partial^2 U/\partial P^2$ [20], where U denotes the elastic deformation energy due to cracking, and P is the element's nodal load.

Considering an open crack in the beam element, the elastic deformation energy can be expressed as [20]

$$U = \frac{1 - \nu^2}{E}\int_{S_c} K_I^2 dS_c \tag{10}$$

where ν is the Poisson's ratio, S_c is the cracked area, and K_I is a stress intensity factor corresponding to mode I of the crack, which can be represented by

$$K_I = \frac{6M}{bh^2}\sqrt{\pi\alpha f\left(\frac{\alpha}{h}\right)} \tag{11}$$

where b is the base and h the height of the cable diameter, α is the variation in the crack depth, as shown in Fig. 2, M is the bending moment in crack position, and f is a correction function of the mode I stress intensification factor, which can be written as

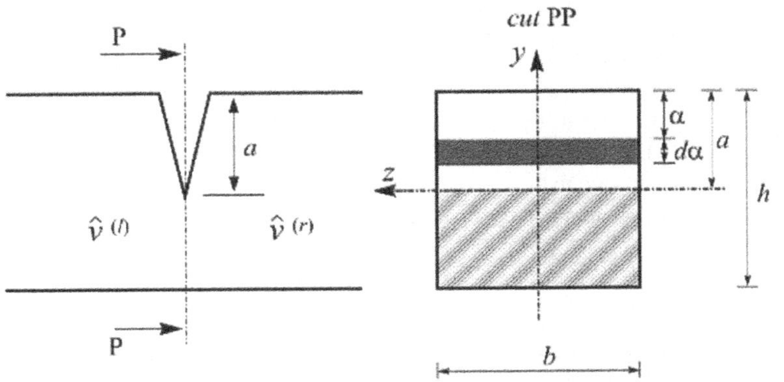

Fig. 2. Cross-section of the cracked beam in the crack position.

$$f\left(\frac{\alpha}{h}\right) = \sqrt{\frac{2h}{\pi\alpha}tan\left(\frac{\pi\alpha}{2h}\right)}\frac{0.923 + 0.199\left(1 - sin\left(\frac{\pi\alpha}{2h}\right)\right)^4}{cos\left(\frac{\pi\alpha}{2h}\right)} \tag{12}$$

The coefficient c that calculates the crack flexibility is expressed as follows

$$c = \frac{72\pi}{bh^2}\int_0^\alpha \alpha f^2(\alpha)d\alpha \tag{13}$$

where $\alpha = a/h$. Finally, the dimensional local flexibility is given by $\theta = EIc/L$.

4 Numerical Results

In the simulations, performed in MATLAB environment, we considered a cable with a length of $L = 5$ m, crack location at $L_I = 0.15$ L, and area of $A = 70 \times 10^{-5}$ mm^2. The material properties are Young's modulus of E = 74GPa, the material density of $\rho = 2700kg/\text{m}^3$, the tensile load of 18.70, 1 870, and 18700 kgf and structural damping factor as $\eta = 0.01$. The structural damping was modelled by a complex elastic modulus, $E = E(1 + i\eta)$. The numerical investigations verified the cable's structural integrity. The damages were simulated considering a non-propagating open crack. The cable has a free-free boundary condition with a unitary excitation at node one (Fig. 1). The receptance is obtained at the same point and along the structure generating a top-view map. The cracked and healthy cables response is compared, aiming to identify changes in the vibration induced by the crack.

Figure 3a–c on top shows the receptance response obtained at node-one for healthy and cracked cable with a crack depth of 1% of the cable's diameter and the tensile load of 18.70, 1 870, and 18700 kgf, respectively. The frequency band of analysis varies from 1 to 100 Hz approaching the first fifth resonance frequencies and mode shapes. Figure 3a–c at the bottom shows the top-view map of the receptance response obtained along the healthy and cracked cable. For a small crack depth, no apparent changes in the cables' vibration are seen. Figure 4a–c shows the receptance response for healthy and cracked cable with crack depth 15% of the cable's diameter obtained at the node-one and over the cable length. Crack depth over 15% induces a slight chance in the vibration response, increasing with the crack depth. The crack causes a shift to the left in the receptance response clear in higher resonance frequencies, which is better seen on the map colour. The tensile load influences the resonance frequencies and the crack effects that affect the first mode shape and as higher is the load.

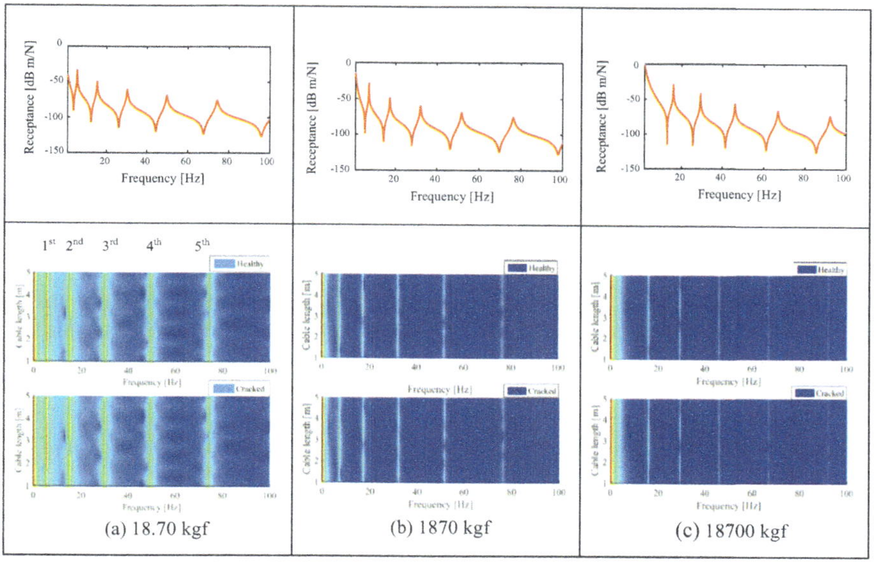

Fig. 3. Receptance response at node-one (top) and the top-view map obtained along the cable (bottom) for healthy and cracked cable with a crack depth of 1% for different tensile loads.

Figure 5a–c shows the receptance response for healthy and cracked cable with crack depth 25% of the cable's diameter obtained at the node-one and over the cable length. Figure 6a–c presents the receptance responses for healthy and cracked cable with a crack depth of 35%. In these cases, the cable vibration changes as the crack increased. For the crack depth with 25% of the cable's diameter, the fourth and fifth modes presented a shift to the left, and the damping increased, leading to a decrease of the modes' magnitude. Changes in the damping are well seen in the colour of the top-view map. For a crack depth of 35%, the five modes changed, including the damping, which is clearly observed in the top-view map.

5 Conclusion

In this paper, a significant issue of vibration and spectral model of a damaged cable was performed. The numerical model of cable structural element with an open crack could induce an effect in the dynamic response. The receptance response and a top-view

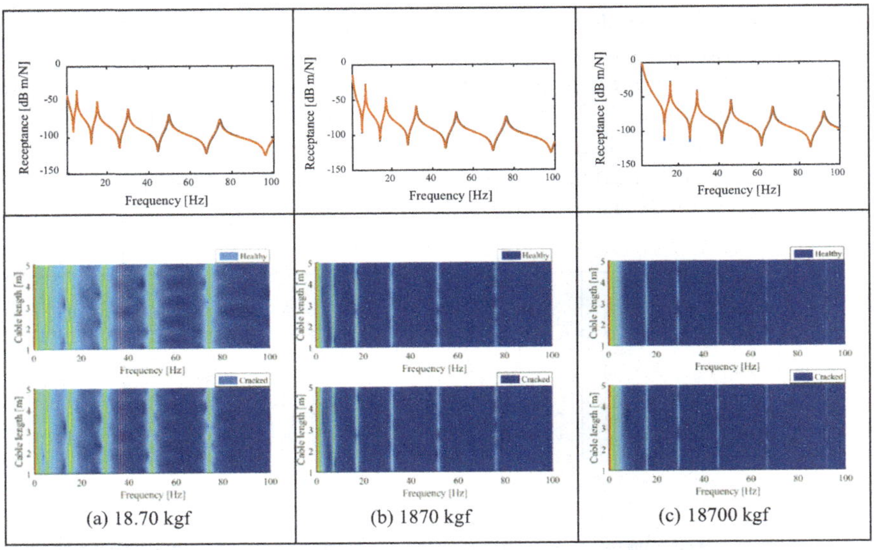

Fig. 4. Receptance response at node-one (top) and the top-view map obtained along the cable (bottom) for healthy and cracked cable with a crack depth of 15% for different tensile loads.

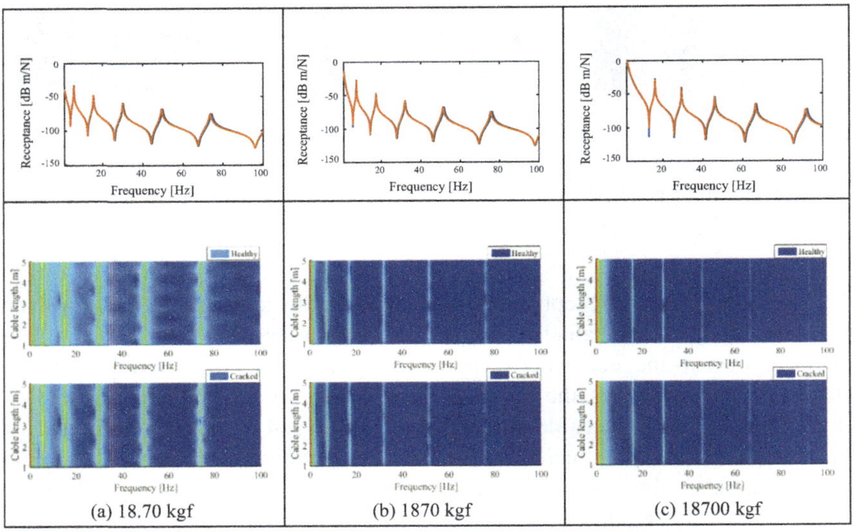

Fig. 5. Receptance response at node-one (top) and the top-view map obtained along the cable (bottom) for healthy and cracked cable with a crack depth of 25% for different tensile loads.

map of the receptance along the cable were presented. The top-view map improves the investigation and accurate characterisation of the cable's vibration. Healthy and damaged cable receptance were compared. A crack depth of 1% of the diameter did not impact the

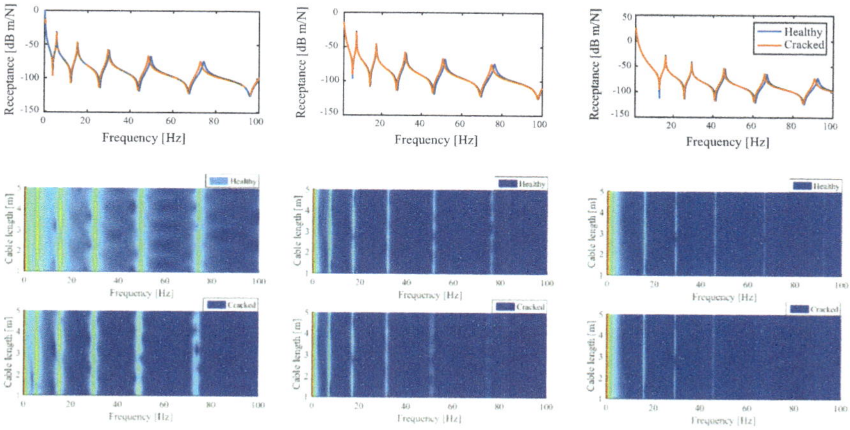

(a) 18.70 kgf (b) 1870 kgf (c) 18700 kgf

Fig. 6. Receptance response at node-one (top) and the top-view map obtained along the cable (bottom) for healthy and cracked cable with a crack depth of 35% for different tensile loads.

receptance, as the crack depth increased, the effect is gradually seen. The crack induced a shift in the mode shapes and increased the damping, leading to decreased modes' magnitude. The vibration map improved the overview of the cable receptance helping the search for the damage.

References

1. Yamaguchi, H., Alauddin, M., Poovarodom, N.: Dynamic characteristics and vibration control of a cable system with substructural interactions. Eng. Structures **23**, 1348–1358 (2001)
2. Barbieri, N., De Souza Junior, O. H., Barbieri, R.: Dynamical analysis of transmission line cables part 1—linear theory. Mech. Syst. Sig. Process. **18**(3) (2004) 659–669.
3. Barbieri, N., De Souza Junior, O. H., Barbieri, R.: Dynamical analysis of transmission line cables part 2 damping estimation. Mech. Syst. Sig. Process. **18**(3) (2004) 671–681.
4. McClure, G., Lapointe, M.: Modeling the structural dynamic response of overhead transmission lines. Comput. Struct. **81**, 825–834 (2003)
5. Taut-helical strand bending stiffness. http://imechanica.org/files/Cardou_art_2006.pdf
6. Spak, K., Agnes G., Inman, D.: Cable modeling and internal damping developments. Appl. Mech. Rev. **65**(1) 10801–10811 (2013)
7. Spak, K., Agnes, G., Inman, D.: Parameters for modeling stranded cables as structural beams. Exp. Mech. **54**(9) 1613–1626 (2014)
8. Spak, K., Agnes, G., Inman, D.: Modeling vibration response and damping of cables and cabled structures. J. Sound Vib. **336**, 240–256 (2015)
9. Dutkiewicz, M., Machado, M. R.: Dynamic response of overhead transmission line in turbulent wind flow with application of the spectral element method. In: IOP Conference Series: Materials Science and Engineering, vol. 471, p. 052031 (2019)
10. Dutkiewicz, M., Machado, M.R.: Measurements in Situ and spectral analysis of wind flow effects on overhead transmission lines. Sound Vib. **53**, 161–175 (2019)

11. Dutkiewicz, M., Machado, M.R.: Spectral approach in vibrations of overhead transmission lines. In: IOP Conference Series: Materials Science and Engineering, vol. 471, p. 052029 (2019)
12. Dutkiewicz, M., Machado, M.R.: Spectral element method in the analysis of vibrations of overhead transmission line in damping environment. Struct. Eng. Mech. **71**, 291–303 (2019)
13. Machado, M.R., Dutkiewicz, M., Matt, C.F.T., Castello, D.A.: Spectral model and experimental validation of hysteretic and aerodynamic damping in dynamic analysis of overhead transmission conductor. Mech. Syst. Sig. Process. **136**, 106483 (2020)
14. Palacz, M., Krawczuk, M.: Analysis of longitudinal wave propagation in a cracked rod by the spectral element method. Comput. Struct. **80**(24), 1809–1816 (2002)
15. Krawczuk, M., Palacz, M., Ostachowicz, W.: The dynamic analysis of a cracked Timoshenko beam by the spectral elemet method. J. Sound Vib. **264**, 1139–1153 (2002)
16. Machado, M.R., Adhikari, S., Dos Santos, J.M.C.: Spectral element-based method for a onedimensional damaged structure with distributed random properties. J. Braz. Soc. Mech. Sci. Eng. **40**, 214–226 (2018)
17. Doebling, S., Farrar, C., Prime, M., Shevitz, D.: Damage identification and health monitoring of structural and mechanical systems from changes in their vibration characteristics: a literature review. Los Alamos, NM: Technical Report LA-13070-MS, Los Alamos National Laboratory, (2018)
18. Sohn, H., Farrar, C., Hunter, N., Worden, K.: A review of structural health monitoring literature: 1996–2001. Los Alamos, NM: Los Alamos National Laboratory report (LA-13976-MS), (2001)
19. Carden, E., Fanning, P.: Vibration based condition monitoring: a review. Struct. Health Monit. **3**(4), 355–377 (2001) https://doi.org/10.1177/1475921704047500
20. Tada, H., Paris, P.C., Irwin, G.R.: The stress analysis of cracks. Del Research Corporation, Handbook (1973)
21. Rao S.S.: Mechanical Vibration, Person Prentice-Hal, (2008)

Aspects Regarding 3D Modeling and Finite Element Analysis for the Work Equipment of a Forklift

Aurora Potîrniche(✉) and Gigel Căpățână

Research Center for Mechanics of Machines and Technological Equipments, Engineering and Agronomy Faculty in Brăila, "Dunărea de Jos" University of Galați, Braila, Romania
{aurora.potirniche,gcapatana}@ugal.ro

Abstract. Advanced industrialization of production processes requires flexible automation, in which the industrial manipulators and robots play a key role. Due to the flexible and versatile nature of industrial robots, their use leads to a number of economic and social benefits (increased labor productivity, replacing humans in hazardous workplaces, raising product quality, and rapid recovery of investment). Thus, the introduction of manipulators and industrial robots led to the transformation of production systems from the human–machine system to the human–robot–machine system. There are a multitude of applications in which industrial robots are present and, from these, the main application is the handling of loading units, manually, with the help of hand-operated pallet trucks. This paper presents aspects related to the generation in the Autodesk Mechanical Desktop software of the 3D model for the work equipment of a forklift, equipment consisting of forks and fastening system. The model thus realized was introduced in the Solid Edge software and a dynamic analysis was performed following which the eigenmodes of vibration of the equipment were highlighted, which led to the knowledge of its eigenfrequencies.

1 Introduction

In the process of advanced industrialization of production processes, industrial manipulators and robots have a decisive role, because they are the basis of automation [1, 6].

The handling, storage, and internal transport of materials are components without which the production of material goods cannot be initiated, continued, and completed [3].

Warehousing is a very important activity in the internal transport technology and consists of storing goods for a shorter or longer period of time, in covered or uncovered warehouses [6].

This paper proposes a frequency analysis for the work equipment of a forklift in order to estimate the range of eigenfrequencies of this type of equipment. The knowledge of the eigenfrequencies domain is used by specialists in the domain of construction equipment, they can take measures both to avoid the resonance phenomenon and to improve the working performance of such equipment.

N. Herisanu and V. Marinca (eds.), *Acoustics and Vibration of Mechanical Structures – AVMS-2021*, Springer Proceedings in Physics 274,
https://doi.org/10.1007/978-3-030-96787-1_36

2 Theoretical Approaches

The forklift is a small industrial vehicle that has a platform with a hydraulic fork attached to the front that can be raised or lowered to handle loads. Pallet transport is almost impossible to do without using a forklift. These machines can be handled in narrow spaces, where larger vehicles do not have access [2, 4, 5].

Forklifts are a much more efficient way to handle bulky items, as opposed to using trolleys or pulley systems or labor. If most heavy lifting is performed by a forklift, the risk of an accident is much lower. Forklifts are commonly used in several industries, such as [6]

- Constructions—they can be used on construction sites and with their help the construction materials can be transported (Fig. 1);

Fig. 1. Construction forklifts [6].

- Warehouses—they are mainly used for loading or unloading trucks and for transporting goods (Fig. 2);

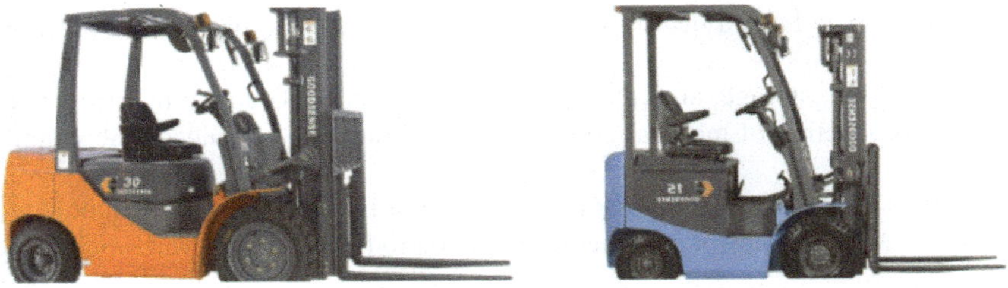

Fig. 2. Warehouse forklifts [6].

- Recycling;
- Shipyards—they are often used in ports, helping to unload or load ships, transport containers;

- Other uses—with the help of a forklift and an attached snow plow, it can be used to clean a snow space (Fig. 3).

Fig. 3. Forklifts for various uses [6].

3 Case Study

In order to achieve the most accurate 3D modeling of the forklift's work equipment, a trip was made to a company that operates in a cash and carry system with a form of self-service sale trade. Given a large number of products on sale, the use of stacking equipment is indispensable. A functional forklift was chosen for which the forks and their support were measured (Fig. 4).

The execution of the 3D model of the lifting arms, called forks, as well as their fastening system was performed in Autodesk Mechanical Desktop software specialized in 3D mechanical engineering, design, visualization, and simulation.

The main commands in Autodesk Mechanical Desktop 2008 that were the basis for making 3D sketches and models were the following: Create basic Work Planes, Profile a Sketch, Extrusion, Fillet.

Since the two forks are identical, the validity of the principle of symmetry was taken into account, and based on this, only the modeling of a fork was needed. The fork clamping system was also modeled. The components resulting from the modeling are shown in Figs. 5, 6, and 7.

With the help of these components, all the work equipment assembly of the forklift was made. The working method has involved switching from Part Menu mode to Assembly Menu mode. In this way, with the help of a Catalog, the fork was attached twice and the fastening system once. After bringing all the necessary components, the 3D Constraints command was used to position them correctly in relation to each other.

The whole work equipment assembly is shown in Fig. 8.

The work equipment assembly was imported in the Solid Edge software, where a modal analysis was performed, which highlighted the first 10 vibration eigenmodes of the forklift's work equipment. Solid Edge software is a 3D modeling CAD software. It provides modeling of solids, modeling of assemblies, and capabilities for the execution of technical drawings for mechanical engineers.

Fig. 4. The real machine used as a model.

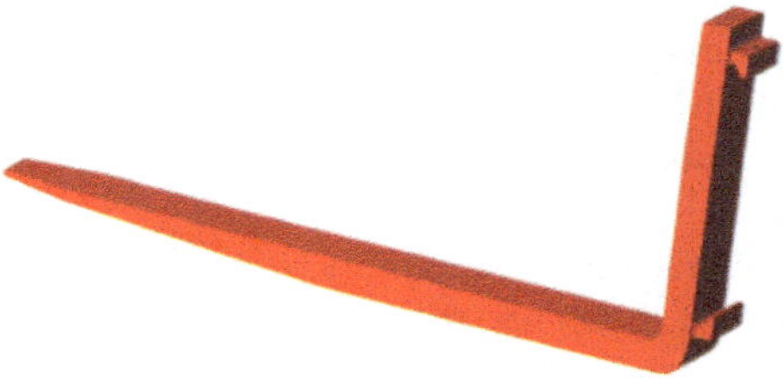

Fig. 5. Fork.

The purpose of performing the frequency analysis was to highlight the behavior of the work equipment of the forklift in terms of displacements of the nodes of the structure in the first 10 eigenmodes of vibration.

The 3D model of the assembly consisting of forks and fastening system was automatically discretized by the Solid Edge software.

After completing the discretization operation, the boundary conditions were specified (in this case, for the frequency analysis, it was only necessary to define a support set). Thus, the support of the assembly was of embedding type at the level of the four cylindrical surfaces that are practiced on the fork clamping system, surfaces that ensure the displacement of the entire work equipment along the mast.

Fig. 6. Fork clamping detail.

Fig. 7. Fork clamping system.

Fig. 8. Realization of the work equipment assembly of the forklift.

In the modal analysis, Solid Edge, like any software that performs analysis of 3D models, provides the user with various results, and the user will choose what he considers suggestive for the analyzed case.

For the present paper, it was interesting to visualize the behavior of the work equipment of the forklift when the frequency response is a displacement. The displacement of the nodes of the structure in the first 10 eigenmodes of vibration is shown in Fig. 9.

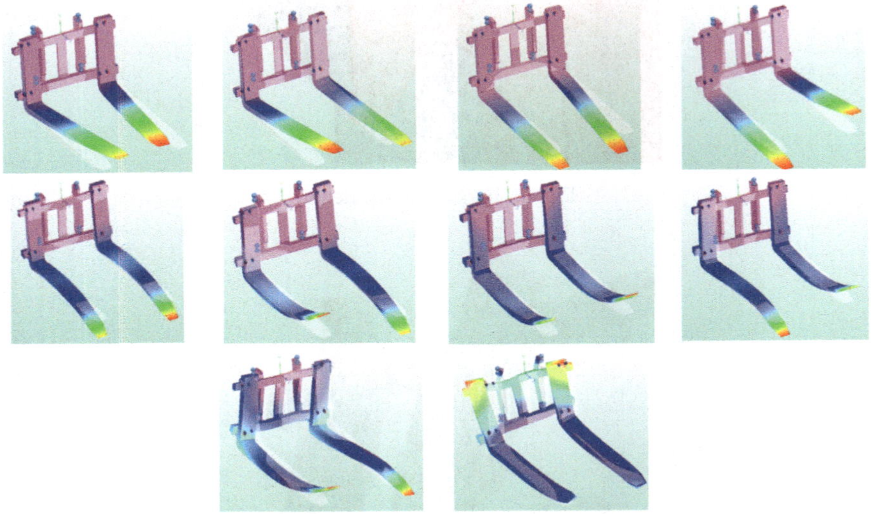

Fig. 9. The displacements of the structure nodes in the first 10 eigenmodes of vibration.

It was also of interest to visualize the strain potential energy accumulated by the system in each of the 10 eigenmodes of vibration for which the analysis was performed, shown in Fig. 10.

In Table 1 are given the values of the eigenfrequencies of the working equipment of the forklift, the values of the maximum displacements of the structure nodes, and also the values of the maximum strain potential energy accumulated by the system in the first 10 eigenmodes of vibration.

Further, in the paper, a graph was made in Microsoft Excel for the variation of the values of the eigenfrequencies of the working equipment of the forklift for the first 10 eigenmodes of vibration in comparison with the maximum values of the displacements of the structure nodes and with the maximum values of the strain potential energy accumulated by the system (Fig. 11).

It can be seen from this graph that moving from eigenmode 1 to eigenmode 10, there is an increase in the values of the system's eigenfrequencies, an increase and then a slight decrease (after the eigenmode number 8) of the maximum displacements values of the structure nodes and a visible increase in the maximum values of the strain potential energy accumulated by the system.

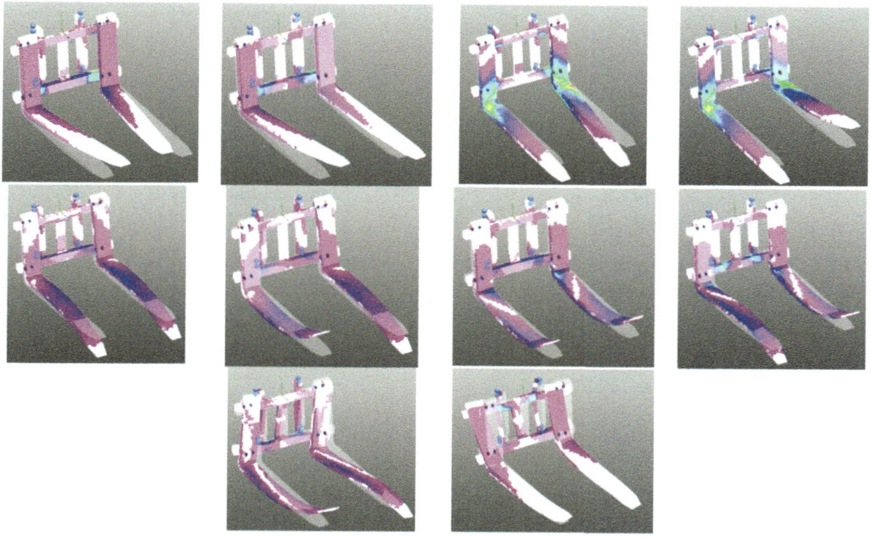

Fig. 10. The strain potential energy accumulated by the system in each of its 10 eigenmodes of vibration.

Table 1. The eigenfrequencies values, the values of the maximum displacements of the structure nodes, and the values of the maximum strain potential energy accumulated by the system for the first 10 eigenmodes of vibration.

Vibration eigen mode	Eigenfrequency value, [Hz]	The values of the maximum displacements of the structure nodes, [mm]	The values of the maximum strain potential energy accumulated by the system, [J]
1	29.09	238	36
2	29.33	236	43.1
3	33.03	260	13.3
4	33.06	261	14.7
5	144.5	346	663
6	146.1	358	714
7	195.6	381	948
8	198	354	945
9	233.1	285	1950
10	263.2	202	2380

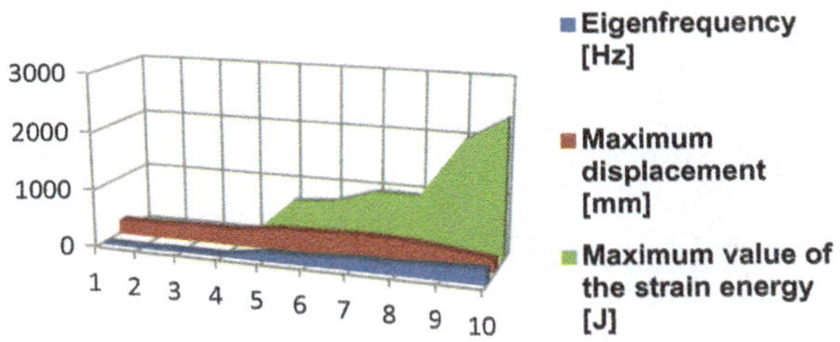

Fig. 11. The variation graph between frequencies—maximum displacements of the structure nodes—maximum value of the strain potential energy for the first 10 eigenmodes of vibration.

4 Conclusions

The analysis of the eigenmodes of the work equipment of the forklift was performed in order to make an estimate of the eigenfrequency range of this constructive type of equipment. The information obtained from this analysis is useful because it highlights the possible occurrence of the resonance phenomenon.

It is well known that resonance is the tendency of a system to oscillate with a higher amplitude at some frequencies than at others. The frequencies at which the amplitude is maximum are called resonant frequencies [7].

When designing a structure, engineers must ensure that the mechanical resonant frequencies of the component parts of the structure are not equal to the oscillating frequencies of the oscillating components, a phenomenon known as destructive resonance [7]. Avoiding destructive resonance is considered when building any type of structure.

Moreover, the structures are made so that resonances occur at hard-to-reach frequencies. For example, buildings located in seismic areas are generally constructed in such a way that there are no resonances at expected frequencies in the event of an earthquake.

References

1. Anghelache, G.D., Goanţă, A.M.: Diging simulation of a narrow trench, MODTECH International Conference—Modern Technologies In Industrial Engineering IV, Book Series: IOP Conference Series—Materials Science and Engineering, vol. 145, Article 042016, (2016) https://doi.org/10.1088/1757-899X/145/4/042016. WOS:000396437600068
2. Goanţă, A.M.: Modern instruments of transfer of geometrical patterns in designing building machines and equipments. In: Proceedings of The 13th International Conference Iaşi&Chişinău - Modtech 2009, pp. 271–274 (2009). ISSN 2066–3919, WOS:000274641800066
3. Haraga, G., Goanţă, A. M.: FEA analysis and design optimization for a multifunctional piece of furniture. In: International Conference Innovative Manufacturing Engineering and Energy IManE&E 2017, Iaşi, România, MATEC Web of Conferences vol. 112, 06009 (2017) eISSN: 2261–236X, WOS:000579349600098. https://doi.org/10.1051/matecconf/201711206009, Proceedings paper ISBN: 978–1–5108–4430–8, pp. 605–611
4. Muscă (Anghelache), G.D.: Static and dynamic stability for floating cranes. In: MODTECH International Conference—Modern Technologies in Industrial Engineering VI, Book Series: IOP Conference Series—Materials Science and Engineering, vol. 400, Article Number: 082015 (2018) https://doi.org/10.1088/1757-899X/400/8/082015. WOS:000461147400195

5. Muscă (Anghelache), G.D., Năstac, S.: Dynamic modelling of overhead crane. In: MODTECH International Conference—Modern Technologies in Industrial Engineering VIII, Online edition, Iași, România (2020) ISSN 2286–4369
6. https://www.utilben.ro/article/cele-mai-importante-lucruri-de-stiut-despre-stivuitoare
7. https://ro.wikipedia.org/wiki/Rezonan%C8%9B%C4%83

Assessing the Crack Location in a Cantilever Beam by Analyzing the Natural Frequency Evolution

Mario Ardeljan, Gilbert-Rainer Gillich, Marius-Vasile Pop, and Codruta Oana Hamat(✉)

Babes-Bolyai University, Str. Mihail Kogalniceanu 1, 400084 Cluj-Napoca, Romania
codruta.hamat@ubbcluj.ro

Abstract. In this paper, we present a method by which a defect of an engineering structure can be located by monitoring only two modes of transverse vibration. The method is based on the particular way in which the energy is stored for the considered vibration modes. At the beginning of the paper, we introduce a deduced mathematical relation that allows to calculate the amount of strain energy that is lost due to a defect and how it can be used to calculate the decrease of the natural frequencies. Then, we show how the defect is located, and finally, the location method is subjected to numerical tests to prove its validity.

Keywords: Crack detection · Frequency shift · Finite element · Strain energy

1 Introduction

Structures fail in operation due to defective manufacturing technology, aged materials, and lack of integrity control methods to detect defects in time. Hence, the importance of developing advanced methods for assessing structural integrity, which can autonomously assess the safety of the structure [1]. Vibration-based methods can signalize the appearance of structural changes in real-time even in operating conditions [2]. Most of these methods rely on the monitoring of frequencies because these are easily measured in situ with robust equipment [3]. However, accurate signal processing algorithms are necessary to estimate the frequencies with accuracy [4].

The main idea of the frequency-based methods is that cracks affect the stiffness of the structure, implicitly decreasing its capacity to store energy [5]. Different types of cracks affect structures, depending on the material properties and the loads applied to the structure. Cracks mainly get a transverse direction [6–8] but can also branch and propagate through the material in oblique directions [9]. Former studies have demonstrated that every damage has a distinct signature [10], which makes it possible to find the location of a crack by knowing the frequency decrease for several modes, and the modal energy distribution through the modal curvature evolution along the beam.

We introduce here a relation to calculate the natural frequencies of beams affected by damage and use it to develop a database that characterizes damage of any type. This

N. Herisanu and V. Marinca (eds.), *Acoustics and Vibration of Mechanical Structures – AVMS-2021*, Springer Proceedings in Physics 274,
https://doi.org/10.1007/978-3-030-96787-1_37

database is accessed by the damage assessment method we propose herein, which we demonstrate can locate cracks with high accuracy.

2 The Effect of a Crack on the Frequency Drop

The occurrence of a crack, whether open or closed, does not affect the mass of the beam but the rigidity of a beam segment, even if the crack has no dimension in the longitudinal direction [5]. Therefore, we analyze the effect of reducing the rigidity on the segment of length ΔL located between points *a–b*, marked hatched in Fig. 1. We maintain here a constant distributed mass by increasing the density in this segment. Considering B the width of the beam and H its thickness, the second moment of inertia is $I = BH^3/12$, and that of the reduced section is $I_C = Bh^3/12$. In the last mathematical relation h represents the thickness of the reduced cross section.

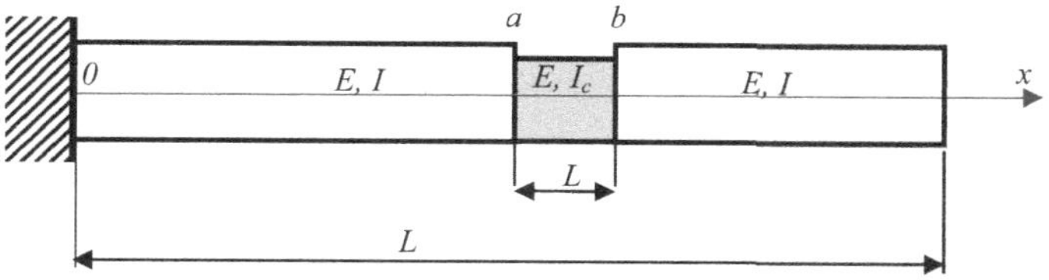

Fig. 1. Cantilever beam with low rigidity in area *a–b*

The bending moment for the beam of the constant section obtained when it is loaded just with the dead mass is proportional to the second-order derivative of the mode shape function. Therefore, the dimensionless bending moment $\overline{\phi}_i(x)$, obtained by normalization, is equal to the normalized second-order derivative.

$$\overline{M}_i(x) = \frac{d^2\overline{\phi}_i(x)}{dx^2} \tag{1}$$

If the cross section of a beam segment is reduced, it can store less energy compared to the intact beam. In order to obtain a deformation of the beam with the reduced section equal to that of the homogeneous beam, an equivalent distributed moment $\overline{M}_{ef-i}(x)$ must be applied to the segment with the reduced section, obviously smaller than the original bending moment. The relation between the two moments, taking into account the moment of inertia of the homogeneous beam I and that of the beam with variable section I(x) is

$$\overline{M}_{ef-i}(x) = \overline{M}_i(x)\frac{I(x)}{I} \tag{2}$$

The decrease in the capacity of the variable section beam to store energy, or in other words the decrease in energy required to deform the beam in a manner similar to that in which the constant section deforms under the effect of the original bending moment, is proportional to the square of the bending moment distributed along with it. The energy stored in the beam with the constant section is

$$\overline{U}_i = \frac{1}{2EI}\int_0^L \left[\overline{M}_i(x)\right]^2 dx \tag{3}$$

and the one stored in the beam with the variable section is

$$\overline{U}_{Ci} = \frac{1}{2EI(x)}\int_0^L \left[\overline{M}_{ef-i}(x)\right]^2 dx \tag{4}$$

The normalized energy ratio calculated for the reduced section beam and the constant section beam becomes

$$\frac{\overline{U}_{Ci}}{\overline{U}_i} = \frac{\frac{1}{2EI(x)}\int_0^L \left[\overline{M}_{ef-i}(x)\right]^2 dx}{\frac{1}{2EI}\int_0^L \left[\overline{M}_i(x)\right]^2 dx} \tag{5}$$

The moment of inertia I_C of the segment between points a and b is constant and smaller than I, then $I_C < I$. Taking into account relation (2), for the segment with reduced rigidity we can write

$$\overline{M}_{Ci}(x \in a, b) = \overline{M}_i \frac{I_C}{I} \tag{6}$$

The distribution of the energy stored in the slices of the beam with stepped rigidity can be deduced from relations (4) and (6), obtaining an expression of the form:

$$\overline{U}_{Ci} = \frac{1}{2EI}\int_0^a \left[\overline{M}_i(x)\right]^2 dx + \frac{1}{2EI_C}\int_a^b \frac{I_C^2}{I^2}\left[\overline{M}_i(x)\right]^2 dx + \frac{1}{2EI}\int_b^L \left[\overline{M}_i(x)\right]^2 dx \tag{7}$$

or

$$\overline{U}_{Ci} = \frac{1}{2EI}\int_0^a \left[\overline{M}_i(x)\right]^2 dx + \frac{1}{2EI}\frac{I_C}{I}\int_a^b \left[\overline{M}_i(x)\right]^2 dx + \frac{1}{2EI}\int_b^L \left[\overline{M}_i(x)\right]^2 dx \tag{8}$$

On the other hand, for the beam with the constant rigidity we have

$$\overline{U}_i = \frac{1}{2EI}\int_0^a \left[\overline{M}_i(x)\right]^2 dx + \frac{1}{2EI}\int_a^b \left[\overline{M}_i(x)\right]^2 dx + \frac{1}{2EI}\int_b^L \left[\overline{M}_i(x)\right]^2 dx \tag{9}$$

We note the integral in the second term of from relations (8) and (9) as follows:

$$\kappa_i^{a-b} = \int_a^b \left[\overline{M}_i(x)\right]^2 dx \tag{10}$$

and we call it the rigidity participation coefficient at the energy of the *a–b* segment of the beam in mode *i*. The same goes for segments 0–*a* and *b–L*.

For simplicity we can write

$$\kappa_i^{0-a} + \kappa_i^{a-b} + \kappa_i^{b-L} = \kappa_i^{0-L} \tag{11}$$

The rigidity participation coefficients are calculated involving the well-known relation of the normalized bending moment for the cantilever beam, which is

$$\overline{M}_i(x) = 0.5\left\{\cosh(\alpha x) + \cos(\alpha x) - \frac{\cos\lambda + \cosh\lambda}{\sin\lambda + \sinh\lambda}\cdot[\sinh(\alpha x) + \sin(\alpha x)]\right\} \tag{12}$$

From here it results that the coefficients κ can be easily calculated. These are in fact the areas under the curves in Fig. 2. The distribution of the potential energy stored in the beam with continuous rigidity is shown in the left column in Fig. 2, while the potential energy stored in the beam with variable rigidity is shown in the right column in the same figure.

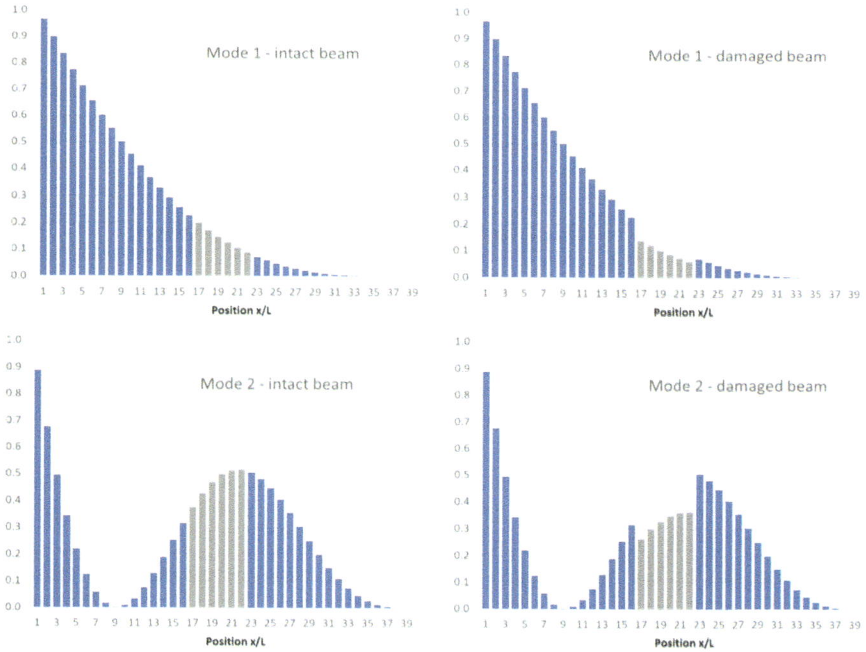

Fig. 2. Normalized strain energy distribution for the first two vibration modes

We know that there is a relation of proportionality between the natural frequencies and the deformation energy

$$f_i \approx \sqrt{U_i} \text{ respectively } f_{Ci} \approx \sqrt{U_{Ci}} \tag{13}$$

In consequence, it results in the ratios between the natural frequencies of the stepped and intact beam and those of the square roots of the energies stored in these beams is equal, thus:

$$\mathrm{f_{cn}} = \mathrm{f_n}\sqrt{\frac{\overline{\mathrm{U}}_{\mathrm{cn}}}{\overline{\mathrm{U}}_{\mathrm{n}}}} = \mathrm{f_n}\sqrt{1 - 4\tau_{\mathrm{n}}\frac{\mathrm{I} - \mathrm{I_c}}{\mathrm{I}}} \tag{14}$$

Therefore, substituting (8) and (9) in (14), after performing simplification, we can write the following:

$$\mathrm{f_{cn}} = \mathrm{f_n}\sqrt{\frac{\overline{\mathrm{U}}_{\mathrm{cn}}}{\overline{\mathrm{U}}_{\mathrm{n}}}} = \mathrm{f_n}\sqrt{1 - 4\tau_{\mathrm{n}}\frac{\mathrm{I} - \mathrm{I_c}}{\mathrm{I}}} \tag{15}$$

It is easy now to calculate the natural frequency of the beam with stepped rigidity when the natural frequency of the intact beam and the damage location and severity are known. This is made with the relation

$$\mathrm{f_{cn}} = \mathrm{f_n}\sqrt{\frac{\overline{\mathrm{U}}_{\mathrm{cn}}}{\overline{\mathrm{U}}_{\mathrm{n}}}} = \mathrm{f_n}\sqrt{1 - 4\tau_{\mathrm{n}}\frac{\mathrm{I} - \mathrm{I_c}}{\mathrm{I}}} \tag{16}$$

One can observe that the effect of the damage location and severity is reflected by the rigidity participation coefficients, i.e., the areas under the curves in Fig. 2 in the right column.

3 The Assessment Method for the Position of the Defect

Let the cantilever beam be the same as in Fig. 1. This beam is now plotted as a brown rectangle in Fig. 3 where it is divided into N finite elements equal in length. If it is intact, all segments have the same rigidity and will store strain energy according to relation (9). This energy evolution with the position on the beam is plotted with a red line. If one element has a crack, the amount of energy that will be stored in the damaged element is reduced by the I_C / I ratio. Figure 3 shows with blue bars the amount of energy stored in each element in the first vibration mode when the 5th element, located between points between a and b, has a crack. The damaged element is plotted with red color to distinguish it from the intact elements plotted with brown color.

Regarding the energy distribution, we can calculate the energy stored in the intact beam as the sum of the rectangles representing the energy stored in each element (or the area under the red curve), which is κ_i^{0-L}. If the number of elements is big enough, the accuracy in calculating this coefficient is good and we obtain the value 0.25 for all vibration modes. So, the normalized energy U_i for the intact beam, expressed with help of the energy contribution coefficient, is known from before.

We can put relation (16) in the form

$$\left(\frac{f_{Ci}}{f_i}\right)^2 \kappa_i^{0-L} = \kappa_i^{0-a} + \frac{I_C}{I}\kappa_i^{a-b} + \kappa_i^{b-L} = \kappa_i^{0-L} - \frac{I - I_C}{I}\kappa_i^{a-b} \tag{17}$$

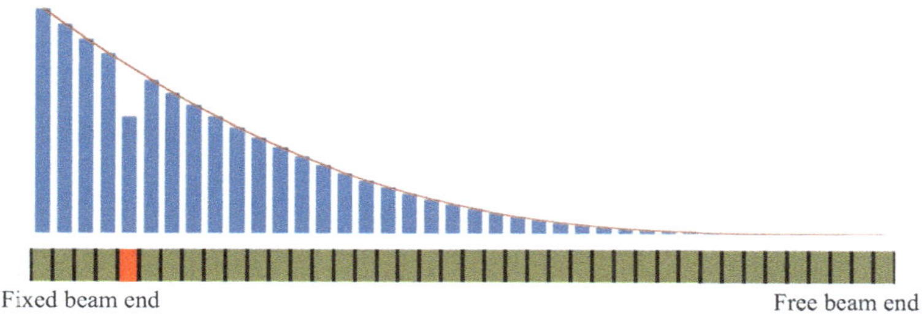

Fig. 3. The beam with a crack in element 5 and the resulted energy distribution

and in consequence

$$\left(1-\frac{I_C}{I}\right)\kappa_i^{a-b}=\kappa_i^{0-L}-\left(\frac{f_{Ci}}{f_i}\right)^2\kappa_i^{0-L} \tag{18}$$

or

$$\left(1-\frac{I_C}{I}\right)\kappa_i^{a-b}=\kappa_i^{0-L}\left[1-\left(\frac{f_{Ci}}{f_i}\right)^2\right] \tag{19}$$

It is convenient to use n, thus the number of the damaged element, instead of the interval a-b. Let us consider the first two bending vibration modes. From relation (18), and considering n instead of a-b, we can calculate the ratio

$$\frac{\left(1-\frac{I_C}{I}\right)\kappa_1^n}{\left(1-\frac{I_C}{I}\right)\kappa_2^n}=\frac{\kappa_1^{0-L}\left[1-\left(\frac{f_{C1}}{f_1}\right)^2\right]}{\kappa_2^{0-L}\left[1-\left(\frac{f_{C2}}{f_2}\right)^2\right]}\ \text{or}\ \frac{\kappa_1^n}{\kappa_2^n}=\frac{1-\left(\frac{f_{C1}}{f_1}\right)^2}{1-\left(\frac{f_{C2}}{f_2}\right)^2}=\varepsilon_{1/2}^n \tag{20}$$

One can observe that the ratio ε in relation (19) does not depend on the severity of the damage given by the stiffness decrease, thus the crack is characterized just by its position. Since the frequencies f_i and f_{Ci} are known from measurements made on the real structure, we can easily find the ratio ε.

Because we know the rigidity participation coefficients, we can calculate the ratios $\varepsilon_{1/2}^n$ for $n = 1 \ldots N$, resulting in N damage scenarios. To locate the damaged element, it is sufficient to find the best fit between the ratios ε calculated for the N scenarios and the measured ratio ε. This can be made by calculating the anti-distance

$$DI=\max\left\{\left[\left(\frac{\kappa_1^n}{\kappa_2^n}\right)^2-\left[\left(1-\left(\frac{f_{C1}}{f_1}\right)^2\right)\Big/\left(1-\left(\frac{f_{C2}}{f_2}\right)^2\right)\right]^2\right]^{-2}\right\}\quad \text{for}\ \ n=1\ldots N \tag{21}$$

The graphical representation made with relation (21) shows the position of the damage by its global maximum. Further, knowing the damage position we can calculate the severity involving a procedure described in [7].

4 Numerical Verification

To prove the validity of the method, we perform numerous simulations. In this section we present two characteristic cases: (*i*) the crack is present in an element close to the fixed end, and (*ii*) the crack is closer to the free end.

Case 1: Simulation performed in ANSYS for a cantilever beam with $L = 1$ m, $B = 50$ mm, and $H = 5$ mm, made of ANSI steel, have shown that the first two frequencies for the intact state are $f_1 = 4.0889$ Hz and $f_2 = 25.6265$ Hz. Reducing the rigidity of the seventh element by 5%, which we realized by multiplying the Young modulus with 0.95, we obtained the frequencies $f_{C1} = 4.0642$ Hz and $f_{C2} = 25.6115$ Hz.

Case 2: For the same intact beam we generated a damage in element 34, by reducing the rigidity of this element by 10% with the same method as for case 1. The frequencies of the damaged beam resulted in $f_{C1} = 4.0642$ Hz and $f_{C2} = 25.6115$ Hz (Fig. 4).

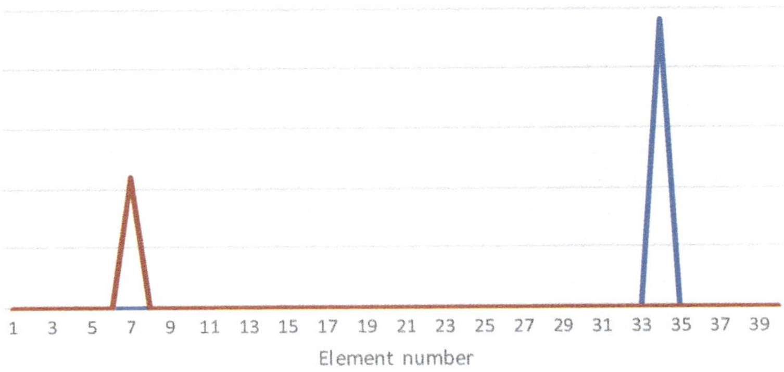

Fig. 4. The damage indicator *DI* for the 2 damage scenarios

The damage indicator DI, calculated with relation (21), reveals the correct damage location for the two cases: with red line is represented case 1, with blue line case 2.

5 Conclusion

The damage detection method we propose in this paper allows finding the location of damage by monitoring just the first two out-of-plane vibration modes. The method base on the fact that the ratio of two frequency shifts is not influenced by the damage severity, therefore being possible to find the crack location in the absence of information about its depth. The disadvantage of the method is that the crack should be placed relatively close to the middle of the element, else it is the risk to get a false crack position. The accuracy of the method can be improved if we take into consideration more vibration modes.

References

1. Dahak, M., Touat, N.: Kharoubi, M: Damage detection in beam through change in measured frequency and undamaged curvature mode shape. Inverse Prob. Sci. Eng. **27**, 89–114 (2019)
2. Mesquita, E., Antunes, P., Coelho, F., André, P., Arêde, A., Varum, H.: Global overview on advances in structural health monitoring platforms. J. Civ. Struct. Heal. Monit. **6**(3), 461–475 (2016)
3. Khatir, S., Dekemele, K., Loccufier, M., Khatir, T., Wahab, M.A.: Crack identification method in beam-like structures using changes in experimentally measured frequencies and particle swarm optimization. C.R. Mec. **346**(2), 110–120 (2018)
4. Gillich, G.R., Mituletu, I.C., Praisach, Z.I., Negru, I., Tufoi, M.: Method to enhance the frequency readability for detecting incipient structural damage. Iran. J. Sci. Technol. Trans. Mech. Eng. **41**, 233–242 (2017)
5. Gillich, G.R., Abdel Wahab, M., Praisach, Z.I., Ntakpe, J.L.: The influence of transversal crack geometry on the frequency changes of beams. In: Sas, P., Moens, D., Denayer, H. (eds.) Proceedings of International Conference on Noise and Vibration Engineering (ISMA2014) and International Conference on Uncertainty in Structural Dynamics (USD2014), 485–98 (2014)
6. Bovsunovsky, A., Surace, C.: Non-linearities in the vibrations of elastic structures with a closing crack: a state of the art review. Mech. Syst. Signal Process. **62–63**, 129–148 (2015)
7. Praisach, Z.I., Gillich, G.R., Protocsil, C., Muntean, F.: Evaluation of crack depth in beams for known damage location based on vibration modes analysis. Appl. Mech. Mater. **430**, 90–94 (2013)
8. Gillich, G.R., Maia, N.M.M., Mituletu, I.C., Tufoi, M., Iancu, V., Korka, Z.I.: A new approach for severity estimation of transversal cracks in multi-layered beams. Lat. Am. J. Solids Struct. **13**(8), 1526–1544 (2016)
9. Ravi, J.T., Nidhan, S., Muthu, N., Maiti, S.K.: Analytical and experimental studies on detection of longitudinal, L and inverted T cracks in isotropic and bi-material beams based on changes in natural frequencies. Mech. Syst. Signal Process. **101**, 67–96 (2018)
10. Gillich, G.R., Praisach, Z.I.: Modal identification and damage detection in beam-like structures using the power spectrum and time-frequency analysis. Signal Processing 96(A), 29–44 (2014).

Experimental Results from Testing a Rolling Pendulum Base Isolation System at Bridge Structures Subjected by Dynamic Shocks and Vibrations

Fanel Scheaua(✉)

Engineering and Agronomy Faculty of Braila, "Dunarea de Jos" University of Galati, MECMET Research Center, Galati, Romania
fanel.scheaua@ugal.ro

Abstract. The isolation concept of vibrations produced by various dynamic actions on construction structures is not new, while there are multiple concerns worldwide for solutions development meant to solve the vibrational problems within the structures where different mechanical systems are mounted for vibration isolation and better stability acquisition. These systems are strategically positioned for avoiding the transmission of vibration energy from the foundation to the insulated superstructure. For bridge structures, base insulation systems can be used composed of elastomeric elements and sliding friction systems in order to obtain good stability both at dynamic loads resulting from traffic and also for the loads dictated by the seismic actions requesting the structure at a certain moment of time. An isolation system working on the principle of a rolling pendulum was made at a reduced scale and tested in laboratory conditions on a bridge beam, in order to highlight the specific values of acceleration that can be obtained at the level of the structural frames where mounted. For a bridge-type structure containing the support pier and superstructure, the insulating system is interposed between these structural frames in order to decouple and isolate the superstructure from the efforts coming from the foundation and the support pier when a dynamic action occurs. The rolling pendulum solution using a spherical pivot is proposed in this study in order to highlight the behavior of the superstructure during dynamic vibratory motions, and the obtained results are compared between the support pier and the isolated superstructure.

1 Introduction

The bridge or viaduct infrastructure elements present maximum strategic importance within a community area because they ensure the vital connection between human communities. That is why these structures must withstand over time the various composed requests to which they are subjected related to not only the traffic demands but also the seismic actions.

Various solutions have been developed and applied over time that aim to isolate the structure from the effects of dynamic actions, as well as energy dissipation through its

N. Herisanu and V. Marinca (eds.), *Acoustics and Vibration of Mechanical Structures – AVMS-2021*, Springer Proceedings in Physics 274,
https://doi.org/10.1007/978-3-030-96787-1_38

consumption at the level of a dissipative system to ensure improved behavior of the structure to seismic events that require the structure strength in vibratory terms.

For a bridge or viaduct structure type, the base isolation procedure plays an important role and in order to apply this method it is used with specific assemblies made of elastomer supports in combination with a friction pendulum system to ensure the structure stability. These types of systems have been also used separately, but in combination they ensure an increase of the structure isolation degree.

In order to isolate a bridge-type structure against vibratory actions, a pendulum-type system is proposed, but it uses a spherical pivot with the possibility of making the rolling movement on the concave surface of the support. This system is positioned between the bridge support pier and the superstructure [1–5].

The aim is to reduce the coefficient of friction between the spherical pivot and the concave surface of the support which can have the effect of greater freedom of movement of the ground with the structure foundation, while the superstructure must remain in a state of relative equilibrium.

The experimental tests were conducted using dynamic actions generated by a special device capable to provide induced vibrations to the bridge structure up to a maximum value of 4.38 J. The values recording was made by means of tri-axial accelerometer sensors which were mounted at the bridge pier and a suprastructure level capable to record the specific acceleration values.

Improved values of acceleration are presented at the superstructure level which, compared to those recorded at the level of the support pier, highlights the insulating character of the system used [6–11].

2 Modeling Aspects for the Rolling Isolation System Operation

A rolling pendulum isolation system is composed of two main plates whose inner surface is in contact with the rolling pivot (Fig. 1).

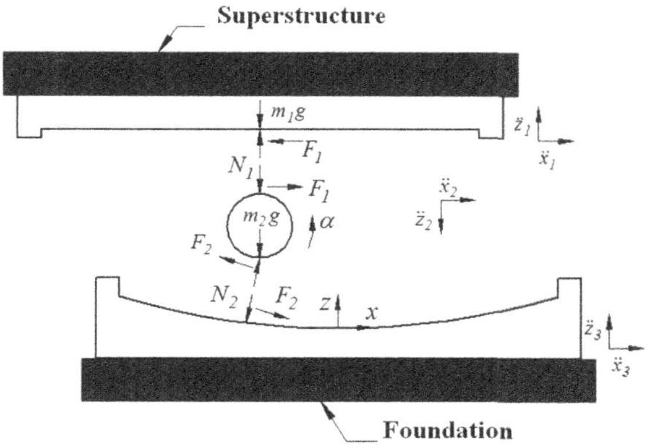

Fig. 1. The mathematical model for the insulation model

The lower plate is attached to the foundation while the upper plate is to the superstructure.

The rolling movement of the spherical pivot is performed without sliding and the friction coefficient between the pivot and the main plates has a lower value in the case of rolling compared to the sliding friction.

The lower plate has a concave surface on which the pivot makes a translational movement combined with a vertical lift.

When the rolling motion is applied at the pivot, a translational displacement of the superstructure is obtained, which is double in value relative to the pivot and due to the geometry of the concave rolling surface, the structure re-centering is ensured at the initial position due to gravitational conditions.

The safe operation of the isolation system involves ensuring permanent contact between the components, which means that the normal forces are in a positive value [12–15].

The excitation components are represented by accelerations in horizontal ($\ddot{x}$) and vertical planes ($\ddot{z}$), and the dynamic equilibrium corresponding to the superstructure on the main directions is described by the relation [16–20]:

Horizontal direction:

$$m_1(\ddot{x}_1 + \ddot{x}_3) + F_1 = 0 \tag{1}$$

Vertical direction:

$$m_1(\ddot{z}_1 + \ddot{z}_3) - N_1 + m_1 g = 0 \tag{2}$$

The dynamic equilibrium for the spherical pivot is written as follows for the main directions of motion:

Horizontal direction:

$$m_2(\ddot{x}_2 + \ddot{x}_3) - F_1 + F_2 \cos\beta - N_2 \sin\beta = 0 \tag{3}$$

Vertical direction:

$$m_2(\ddot{z}_2 - \ddot{z}_3) - N_1 + F_2 \sin\beta + N_2 \cos\beta - m_2 g = 0 \tag{4}$$

3 Experimental Model of Rolling Pendulum Isolation System

An experimental model of a rolling pendulum isolation system was developed (Fig. 2). This system is used for experimental research materialized in the insulation of a bridge model (beam) which was equipped with 4 insulation systems positioned symmetrically at the ends of the beam (Fig. 3) [16–20].

This system has the ability to provide freedom of movement for the supporting pier, while the superstructure tends to remain in a state of dynamic equilibrium when a shock or seismic movement of the ground is recorded.

The beam was modelled by a homogeneous element with a constant geometry of length 2 m, width 0.36 m and thickness 0.017 m, the material being OSB.

Fig. 2. Experimental model of rolling pendulum system (RPS)

Fig. 3. The experimental stand with rolling pendulum system mounted

The loading was performed with a uniformly distributed concentrated load having a value of 30 kgf, so as to avoid accidental contact loss between the rolling elements and the corresponding pendulum supports when applying the dynamic load. The induced vibrations at the insulated structure level were transmitted by means of a pendular exciter system with a constant mass of 1,135 kg which has the possibility to provide impact energy of up to 4.38 J depending on the angle of fall [16–20].

Acceleration values were recorded at the support pier and superstructure level by means of mounted tri-axial accelerometers (Fig. 4).

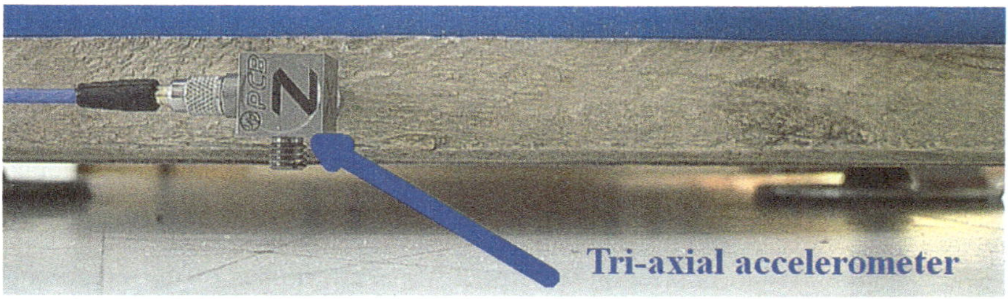

Fig. 4. Isolated experimental model with accelerometer mounted for signal recording

4 Experimental Research Results

The experimental tests performed on the scaled model of the isolated beam by means of rolling pendulum isolation system (RPS) with spherical rolling pivot are presented that revealed specific result values as presented below.

The excitation that simulates the seismic shock had as application point the upper area of the support pile (beam support pier), in the vicinity of the support area, so that the excitation force is acting on the lower plate of the isolation device.

The obtained experimental results are presented in terms of accelerations in time and acceleration amplitude in relation to excitation frequency on the transversal main orthogonal direction of movement for both structural elements: support pier and superstructure both with tri-axial accelerometers mounted (Table 1) [16–20].

Figure 5 explicitly shows the acceleration values in time and frequency obtained both for the foot and for the beam.

Thus, the lower values are highlighted due to the use of the rolling isolation system.

Significant differences are registered between the structural elements the isolation character of the rolling pendulum system (RPS) being highlighted which has the possibility to interrupt the mainstream transmission of motion components on vertical direction protecting the superstructure at dynamic stresses.

5 Conclusion

The base isolation principle on construction structures is one of the basic techniques in use not only for protection mainly against seismic actions but also for other vibrational actions that may occur over time in bridge or viaduct structures.

This paper presents an isolation solution that involves a rolling pendulum-type system (RPS) with a spherical central pivot that has the ability to roll without sliding on a concave surface during the occurrence of an earthquake or vibrational shock. Through this system, a disconnection is introduced between the structural frames of the bridge, being interrupted the efforts transmission main route on vertical direction from the ground to the foundation, support pier and superstructure. In this idea, it is expected that the bridge superstructure or deck will experience a much attenuated movement during an earthquake of high magnitude precisely as a result of the action of the insulating system.

The starting point for this system is the friction pendulum system, which uses a spherical pivot but uses dry friction force to limit the relative displacement between the insulating system assembly elements each connected to different structural frames.

An experimental model of the rolling pendulum system (RPS), which contains a steel pivot with the possibility of rolling on a spherical steel surface, was made and tested in laboratory conditions on a bridge beam made on a small scale, in order to identify the structure behavior while required by shocks made through a mechanical exciter that simulates the occurrence of an earthquake.

The isolation system is positioned between the main frames of the experimental structure.

The obtained results presented for the transversal direction of motion show eigenvalues of the acceleration as a function of time and magnitude of the acceleration as a function of frequency for both the bridge pier and superstructure.

Table 1. The experimental results obtained for pier and superstructure isolated with rolling pendulum system

Pier

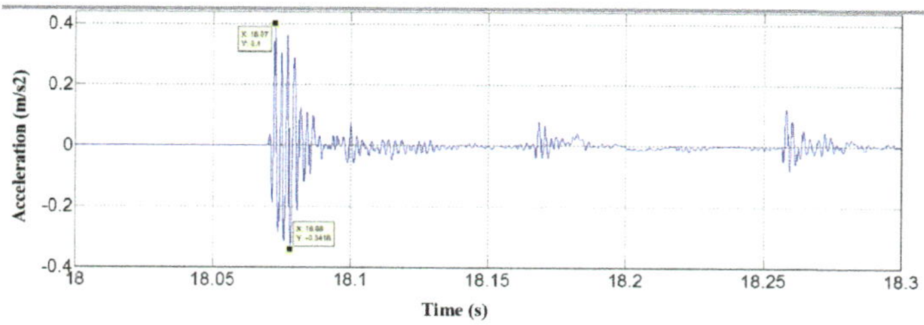

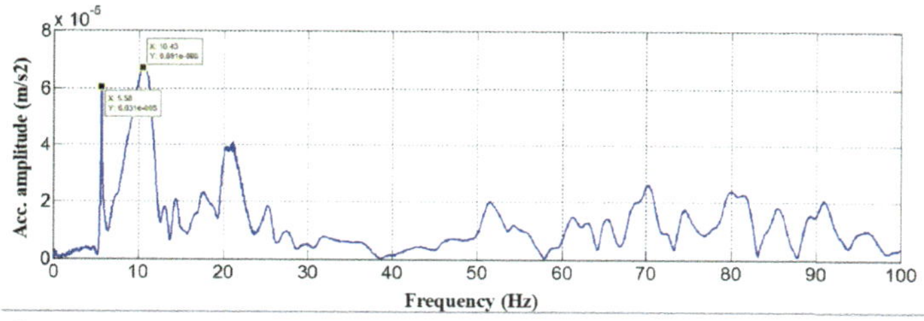

Superstructure

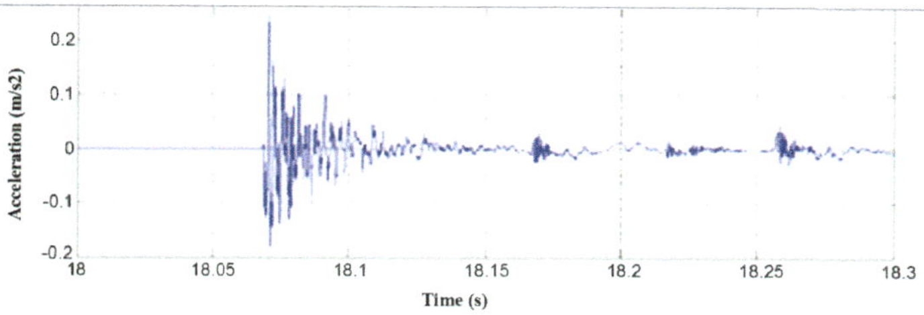

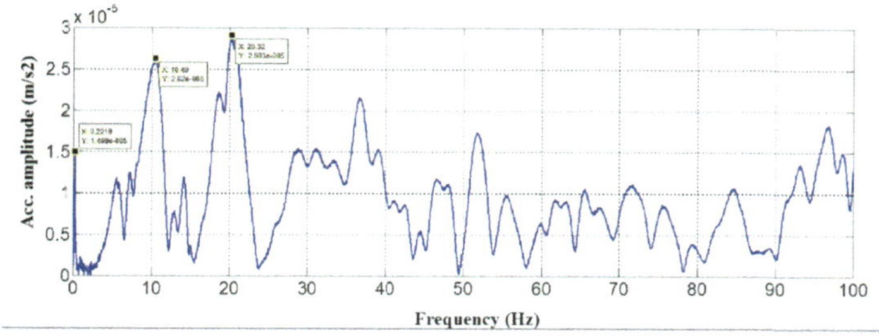

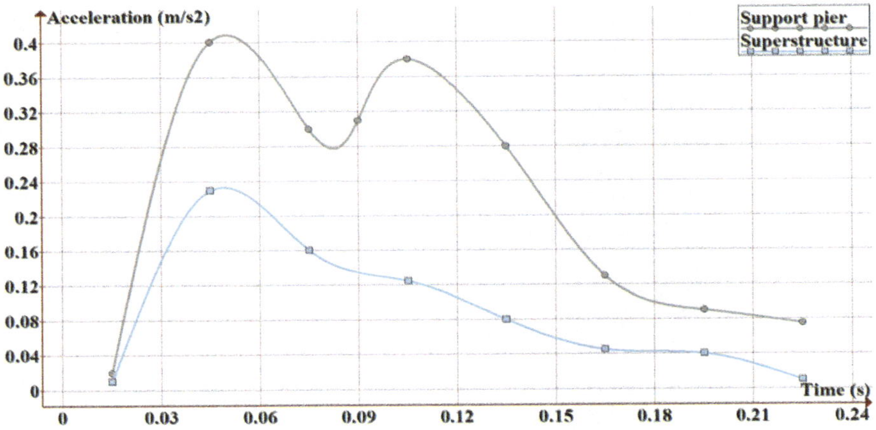

(a) Acceleration in time

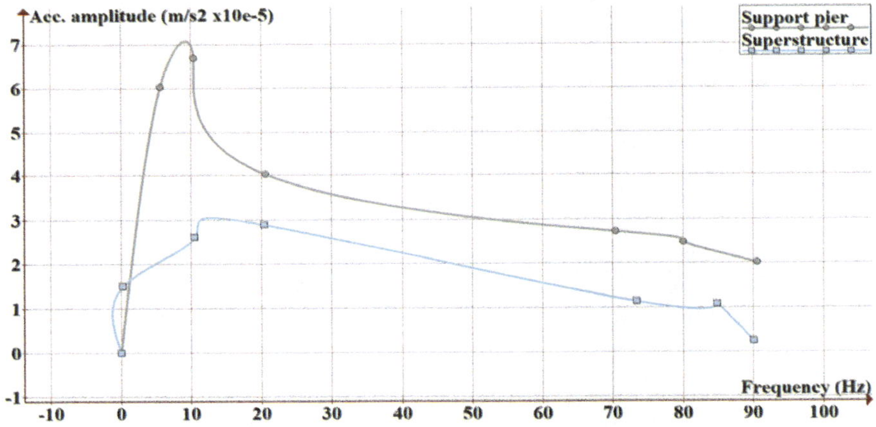

(b) Acceleration amplitude function of frequency values

Fig. 5. Specific proper values recorded for the support pier and superstructure

The lower values recorded at the level of the superstructure are compared to those recorded at the level of the support pier, thus highlighting the insulating character of the system used to equip the structure.

References

1. Harvey, P.S., Kelly, K.C.: A review of rolling-type seismic isolation: historical development and future directions. Eng. Struct. **125**, 521–531 (2016)
2. Jeon, B.G., Chang, S.J., Kim, S.W., Kim, N.S.: Base isolation performance of a cone-type friction pendulum bearing system. Struct. Eng. Mech. **53**(2), 227–248 (2015)
3. Wang, S.J., Sung, Y.L., Yang, C.Y., Lin, W.C., Yu, C.H.: Control performances of friction pendulum and sloped rolling-type bearings designed with single parameters. Appl. Sci. **10**(20), 7200 (2020)
4. Lee, G.C., Ou, Y.C., Song, J., Niu, T., Liang, Z.: A roller seismic isolation bearing for highway bridges. In: The 14-th International Conference on Seismic Engineering, Beijing, China (2008)
5. Tsai, M.H., Chang, K.C., Wu, S.Y.: Seismic isolation of a scaled bridge model using rolling-type bearings. In: a- 4-a Conferinţă Internaţională în Ingineria Seismică, Taipei, Taiwan (2006)
6. Matta, E., Greco, R.: Modeling and design of tuned mass dampers using sliding variable friction pendulum bearings. Acta Mech. **231**(12), 5021–5046 (2020)
7. Ciornei, M.C., Alaci, S., Ciornei, F.C., Romanu, I.C.: (2017, August). A method for the determination of the coefficient of rolling friction using cycloidal pendulum. In: IOP Conference Series: Materials Science and Engineering, vol. 227, No. 1, p. 012027. IOP Publishing (2017)
8. Tsai, C.S., Lin, Y.C., Chen, W.S., Su, H.C.: Tri-directional shaking table tests of vibration sensitive equipment with static dynamics interchangeable-ball pendulum system. Earthq. Eng. Eng. Vib. **9**(1), 103–112 (2010)
9. Ştefănescu, T.M., Volintiru, O.N., Scurtu, I.C.: The variation of the forces acting in a helical cylindrical gear according to its operating modes. In: Journal of Physics: Conference Series, vol. 1122, No. 1, p. 012027. IOP Publishing (2018)
10. Rahmani, M., Scurtu, I.C., Petrudi, A.M.: Analytical and dynamic study of pulled mass nonlinear vibration by two cables using Newton's harmonic balance method. Technium Rom. J. Appl. Sci. Technol. **2**(2), 79–86 (2020)
11. Fathi, P., Petrudi, A.M., Scurtu, I.C.: Investigation of the effects of drilling on mechanical joints in composite structures. In: IOP Conference Series: Earth and Environmental Science, vol. 635, No. 1, p. 012010. IOP Publishing (2021)
12. Goanta, A. M. (2011). E-LEARNING METHOD APPLIED TO TECHNICAL GRAPHICS SUBJECTS. J. Ind. Des. Eng. Graph. **6**(2) (2011)
13. Goanţă, A.M.: Considerations on design controls intended to achieve thin walled bodies. Analele Universităţii" Dunărea de Jos" din Galaţi. Fascicula XIV, Inginerie mecanică= Annals of "Dunarea de Jos "University of Galati. Fascicle XIV, Mechanical Engineering, **20**(2), 9–12 (2013)
14. Drăgan, N.: Dynamic analisys of the parameters of the mechanical systems with structural damping. Viscoelastic SLS model. Part 2: Transmissibility factor and isolation degree. Analele Universităţii" Dunărea de Jos" din Galaţi. Fascicula XIV, Inginerie mecanică= Annals of "Dunarea de Jos "University of Galati. Fascicle XIV, Mechanical Engineering, **23**(2), 13–16 (2016)
15. Drăgan, N.: Analiza dinamică a podurilor din grinzi de beton armat-determinarea modurilor proprii de vibraţie. Sinteze de mecanică teoretică şi aplicată, Volumul, 9 (2018)

16. Scheaua, F.D.: Behavior analysis of the dry friction dissipating systems at dynamic actions, PhD thesis, "Dunarea de Jos" University of Galati (2013)
17. Scheaua, F., Nastac, S.: On dynamics of a dual vibration insulation device based on Elastomeric Kernel and rolling friction dissipation. In: Acoustics and Vibration of Mechanical Structures—AVMS-2017, pp. 397–403. Springer, Cham (2018)
18. Scheaua, F.: Aspects regarding rolling pendulum seismic isolation systems. Annals of "Dunarea de Jos "University of Galati. Fascicle XIV, Mechanical Engineering **22**(1), 5–8 (2015)
19. Şcheaua, F.: Experimental dynamics of a bridge deck isolated with rolling pendulum system (RPS), Annals of "Dunarea de Jos "University of Galati. Fascicle XIV, Mechanical Engineering **20**(2), 81–84 (2013)
20. Scheaua, F.: Modelling of the visco-elastic pendular hybrid system with dissipative rolling elements. In: IOP Conference Series: Materials Science and Engineering, vol. 591, No. 1, p. 012029. IOP Publishing (2019)

Study on the Behavior of the Isolated Structures with Friction Pendulums and a Counterweight

Tatian-Cristian Malin, Gilbert-Rainer Gillich(✉), and Dorian Nedelcu

Babeş-Bolyai University, Str. Mihail Kogalniceanu 1, 400084 Cluj-Napoca, Romania
gilbert.gillich@ubbcluj.ro

Abstract. The paper presents a study regarding the behavior of the isolated structures with friction pendulums and a counterweight, as well as the influence of the spring constant to assure efficient isolation. To determine the structural behavior, we created a model of a six-story building in SolidWorks software with the help of which we made the simulations. We found that the acceleration and the displacement of the structure decreased if we have a higher value for the spring constant and a second friction contact between the structure and the counterweight.

Keywords: Friction pendulum · Structural response · SolidWorks simulation

1 Introduction

Our society has been and still is affected by one of the deadliest and costly natural catastrophes that are earthquakes. The two main objectives of earthquake safety are life safety and collapse prevention, so the desire for high-performance buildings has increased. To avoid harmful effects, the structures are isolated by inserting them between the ground and the superstructure devices that reduce the effect of inertial forces. Description of these devices and the expected structural response are presented in detail in the literature; see, for example, [1–4]. A spread method to protect structures is the insertion of reinforced elastomeric or natural rubber elements between the ground and the base of the structure [5–9]. Other seismic isolation devices based on the energy dissipation by friction, which has been introduced in 1985, are friction pendulums (FP). In the meantime, a lot of advanced friction pendulum-type bearings were designed and optimized. The FP can have one, two, and three sliding surfaces [10–13] of which some can have a polynomial surface [14].

2 Materials and Methods

In this section, we present the study on the behavior of the isolated structure with a friction pendulum. The research was performed with the help of the SolidWorks program involving the Motion module. In Fig. 1 is presented the test structure generated in SolidWorks as an assembly with four parts: 1—the rigid structure, made of steel bars

N. Herisanu and V. Marinca (eds.), *Acoustics and Vibration of Mechanical Structures – AVMS-2021*, Springer Proceedings in Physics 274,
https://doi.org/10.1007/978-3-030-96787-1_39

with the dimensions 1200 × 400 × 200 mm; 2—the base plate that is fixed and is used as a reference, with the dimensions 600 × 200 × 10 mm; 3—the shaking plate reproducing the horizontal trepidation of the ground and has the same dimensions as the base plate; 4—the counterweight, with the dimensions 100 × 150 mm and with the weight of 17.784 kg, which is fastened with 2 springs in the upper part of the structure and has the role of stabilizing the structure subjected to the horizontal trepidation of the earth.

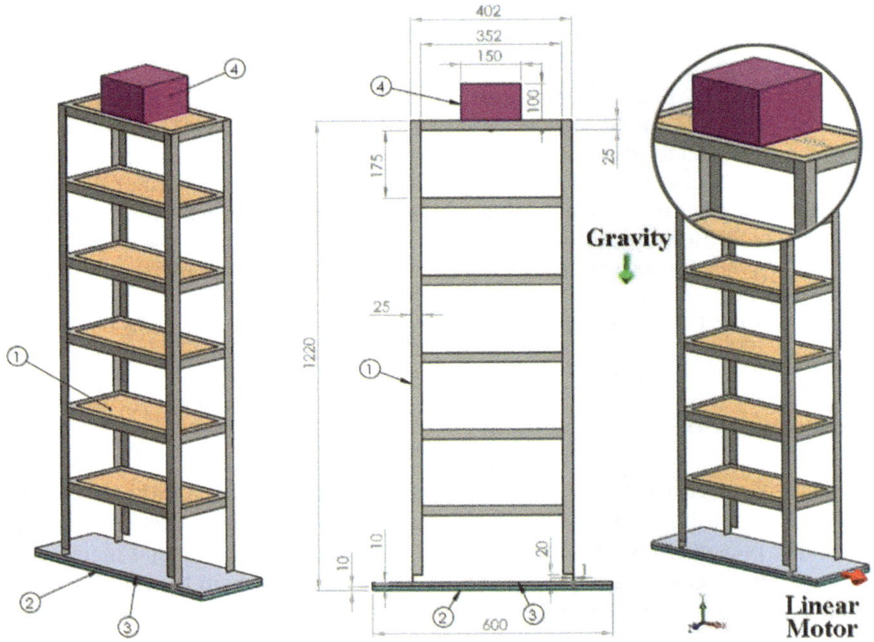

Fig. 1. The test structure

The simulations were performed in the SolidWorks Motion module, for the following conditions:

- The base plate is fixed and is used as a reference;
- The shaking plate, simulating the horizontal trepidation of the earth, is moved in the X direction with the help of the Linear Motor, a feature of the SolidWorks Motion module. It imposes displacement with the following parameters: Oscillating motion, Max Displacement A = 20 mm, Frequency $f = 1$ Hz, and Shift = 0 deg.
- The gravitational acceleration is g = 9806.65 mm/s2 and is oriented in the Y direction.
- A *SolidBody Contact* with friction is imposed between the structure and the sliding surface, the friction coefficient μ being dependent on the chosen materials. The contact between the structure and the shaking plate was simulated considering the static and dynamic friction coefficients μ_D and μ_S, respectively, the dynamic and static velocity coefficients v_D and v_S presented in Table 1.

- For this study, the time of analysis is imposed as 30 s.
- The counterweight and the structure are caught by the edge of the steel bars, respectively, the edge of the shaking plate by four springs. The spring parameters (k—spring constant, coil diameter, number of coils, and wire diameter) are presented in Fig. 2.

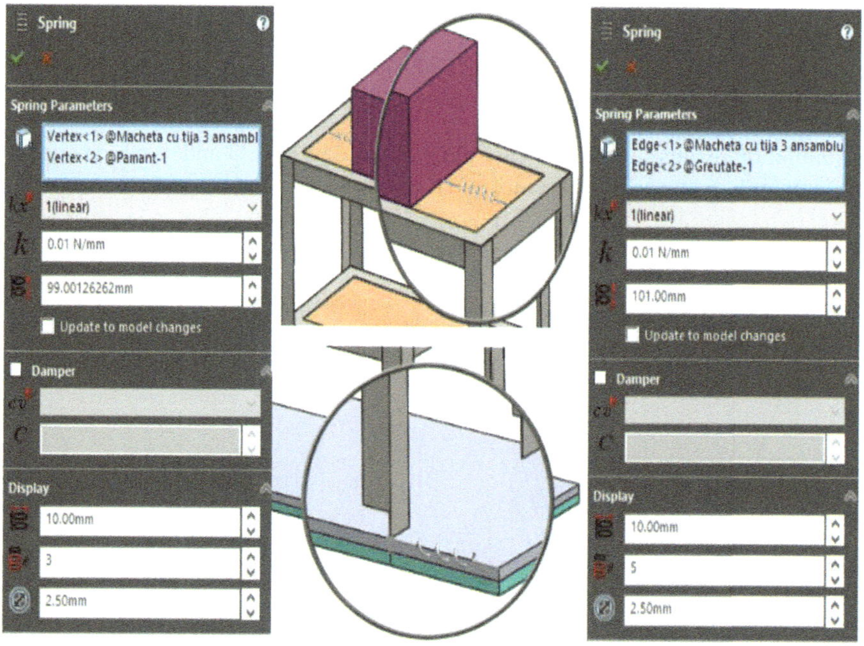

a) Structure springs b) Counterweight springs

Fig. 2. The spring position and parameters

Table 1. Contact condition based on friction coefficients.

Contact case	Components	Contact type	μ_D [−]	ν_D [mm/s^2]	μ_S [−]	ν_S [mm/s^2]
1	Structure	Steel (Greasy)	0.05	10.16	0.08	0.1
	Shaking plate	Steel (Greasy)				

In the first case of the research, the contact between the structure and the shaking plate was simulated considering the friction coefficients presented in Table 1, but without contact between the structure and the counterweight. In this case, the constant k of the springs with which the structure is caught with the shaking plate remains unchanged ($k = 0.01$ N/mm), while the constant k of the spring with which the counterweight is caught is modified with each simulation ($k = 0.004 \div 0.01$ N/mm).

The simulations in the second case are made with the same spring parameters and conditions from the first case, both for the structure and the counterweight, but in this case, we also added contact with friction between the structure and the counterweight. The contact structure-shaking plate and structure-counterweight was simulated considering the static (μ_S) and the dynamic (μ_D) friction coefficients presented in Table 1.

3 Results and Discussions

The simulation results for the first case are presented in Figs. 3 and 4, where contact is considered only between the structure and the shaking plate.

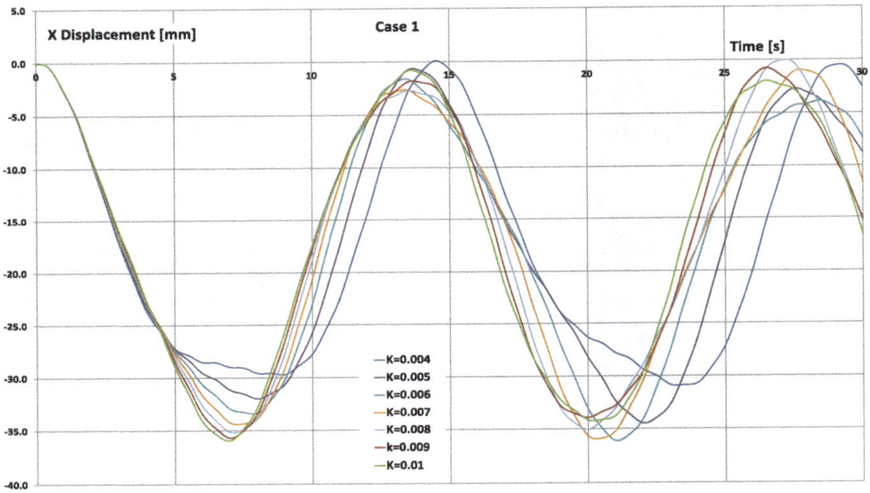

Fig. 3. Case 1—Displacement between the structure and the base plate

One can observe that the acceleration and the displacement of the structure have been decreased as the spring constant k increased.

In Figs. 5 and 6 are presented the results obtained for the second case, considering the friction coefficient of the contacts structure-base plate and structure-counterweight.

From Figs. 5 and 6, we can observe that the acceleration and the displacement of the structure decreased, even more than in the first case, when we added the second friction contact between the structure and the counterweight. This means that to have better insulation, both contacts (structure—base plate and structure—counterweight) must be considered.

In Figs. 7 and 8 is present a comparison between the results obtained for Case 1 and Case 2. The analysis is made to find the best constructive solution in regard to the surface friction coefficient and the added mass.

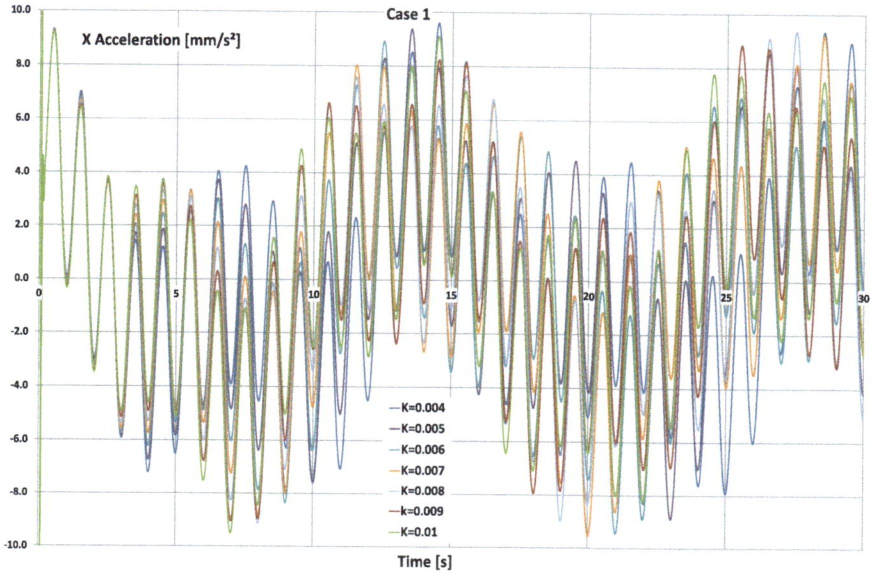

Fig. 4. Case 1—The acceleration of the structure

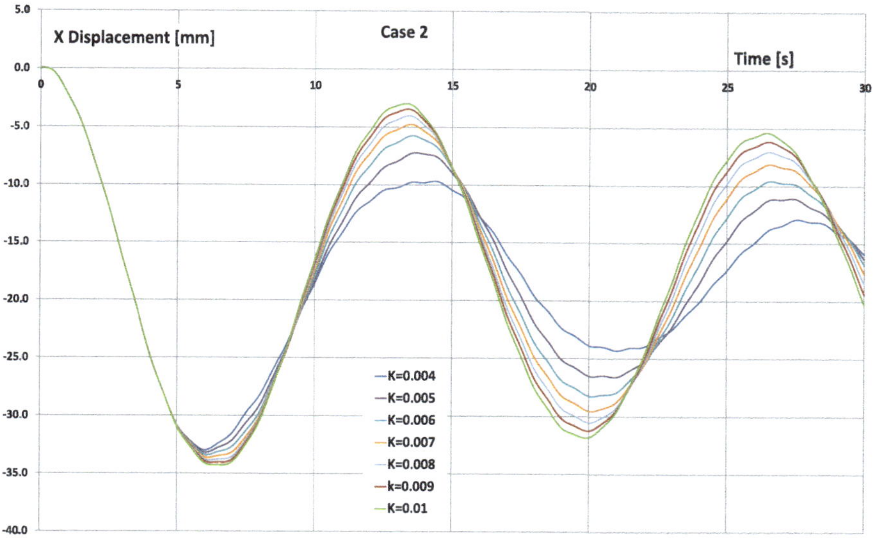

Fig. 5. Case 2—Displacement between the structure and the base plate

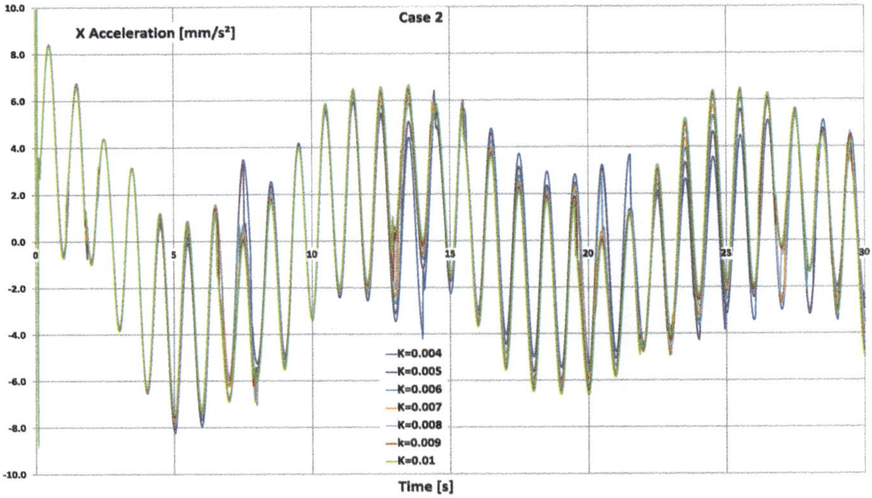

Fig. 6. Case 2—The acceleration of the structure

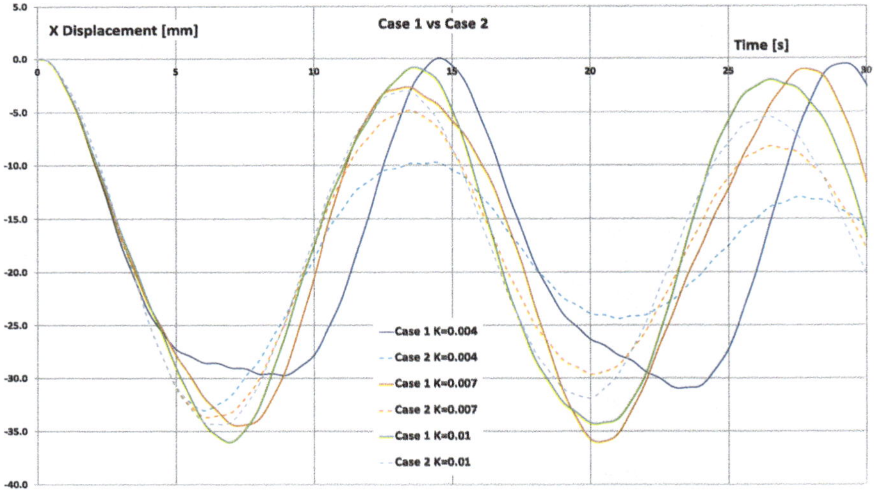

Fig. 7. Displacement—Case 1 versus Case 2

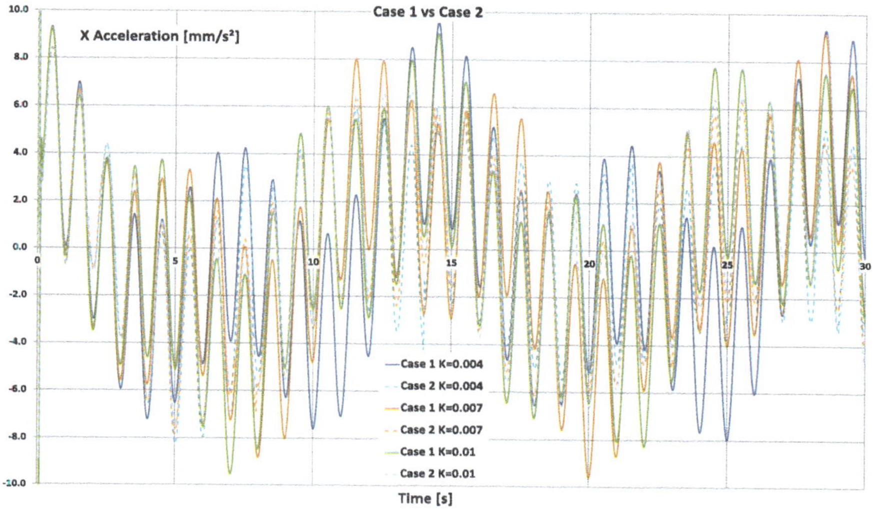

Fig. 8. Acceleration—Case 1 versus Case 2.

4 Conclusion

The aim of the study was to identify the behavior of the isolated structures with a friction pendulum and a counterweight, as well as the influence of the spring constant k to assure efficient isolation. It was found that the best isolation is achieved if the spring constant k, with which the mass is caught, is higher and if we have two surface contacts. The two constructive parameters, namely the spring constant and the friction coefficient, must be carefully adapted to achieve efficient isolation.

References

1. Skinner, R.I., Robinson, W.H., McVerry, G.H.: An introduction to seismic isolation. John Wiley and Sons, London (1993)
2. Fenz, D.M., Constantinou, M.C.: Spherical sliding isolation bearings with adaptive behavior: experimental verification. Earthquake Eng. Struct. Dynam. **37**(2), 185–2015 (2008)
3. Tsai, C.S.: Advanced base isolation systems for light weight equipment's. In: Earthquake-Resistant Structures-Design, Assessment and Rehabilitation, Intech, p. 79–130 (2012).
4. Wilde, K., Garboni, P., Fujino, Y.: Base isolation system with shape memory alloy device for elevated highway bridges. Eng. Struct. **22**(3), 222–229 (2000)
5. Taylor, A., Lin, A., Martin, J.: Performance of elastomers in isolation bearings: a literature review. Earthq. Spectra **8**(2), 279 (1992)
6. Iancu, V., Gillich, G.R., Iavornic, C.M., Gillich, N.: Some models of elastomeric seismic isolation devices. Appl. Mech. Mater. **430**, 356–361 (2013)
7. Kelly, J.M.: Earthquake-resistant design with rubber, 2nd edn, Springer-Verlag, London (1997)
8. Kelly J.M., Konstantinidis D.: Mechanics of rubber bearings for seismic and vibration isolation, Wiley (2011)

9. Robinson, W.H.: Lead-rubber hysteretic bearings suitable for protecting structures during earthquakes. Earthquake Eng. Struct. Dynam. **10**(4), 593–604 (1982)
10. Constantinou, M.C.: Behavior of the double concave Friction Pendulum bearing. Earthquake Eng. Struct. Dynam. **35**(11), 1403–1424 (2006)
11. Gillich, G.R., Nedelcu, D., Malin, T.C., Iancu, V., Hamat, C.A., Gillich, N.: The effect of the friction coefficient and the pendulum radius on the behavior of structures isolated with simple friction pendulums. Rom. J. Acoust. Vib. **15**(2), 130–135 (2019)
12. Malin, T.C., Nedelcu, D., Gillich, G.R., Petrica, A., Padurean, I.: Comparison of the performance of friction pendulums with uniform and variable radii. Vibroengineering Procedia **23**, 81–86 (2019)
13. Nedelcu, D., Gillich, G.R., Iancu, V., Mălin, C.T.: Study on the effect of a simple friction pendulum radius on the response of isolated structures. ICMSAV 2018 & COMAT 2018 & Emech 2018, Brasov, Romania (2018)
14. Minda, A.A., Gillich, G.R., Iavornic, C.M., Minda, P.F.: Analytical and finite element study for friction pendulum with parameterized sliding surfaces. In: Proceedings of the World Congress on Engineering 2012, vol. 3. WCE 2012, London, U.K. (2012)

The manufacturer's authorised representative in the EU is Springer Nature Customer Service Centre GmbH, Europaplatz 3, 69115 Heidelberg, Germany. If you have any concerns regarding our products, please contact ProductSafety@springernature.com

Printed and bound by CPI Group (UK) Ltd, Croydon, CR0 4YY

30/09/2026

02225228-0015